Gender in Agriculture

Agnes R. Quisumbing • Ruth Meinzen-Dick
Terri L. Raney • André Croppenstedt
Julia A. Behrman • Amber Peterman
Editors

Gender in Agriculture

Closing the Knowledge Gap

Springer

FAO FIAT PANIS

IFPRI

Editors
Agnes R. Quisumbing
Poverty, Health, and Nutrition Division
International Food Policy Research Institute (IFPRI)
Washington, DC, USA

Ruth Meinzen-Dick
Environment and Production Technology Division
International Food Policy Research Institute (IFPRI)
Washington, DC, USA

Terri L. Raney
The State of Food and Agriculture
Agricultural Development Economics Division, Economic and Social Development Department
Food and Agriculture Organization of the United Nations (FAO)
Rome, Italy

André Croppenstedt
Agricultural Development Economics Division, Economic and Social Development Department
Food and Agriculture Organization of the United Nations (FAO)
Rome, Italy

Julia A. Behrman
Department of Sociology
New York University, New York
NY, USA

Amber Peterman
Department of Public Policy
University of North Carolina, Chapel Hill
Chapel Hill, NC, USA

Published by
The Food and Agriculture Organization of the United Nations
and
Springer Science + Business Media B.V.,
Dordrecht 2014

FAO hardback ISBN 978-92-5-107439-8
FAO paperback ISBN 978-92-5-107182-3
Springer hardback ISBN 978-94-017-8615-7
Springer paperback ISBN 978-94-017-8636-2

ISBN 978-94-017-8615-7 ISBN 978-94-017-8616-4 (eBook)
DOI 10.1007/978-94-017-8616-4
Springer Dordrecht Heidelberg New York London

Library of Congress Control Number: 2014936734

Printed on acid-free paper

Springer is part of Springer Science+Business Media (www.springer.com)

Foreword

The publication of *The State of Food and Agriculture 2010–11, Women in agriculture: Closing the gender gap for development*, called for renewed attention to the importance of the gender gap in agriculture. This issue had not been addressed in *The State of Food and Agriculture 2010–11, Women in agriculture: Closing the gender gap for development* since the 1983 special section on "Women in Developing Agriculture." In almost three decades there have been vast changes owing to the structural transformation of the global economy and changes in social and political institutions, yet fundamental social, economic, and political gender inequalities have continued to persist. Clearly, a knowledge gap needed to be closed in order to give policymakers, development practitioners, and civil society tools with which to attempt to close the gender gap.

This volume represents the culmination of a fruitful collaboration between the Food and Agriculture Organization of the United Nations (FAO) and the International Food Policy Research Institute (IFPRI). As the premier global policy advisory body on food and agriculture, FAO is well placed to advise governments on the design and implementation of agricultural policies that reflect the latest advances from empirical research. Since the 1980s, IFPRI has undertaken pathbreaking empirical research that challenged the dominant paradigm that men and women within households behave "as one." Its findings that putting more resources into the hands of women has payoffs in terms of agricultural productivity, health, and nutrition have been influential in the design of a new generation of development projects. In the process leading up to *The State of Food and Agriculture 2010–11*, FAO commissioned many background papers on gender issues in agriculture, and undertook consultations with researchers and policymakers all over the world. The sheer volume of new evidence meant that all this new knowledge could not be presented in *The State of Food and Agriculture 2010–11*. Because of IFPRI's contributions to research on gender and development policy, FAO commissioned IFPRI to edit a volume comprising the background papers, and to commission additional papers on specific aspects of gender in agriculture.

The chapters in this book provide evidence to support the main message of *The State of Food and Agriculture 2010–11*: closing the gender gap in agriculture will generate significant gains for the agricultural sector and for society. Gender equity in access to and control of productive resources—on the farm, in labor markets, and in the nonfarm economy, as well as in the institutions that support agricultural research, development, and extension—is not merely an issue of political correctness, but an essential aspect of development that can benefit women, men, and their families, and society.

Jomo Kwame Sundaram
Assistant Director General
Economic and Social Development Department
Food and Agriculture Organization of the UN
Shenggen Fan
Director General
Director General's Office
International Food Policy Research Institute

List of Editors

Agnes R. Quisumbing is a Senior Research Fellow, Poverty, Health, and Nutrition Division, International Food Policy Research Institute.

Ruth Meinzen-Dick is a Senior Research Fellow, Environment and Production Technology Division, International Food Policy Research Institute.

Terri L. Raney is a Senior Economist, and Editor of The State of Food and Agriculture, Agricultural Development Economics Division, Economic and Social Development Department, Food and Agriculture Organization of the United Nations.

André Croppenstedt is an Economist, Agricultural Development Economics Division, Economic and Social Development Department, Food and Agriculture Organization of the United Nations.

Julia A. Behrman, formerly a Research Analyst at the International Food Policy Research Institute, is currently a doctoral student at New York University.

Amber Peterman is an Assistant Research Professor of Public Policy at the University of North Carolina at Chapel Hill.

Contents

Abbreviations

AIDS	Acquired Immunodeficiency Syndrome
ASTI	Agricultural Science and Technology Indicators
ATMA	Agricultural Technology Management Agency
AWARD	African Women for Agricultural Research and Development
BRAC	Bangladesh Rural Advancement Committee
CBAs	Collective Bargaining Agreements
CBOs	Community-Based Organizations
CGIAR	Consultative Group on International Agricultural Research
CSR	Corporate Social Responsibility
EADD	East African Dairy Development
FAO	Food and Agriculture Organization of the United Nations
FFS	Farmer Field Schools
GAAP	Gender, Assets, and Agricultural Program
GAD	Gender and Development
GDP	Gross Domestic Product
HIV	Human Immunodeficiency Virus
ICRW	International Center for Research on Women
ICTs	Information Communication Technologies
IFAD	International Fund for Agricultural Development
IFPRI	International Food Policy Research Institute
ILO	International Labour Organization
IOM	International Organization for Migration
MDGs	Millennium Development Goals
MFI	Microfinance Institution
NAADS	National Agricultural Advisory Services (Uganda)
NARS	National Agricultural Research Systems
NGOs	Nongovernmental Organizations
NTAE	Nontraditional Agricultural Export
OECD	Organisation for Economic Co-operation and Development
PRA	Participatory Rural Appraisal

R&D	Research and Development
RD&E	Research, Development, and Extension
S&T	Science and Technology
SEWA	Self-Employed Women's Association
UNDP	United Nations Development Programme
UNESCO	United Nations Educational, Scientific, and Cultural Organization
UNIFEM	United Nations Development Fund for Women
WID	Women in Development

Contributors

Isabelle Baltenweck International Livestock Research Institute, Nairobi, Kenya

Julia A. Behrman Department of Sociology, New York University, New York, NY, USA

Nienke Beintema Agricultural Science and Technology Indicators (ASTI) Initiative, Environment and Production Technology Division, International Food Policy Research Institute, Washington, DC, USA

André Croppenstedt Agricultural Development Economics Division, Economic and Social Development Department, Food and Agriculture Organization of the United Nations (FAO), Rome, Italy

Jennie Dey de Pryck Global Forum on Agricultural Research (GFAR), Gender in Agriculture Partnership (GAP), Rome, Italy

Cheryl Doss African Studies and Economics, Yale Center for International and Area Studies, Yale University, New Haven, CT, USA

Diana Fletschner Director of Research, Monitoring and Evaluation, Landesa, Seattle, WA, USA

Delia Grace International Livestock Research Institute, Nairobi, Kenya

Jody Harris Poverty, Health, and Nutrition Division, International Food Policy Research Institute, Washington, DC, USA

Ruth Vargas Hill Senior Economist, World Bank, Washington, DC, USA

Nancy Johnson IFPRI, Agriculture for Nutrition and Health (A4NH), Washington, DC, USA

Lisa Kenney PhD Student, Department of Geography, University of Calgary, Calgary, Alberta, Canada

Patti Kristjanson CGIAR Program on Climate Change, Agriculture and Food Security (CCAFS), World Agroforestry Centre, Nairobi, Kenya

Susana Lastarria-Cornhiel Emeritus, Department of Urban and Regional Planning, Affiliate, Land Tenure Center and The Women's Studies Program, University of Wisconsin-Madison, Madison, WI, USA

Susan MacMillan International Livestock Research Institute, Nairobi, Kenya

Cristina Manfre Cultural Practice, LLC, Bethesda, MD, USA

Ruth Meinzen-Dick Environment and Production Technology Division, International Food Policy Research Institute (IFPRI), Washington, DC, USA

Jemimah Njuki Senior Program Officer, International Development Research Centre, Nairobi, Kenya

Lauren Pandolfelli Statistician, Evidence and Data for Gender Equality (EDGE), United Nations Statistics Division, New York, NY, USA

Stony Brook University, Stony Brook, NY, USA

Amber Peterman Department of Public Policy, University of North Carolina, Chapel Hill, Chapel Hill, NC, USA

Agnes R. Quisumbing Poverty, Health, and Nutrition Division, International Food Policy Research Institute (IFPRI), Washington, DC, USA

Catherine Ragasa Development Strategy and Governance Division, International Food Policy Research Institute, Washington, DC, USA

Terri L. Raney The State of Food and Agriculture, Agricultural Development Economics Division, Economic and Social Development Department, Food and Agriculture Organization of the United Nations (FAO), Rome, Italy

Deborah Rubin Cultural Practice, LLC, Bethesda, MD, USA

Paola Termine Child Labour in Agriculture, Fundamental Principles and Rights at Work (FPRW), International Labour Organization, Genève 22, Switzerland

Annita Tipilda Global Forests and Climate Change Programme, Nature Based Solutions Group, International Union for Conservation of Nature (IUCN), Gland, Switzerland

Marcella Vigneri Independent Development Economist (Specializing in Agricultural Development and research collaborator for IFPRI 's Ghana Strategy Support Programme [GSSP]), Oxford, UK

Elizabeth Waithanji International Livestock Research Institute, Nairobi, Kenya

Ann Waters-Bayer ETC Foundation, Leusden, Netherlands

Part I
Closing the Knowledge Gap on Gender in Agriculture

Chapter 1
Closing the Knowledge Gap on Gender in Agriculture

Agnes R. Quisumbing, Ruth Meinzen-Dick, Terri L. Raney, André Croppenstedt, Julia A. Behrman, and Amber Peterman

Abstract This introductory chapter provides an overview of the book, *Gender in Agriculture: Closing the Knowledge Gap*. The book grew out of collaborative work done for Food and Agriculture Organization of the United Nations (FAO) flagship report, *The State of Food and Agriculture 2010–11, Women in agriculture: Closing the gender gap for development*, highlighting the important and varied roles of women in agriculture, their unequal access to productive resources and opportunities relative to men, and the gains that could be achieved by closing the gender gap in agriculture. This book provides a more thorough treatment of the conceptual and empirical basis of the FAO report, and fills a niche in the literature for a standard reference for the analysis of gender issues in agriculture. This chapter defines basic concepts related to sex and gender and discusses changes in the way gender issues have

---

A.R. Quisumbing (✉)
Poverty, Health, and Nutrition Division, International Food Policy Research Institute (IFPRI), 2033 K Street, N.W, Washington, DC 20006, USA
e-mail: a.quisumbing@cgiar.org

R. Meinzen-Dick
Environment and Production Technology Division, International Food Policy Research Institute (IFPRI), 2033 K Street, N.W, Washington, DC 20006, USA
e-mail: r.meinzen-dick@cgiar.org

T.L. Raney
The State of Food and Agriculture, Agricultural Development Economics Division, Economic and Social Development Department, Food and Agriculture Organization of the United Nations (FAO), Viale delle Terme di Caracalla, 00153 Rome, Italy
e-mail: Terri.Raney@fao.org

A. Croppenstedt
Agricultural Development Economics Division, Economic and Social Development Department, Food and Agriculture Organization of the United Nations (FAO), Viale delle Terme di Caracalla, 00153 Rome, Italy
e-mail: Andre.Croppenstedt@fao.org

A.R. Quisumbing et al. (eds.), *Gender in Agriculture: Closing the Knowledge Gap*, DOI 10.1007/978-94-017-8616-4_1,

been conceptualized in agriculture from the work of Ester Boserup, to the Women in Development (WID) and Gender and Development (GAD) debate, to current approaches that recognize the importance of both women and men and the interplay between the two in agriculture. It traces how gender issues have been addressed institutionally and discusses shifting paradigms in the economic analysis of the household, including how demographic processes surrounding household formation and dissolution, gender differences across the life cycle, and migration have implications for the gender gap in agriculture. It then provides a summary of each of the chapters, suggests areas for future research, and explores implications for development policy and practice.

Keywords Gender gaps • Agriculture • Data and methods • Assets • Research, development and extension (R D & E)

The Food and Agriculture Organization (FAO) of the United Nation's *The State of Food and Agriculture 2010–11, Women in agriculture: Closing the gender gap for development* had a clear message: agriculture is underperforming because half of its farmers—women—do not have equal access to the resources and opportunities they need to be more productive. Arriving at this message was not a foregone conclusion. Almost three decades have passed since the 1983 *The State of Food and Agriculture: Women in Developing Agriculture.* During this period, the structural transformation of many economies meant a decline in the contribution of the agricultural sector to Gross Domestic Product (GDP) and a rise in industry and services, with consequences for the structure of employment and rural–urban migration. Many rural households have diversified their livelihoods to include nonagricultural income sources. There has been considerable discussion about consequent changing roles: whether male migration and nonfarm occupations were causing a "feminization of agriculture." This period was also marked by an increase in attention to issues of gender inequality on a global scale, starting from the United Nation's (UN) 1975 Year of Women and World Conference on Women, which led to the establishment, among other things, of the United Nations Development Fund for Women (UNIFEM), and more recently, UN Women, an umbrella organization that coordinates the efforts of 25 UN organizations

J.A. Behrman
Department of Sociology, New York University,
295 Lafayette Street, 4th floor, New York, NY 10012, USA
e-mail: jab965@nyu.edu

A. Peterman
Department of Public Policy, University of North Carolina, Chapel Hill,
Chapel Hill, NC, USA
e-mail: amberpeterman@gmail.com

to promote gender equity and support the full realization of women's rights and opportunities. Over the course of the next decades, the issue continued to gain momentum—notably in the Beijing Platform for action, a global agenda for women's empowerment, which developed out of the 1995 Fourth World Conference on Women. More recently, the adoption of gender equality and women's empowerment as one of the eight United Nations Millennium Development Goals (MDGs) signaled that gender issues continue to gain traction within the international community.

In spite of the attention to gender accorded by the United Nations and other key international actors, the agricultural research and development community did not readily grasp the importance of gender issues in rural development. By the end of the first decade of the twenty-first century, the evidence base on gender in agriculture remained woefully outdated. Arguments used to support attention to gender in agriculture relied on stylized facts that have attained mythological status, such as the oft-quoted "Women produce 60–80 % of food in the developing world." The editorial team drafting *The State of Food and Agriculture 2010–11*, therefore, had to commission new work, compile existing statistical evidence, and extensively review the empirical literature that had been written since 1983. The volume of new material meant that *The State of Food and Agriculture 2010–11* itself would not be able to accommodate all this new information in its necessarily succinct format. Thus, the FAO commissioned the International Food Policy Research Institute (IFPRI) to edit a book based on the background studies for *The State of Food and Agriculture 2010–11*. This volume brings together several of the studies commissioned for *The State of Food and Agriculture* focusing on gender gaps in key agricultural inputs, plus additional material on data and methods, the gender asset gap, and institutional changes toward more responsive agricultural research, development, and extension systems. In short, writing the volume on closing the gender gap was predicated on an attempt to close the knowledge gap on gender in agriculture.

This book represents a compendium of what we know now about gender gaps in agriculture, written for a nontechnical audience. This chapter begins with a discussion of the two motivations for closing the gender gap—equality and efficiency—and why these motivations may not be mutually exclusive, but rather, reinforcing. It then presents a summary of the main messages of *The State of Food and Agriculture 2010–11*, and a brief overview of how gender has been conceptualized in agriculture and how these concepts have changed in the past three decades. It also discusses the complexities surrounding gender relations as they are mediated by demographic phenomena such as household structure, age, and migration.

The remainder of the chapter consists of an overview of the four major sections of the book. Part II of the book is on data and methods for understanding gender issues in agriculture; Part III brings together the background studies documenting gender gaps in assets and key agricultural inputs; Part IV looks beyond the farm to consider gender roles in markets and value chains; Part V examines how the research, development, and extension system can be made more responsive to the needs of both male and female farmers. After reviewing each of these sections of the book, we conclude this chapter by considering future actions to close the knowledge gap on gender in agriculture and to better inform gender-sensitive agricultural policies and programs.

1.1 Closing the Gender Gap: Why Is It Important?

The concepts of "sex" and "gender" can be confusing to researchers and practitioners alike, particularly because they are often used inconsistently and interchangeably, when, in fact, they refer to two distinct concepts. Sex refers to the innate biological categories of male or female and is thus a fixed category rooted in biological differences. On the other hand, gender refers to the social roles and identities associated with what it means to be a man or a woman in a given society or context. Gender roles may be shaped by ideological, religious, ethnic, economic, and cultural factors and are a key determinant of the distribution of responsibilities and resources between men and women (Moser 1989). Because gender roles are socially, rather than biologically, determined, they are fluid and subject to change based on changing norms, resources, policies, and contexts. Every society is marked by gender differences, but these vary widely by culture and can change dramatically—within or between cultures—over time. A further complication is that "gender" and "women" are often used interchangeably. Gender refers to relations between men and women, not an exclusive focus on women. In much of agricultural development, the focus has been on men, so achieving gender equality requires rebalancing by paying greater attention to women. However, the importance of relations between women and men, as well as the differential roles, resources, and responsibilities of women and men of different ages, ethnicity, and social class need to be kept in mind in both analysis and programming.

Gender equality is a basic human right—with value in and of itself. The importance of gender equality is highlighted in its prominence in the United Nations Millennium Development Goals (MDGs), which have been commonly accepted as a framework for measuring development progress. Of the eight goals, four are directly related to gender: achieving universal primary education, promoting gender equality and the empowerment of women, reducing infant and child mortality, and improving maternal health. Closing gender gaps—which tend to favor males—has also been seen to contribute to women's empowerment. However, the term "empowerment" is a broad concept that gets used differently by various writers, depending on the context or circumstance. In an attempt to come to a common understanding applicable across multiple domains and disciplines, Kabeer (2001) defines empowerment as expansion of people's ability to make strategic life choices, particularly in contexts where this ability had been denied to them.[1]

[1] There is a growing literature on the measurement of empowerment (see Kabeer 1999; Alsop and Heinsohn 2005); the most recent studies attempt to develop multidimensional indices because empowerment is a multidimensional concept. See, for example, Ibrahim and Alkire (2007).

Another argument for reducing the gender gap revolves around improving productivity and increased efficiency, a strong message of *The State of Food and Agriculture 2010–11* (see next section), as well as its significant effects on the efficiency and welfare outcomes of project or policy interventions. We argue that the motivations for closing the gender gap are not mutually exclusive: rather, they reinforce each other. Closing the gender gap in assets—allowing women to own and control productive assets—both increases their productivity and increases self-esteem. A woman who is empowered to make decisions regarding what to plant and what (and how many) inputs to apply on her plot will be more productive in agriculture. An empowered woman will also be better able to assure her children's health and nutrition, in no small part because she is able to take care of her own physical and mental well-being (see Smith et al. 2003 and studies reviewed therein). These linkages are among *The State of Food and Agriculture 2010–11*'s key messages; the evidence supporting these messages is found in this volume.

1.2 Key Messages of *The State of Food and Agriculture 2010–11*

The State of Food and Agriculture 2010–11 can be distilled into the following key messages (FAO 2011):

- Women make essential contributions to agriculture in developing countries, but their roles differ significantly by region and are changing rapidly in some areas. Women comprise, on average, 43 % of the agricultural labor force in developing countries, ranging from 20 % in Latin America to 50 % in Sub-Saharan Africa and East Asia. Their contribution to agricultural work varies even more widely, depending on the specific crop and activity.
- Women in agriculture and rural areas have one thing in common across regions: they have less access than men to productive resources and opportunities. This gender gap is found for many assets, inputs, and services—land, livestock, labor, technology, education, and extension and financial services—and it imposes costs on the agricultural sector, the broader economy and society, as well as on women themselves.
- Closing the gender gap in agriculture would generate significant gains for the agricultural sector and for society. If women had the same access to productive resources as men, they could increase yields on their farms by 20–30 %. This could raise total agricultural output in developing countries by 2.5–4 %, which could, in turn, reduce the number of hungry people in the world by 12–17 %. The potential gains would vary by region, depending on how many women are currently engaged in agriculture, how much production or land they control, and how wide a gender gap they face.
- Policy interventions can help close the gender gap in agriculture and rural labor markets. Priority areas for reform include

- eliminating discrimination against women in access to agricultural resources, education, extension, and financial services and labor markets;
- investing in labor-saving and productivity-enhancing technologies and infrastructure to free women's time for more productive activities; and
- facilitating the participation of women in flexible, efficient, and fair rural labor markets.

1.3 Reconceptualizing Gender in Agriculture

The State of Food and Agriculture 2010–11 messages embody lessons from research and practice over the past three decades. They also reflect changes in the way gender relations in agriculture have been conceptualized. These are reflected in (1) changes in approaches to understanding gender relations in agriculture, from the work of Ester Boserup, to Women in Development (WID) and Gender and Development (GAD) approaches, to approaches that recognize the importance of both women and men and the interplay between the two in agriculture; (2) changes in the way gender issues are addressed institutionally, whether through stand-alone approaches, gender mainstreaming, or gender integration; and (3) shifting paradigms in economic analysis of the household, from the unitary to the collective model of the household.

1.3.1 Conceptualizing Gender Relations in Agriculture

Gender dimensions remained largely absent from mid-twentieth century discussions of agriculture, rural economies, and development. However, Ester Boserup (1970) brought gender to the forefront of the discussion with her 1970 book on *Woman's Role in Economic Development.* In this groundbreaking publication, Boserup draws explicit attention to the gendered division of labor that arises in both "traditional" and "modern" agricultural systems and to the fact that—for better or for worse—men and women experience the transition to modernity in different ways. Boserup was an early critic of the notion that gender differences in the labor market were due to biological, as opposed to socially constructed, differences. She went on to argue that economic development could not be fully evaluated without the recognition of the myriad "hidden contributions" of women throughout the world, particularly in the form of unpaid work. In this and other publications, Boserup illuminated the complexity of women's work, an area that had traditionally been downplayed or ignored.

Further research by anthropologists and then economists have demonstrated the wide range of roles played by men and women in agriculture, especially in Sub-Saharan Africa, where individual household members may have separate fields and crops, as well as some joint household plots and enterprises. Examples of evidence that men and women may have different production priorities can be found in the work of Dey (1985) on irrigated rice in The Gambia; Jones's (1983) formal model of intrahousehold conflict and husband's and wife's gains from cooperation based on work in the Cameroon; Mukhopadhyay's (1984) decision model of the sexual division of labor for household tasks; Hill's (1963, 1978) description of Fante women's entrepreneurial behavior; and Gladwin's (1975, 1982) models of women's marketing and farming decisions.[2] Studies documenting how assumptions of men as the primary or sole farmers led to adverse outcomes in agricultural projects (e.g., von Braun et al. 1989; Schroeder 1993) and further raised attention to gender in agriculture as more than an academic issue, requiring attention in policy and development programs.

1.3.2 Approaches to Addressing Gender in Development Policy and Institutions

The myriad roles women play in rural development was further taken up by the emergence of the WID movement in the 1970s, which called for development practitioners to pay explicit attention to the needs and preferences of women when designing and implementing development projects, as well as championing for a more general understanding and appreciation of the important roles women play in agriculture and rural development (see, for example, Tinker and Bo Bramsen 1976). The common response was to set up separate women's programs or units. Particularly in the case of agriculture (as compared to, for example, health), the women's programs were often relatively small and under-resourced, often staffed by relatively junior women, who might or might not have had any specific training in gender analysis (but were assumed to have expertise by virtue of being women). These programs were often on the sidelines, focused on gardens, nutrition, or small-scale income-earning activities.

The WID movement was ultimately succeeded by the GAD movement in the 1980s, which shifted the emphasis to looking at the relationships between women and men in rural development processes. The GAD movement critiqued the exclusive focus of the WID movement on women, by pointing out that looking at women in isolation was problematic because it only told part of the story of what transpired in rural contexts. Instead, proponents of the GAD approach argued that it is essential to shift the focus of rural development work and research to understand the gender elements of development; that is, the social roles, norms, and resources ascribed to women and men and how these gender roles shape the opportunities and constraints faced by both women and men.

[2] Implications of these differences for project design and agricultural technology are discussed by Cloud (1983), Doss (2001), and Feldstein and Poats (1989). These and other studies from Sub-Saharan Africa are reviewed in Gladwin and Macmillan (1989).

A response within many development agencies to the isolation and frequent marginalization of specialized women's programs was to "mainstream" gender in development policy or organizations, requiring all programs to address gender (Miller and Razavi 1995). While this was promoted as a transformative agenda, Rao and Kelleher (2005, 58) declared that "gender mainstreaming has become a random collection of diverse strategies and activities, all ostensibly concerned with moving forward a gender equality agenda, but often not working in ways we would have hoped." A major problem with mainstreaming was that declaring gender was to be addressed everywhere often led to gender becoming invisible, and without staff skilled in gender analysis, many programs did not develop significant gender-related programs; without committed leadership from the top to create clear forms of accountability, "mainstreaming" led to invisibility of gender. Furthermore, both women's programs and mainstreaming generally required that gender be retrofitted into a predefined agenda, which often missed the critical gender issues. For example, donor programs on "women in irrigation" in Asia in the 1980s often led to frustration when it appeared that irrigation was a very male-dominated activity. However, redefining the issue from a narrow focus on irrigation to look at the whole range of water uses led to identification of a wide range of gender issues related to different priorities and roles in water management between women and men.

With growing evidence of the importance of addressing gender issues to increase agricultural productivity and ensure that those increases are translated into improved welfare of the children, there is renewed commitment to addressing gender in a wide range of development organizations. The World Bank's program on "Gender Equality as Smart Economics" (World Bank 2008) created the rationale, with a Gender Action Plan to follow up. In the agricultural sector, the World Bank/FAO/IFAD *Gender in Agriculture Sourcebook* (2009) laid out the gender dimensions of a wide range of topics, from land tenure to marketing. *The State of Food and Agriculture 2010–11* has created high-level attention to gender in agriculture, and the 2012 *World Development Report* has similarly raised the profile of gender, and indicated significant ways in which gender can be integrated into development programs, including in agriculture, not as an afterthought, but as a priority for development.

1.3.3 Shifting Paradigms in Economic Analysis

Historically, the field of economics was dominated by a traditional "unitary" model of household behavior in which households were viewed as groups of individuals who have the same preferences and fully pool their resources. According to this view, most economists see the household as a collection of individuals who behave as if they agree on how best to combine time, goods purchased in the market, and goods produced at home to produce commodities that maximize some common welfare index. This approach is appealing because of its relative simplicity and the diversity of issues it can address—such as the determinants of education, health, fertility, divorce, migration, labor supply, home production, land tenure, and crop adoption (see Becker 1973). However, economists have increasingly questioned the unitary model of household behavior and proposed alternative models that bear closer resemblance to reality (see Haddad et al. 1997 for a review of these models).[3] Haddad et al. (1997) argue that using the unitary model of the household as a guideline for policy prescriptions may lead to four types of policy failures.

1. The effect of public transfers may differ, depending on the identity of the income recipient. If this is so, targeting transfers to the household may not result in the desired outcomes, if transfers directed to the husband or the wife have different impacts.
2. The response of nonrecipients of the income transfer must also be considered. If households reallocate resources away from the transfer recipient to compensate for the transfer receipt, the intended effect of the income transfer may not be realized. For example, suppose that public social security schemes are introduced that tax the young and subsidize the old, leaving aggregate income unchanged. This might lead to a reduction in private urban–rural remittances, with consumption by individual members unchanged.
3. At the project level, the unitary model predicts that it does not matter to whom policy initiatives are addressed, since information, like other resources within the household, will be shared. However, numerous examples, many from Sub-Saharan Africa, have shown that targeting one individual rather than the other has led to non-adoption of particular policies or unintended consequences of policies adopted.
4. Adherence to a unitary model of the household disables many policy instruments that could be brought to bear on development problems. The unitary model predicts that household behavior can be changed only by changes in prices and household incomes. In contrast, the collective model (see discussion below) posits that a large range of policies can be used to affect household allocation outcomes, such as changes in access to common property resources, credit, public works schemes, and legal and institutional rights. Policymakers have taken these findings to heart in designing programs that aim to change household behavior by transferring income directly into the hands of women, as illustrated by conditional cash transfer programs in Latin America, most of which designate the mother as the transfer recipient, and microfinance projects that target poor women as credit recipients.

A number of alternatives to the unitary model have emerged that focus on the individuality of household members and the possible differences in their

---

[3] Many of these challenges came from studies in the 1980s that suggested that men and women spend income under their control in systematically different ways. These studies included Guyer (1980), Tripp (1982), Pahl (1983), and studies from different countries (for example, Fapohunda 1988) in the volume edited by Dwyer and Bruce (1988). A series of studies on agricultural commercialization and nutrition in developing countries also found that female-controlled income is more likely than male-controlled income to be spent on food (Kennedy 1994).

preferences. One class of these is the so-called collective models (Chiappori 1988, 1992, 1997) that allow differing preferences and only assume that allocations are made in such a fashion that the outcomes are Pareto optimal or Pareto efficient. That is, an individual within the household can only be made better-off at the expense of another household member.

All collective models have two common features: they allow different decision-makers to have different preferences, and do not assume a single household welfare index or utility function (Chiappori 1992). Both unitary and collective models permit existing intrahousehold resource allocation rules to affect household responses to public policy and public policy to change intrahousehold allocations of a good. However, only the collective model permits public policy to affect the *rules* of intrahousehold allocation. In the collective model as described above, nothing is assumed a priori about the nature of the decision process, that is, the model does not directly address the question of how individual preferences lead to a collective choice. If one is willing to put more structure on the decisionmaking process, two subclasses of collective models emerge, one rooted in cooperative and the other in noncooperative game theory.[4] While a detailed discussion of cooperative and noncooperative household models is outside the scope of this chapter, one key policy-relevant implication is the importance of the roles of "outside options" or "threat points" in influencing the bargaining power of individuals within the household (Manser and Brown 1980; McElroy and Horney 1981). These "outside options"—whatever utility or welfare individuals would have outside the household—affect the gains from being in marriage or within the household. The better the outside options, the more likely that an individual will have higher welfare from being in the household; otherwise, the person would leave. This implies that one way to affect intrahousehold welfare is to improve the exit options of disadvantaged groups—such as women—through improving the skills or assets that they could use to support themselves, outside of the household, or to provide alternative employment opportunities (say, through public works programs). For policies to be successful, however, one must first identify the relevant exit options, which are themselves a function of "extra environmental parameters" such as laws concerning access to common property and prohibitions on women working outside the home (McElroy 1990).

In the case of agriculture, the collective models of the household recognize that men and women of different generations may have separate plots, animals, or production activities, with varying degrees of independent control over the output, and varying degrees of claims on the land, labor, income, or other resources of other household members. This is particularly observed in much of Sub-Saharan Africa, but even in Asia, where there appears to be a stronger pattern of joint family farming, women may have prime responsibility and control over homestead gardens. And in many countries, occupational diversification and migration are increasingly breaking up household joint farming enterprises, as men (or women) move into off-farm or nonfarm activities and earn incomes that may be pooled or held separately to varying degrees. The collective models of the household allow us to see beyond

[4] See Haddad et al. (1997) for a detailed discussion of both types of models.

the image of a single farm unit, to consider how gender differences in the control over farming assets, inputs, and information will affect agricultural productivity, and how the control over assets, production, and income affect welfare outcomes for different members of the household, especially children's nutrition, health, and education. The evidence on many of these issues is explored in subsequent chapters of this book.

Researchers working on gender issues in agriculture have often blamed the lack of sex-disaggregated data for the underestimation of women's contribution to agricultural production (Dixon 1982). A collection of papers on farming systems research in Africa, Asia, and Latin America (Poats et al. 1988) argued convincingly for the collection of sex-disaggregated data and proposed methodologies for collecting such data for farming systems research and extension. Unfortunately, most agricultural data have continued to be at the level of the household, and many analyses of "gender" in agriculture have been limited to comparisons of "male-headed" and "female-headed" households (usually defined in terms of whether there is an adult male present who is assumed to be the head of household). The result has been that even these analyses employ an implicit unitary model of the household, and ignore the role of women (and junior men) within male-headed households (for more detail, see Sect. 1.4.1 below).

The recent "explosion" in both the quantity and quality of household- and individual-level data (Strauss and Thomas 1995) has contributed to the growing empirical literature on intrahousehold allocation. While national statistical agencies have been collecting nationally representative data since after World War II, it is only relatively recently that these datasets have permitted the analysis of intrahousehold and gender issues, and only since the early 2000s that sex-disaggregated data have been collected on agricultural production, especially data that permit more than comparison based on sex of the "head of household." Where programs are designed to affect individual outcomes such as health, nutrition, or education, evaluations of such programs will naturally collect individual-level, and thus, sex-disaggregated, indicators. Multiple survey rounds undertaken for monitoring purposes also yield panel datasets for future analysis. Most of the more completely sex-disaggregated data in agriculture are not based on nationally representative sampling frames, but are based on more limited surveys that can be treated as complementary to large-scale household surveys, due to the detail they provide at the farm and even the plot level. More recent efforts to collect agricultural census data at the plot or enterprise level, rather than the farm level, in Africa would also lay the basis for more comprehensive gender analysis of agricultural production (see FAO 2005).

1.4 Gender and Demographic Phenomena

Going beyond the realization that not all individuals within the household share the same preferences, reconceptualizing gender in agriculture has also meant a greater appreciation of the complexities introduced by demographic processes surrounding household formation and dissolution, gender differences across the life cycle, and migration.

1.4.1 Household Structure, Headship, and Gender

In the absence of data about men and women within the household, the sex of the household head has often been used as a "gender indicator" in many studies (see critical reviews by Doss Chaps. 3 and 4 and Peterman et al. Chap. 7). Comparisons are typically made between male-headed and female-headed households, and interpreted as comparisons between men and women. There are two main reasons why the sex of the household "head" is a misleading proxy for gender relations. First, in many comparisons, "male-headed households" are composed of households in which both spouses are present, while female-headed households are made up mostly of households in which a husband is not present (Fafchamps and Quisumbing 2008, 3232). Doss (Chaps. 3 and 4) points out that since women frequently live in male-headed households, but adult men rarely live in households that are defined as female-headed, this approach conflates measures of household structure and composition with the sex of the head. Moreover, considering only the sex of the head of household renders women living in male-headed households invisible. Household heads may also be defined by age, sex, custom, or economic contributions, among other factors, and differ widely across contexts (Budlender 2003). Second, comparisons between male-and female-headed households do not pay attention to the endogeneity of female headship. As Fafchamps and Quisumbing (2008) point out, not all female heads of households are alike, and the reasons why a woman heads a household vary considerably. A woman could head a household if she does not marry (as in many Western societies), if she postpones marriage, if she is widowed, or if she temporarily heads a household owing to migration, war, etc. One needs to distinguish between *de jure* female-headed households (headed by divorced or widowed women) and *de facto* female-headed households (in which the husband is absent, but may contribute to household finances). The welfare implications of female headship are quite different, depending on the process by which it occurs—by death or divorce, or through migration of the husband (see Joshi (2004) for an example in Bangladesh). Paying attention to headship and the endogeneity of headship is not a mere academic exercise; it has important policy implications. If it is mistakenly assumed that all female-headed households are poor, then resources may not be targeted to the poorest. Using headship as a proxy for gender differences within households may also lead to underestimation of gender differences in agricultural productivity (Peterman et al. 2011).

1.4.2 Differences Across the Life Cycle

A narrow focus on differences between men and women often masks more important differences between women, including those arising from where they are in the life cycle. Many of these needs are biological, owing to different needs for reproduction and childcare obligations (see Chap. 11 by Harris); others arise from differences in men's and women's position in society, depending on their life-cycle stage.

There has been recent interest in both programmatic and research arenas on the role of adolescent girls in agriculture and rural economies, predicated on the hypothesis that investments in the "future generation" of farmers will have spillover effects on households, communities, and the intergenerational transmission of poverty (Kirk et al. 2011; Bertini 2011). The focus on girls is motivated by the lower levels of schooling and resources and the higher levels of domestic work and security concerns as compared to boys of their same socioeconomic status. Although we still know very little about the challenges and opportunities for adolescents in agriculture, as well as how to successfully manage trade-offs between traditional schooling, marriage, and labor force participation, there have been promising interventions ranging from agricultural training, legal rights and inheritance interventions, microfinance, and asset transfers to both girls and boys across diverse countries (Bertini 2011; Bandiera et al. 2010; Catino et al. 2011). For example, an evaluation of the Berhane Hewan project aimed at delay of child marriage through girls' groups and asset transfers (livestock) showed improvements in schooling and delays in age at marriage (Erulkar and Muthengi 2009).

However, youth are not the only important demographic group to consider along the life cycle in agriculture. In Kenya, young Luo women, who first learn to farm under the guidance of their mothers-in-law, defer much of the decisionmaking about their farms to their mothers-in-law and do not obtain the rights to farm independently until they have had children (Potash 1981). This suggests that while older, actively farming women may have more resources to draw upon to better respond to extension messages, interventions that target younger female farmers need to be aware of the differential constraints they may face. Older women—particularly grandmothers—may also be an untapped resource for spreading extension messages owing to their status in a particular society, although this is, of course, context-specific. A study in Guatemala (Kevane and Wydick 2001) also found that gender differences in the ability to expand family-owned enterprises were highly correlated with the life cycle. Young male entrepreneurs were more aggressive in generating employment than older male entrepreneurs, but older women generate more employment than young women or older men. Older women may therefore be good targets for microenterprise funds, not only because they can expand the enterprise, but also because preferences would tend toward welfare of grandchildren. As varying demographic shifts take place across the globe, life-cycle heterogeneities of men and women will become increasingly important considerations in agricultural programming.

1.4.3 Migration

Globally, female migration is now virtually equal to male migration (Donato et al. 2006). Migration creates changes in family structure and sources of incomes and thereby has the potential to affect both the sex composition of the agricultural labor force and gender roles in agriculture. The impacts of migration will depend on the age and sex of the migrant, the purpose of migration (for marriage or for work), the destination of the migrant (internal or international), the duration of the move (temporary, circular, or permanent), and whether migrants make remittances to their origin households. The literature examining the gendered impact of migration is uneven. In a review of the recent sociological migration literature, Curran et al. (2006) bemoan the slow progress of gender analysis in quantitative studies of migration. They also point out the male bias of migration studies (since most surveys interview the male household head) and the failure to observe pre- and post-migration impacts and responses, particularly of nonmigrants, who are often women, in many contexts. Owing to the focus on international migration and the lack of data on internal migration and on the point of origin (Deshinkar 2009), we know even less about the intersection of gender, migration, and agriculture. As Tacoli (2010, 296) points out, less attention has been given to internal migration, even though this type of movement more closely reflects the demographic, socioeconomic, and cultural transformations in low- and middle-income nations. Despite the proliferation of myths regarding the "feminization" of migration, there is no evidence that the proportion of females who are migrating internationally has increased. The proportion of female migrants was 47 % of all international migrants as far back as 1960 and this percentage increased by only two points during the next four decades (Zlotnik 2003, cited in Deshinkar 2009). Prevailing myths that most internal migration is male-dominated and primarily rural–urban have also been debunked by recent studies based on demographic and health surveys (see Tacoli 2010 for a review). Rural-rural migration is the most common type of movement among female migrants, and tends to be highest in Africa, although urban-urban migration predominates in Brazil, Peru, Paraguay, Colombia, and Bolivia. With regard to male migrants, urban-urban migration predominates, although rural-rural migration is also most common in Africa. However, women have increasingly been migrating for economic reasons rather than just family reunification (IOM 2004; see, for example, Elmhirst (2007), on the feminization of rural–urban migration in Indonesia; Piper (2008) on women's migration within the Asian region, and Adepoju (2004) on Sub-Saharan Africa), even if remittances from women who originally migrated for marriage have been documented to play an important role in risk-smoothing (see, for example, Rosenzweig (1993) on India). Agricultural development programming will increasingly need to take into account the gendered nature of migration flows and their impacts on origin and destination labor markets.

1.5 Outline of the Book

This volume draws heavily upon the growing body of data and analyses of differences of gender in agriculture. The following is an overview of the issues and themes taken up in the remaining sections of this book.

1.5.1 Part II: Data and Methods for Gender Analysis in Agriculture

One of the important contributions of *The State of Food and Agriculture 2010–11* is the compilation and review of the available statistical evidence on gender in agriculture. The statistical appendix merits attention, both for what is there, and for the number of countries and variables for which sex-disaggregated data are not available. Making sure that women's and men's roles are counted is an important step to answer questions and to monitor progress in a number of areas.

Three chapters in this section discuss how developments in data collection and analytical methods have helped expand the knowledge of gender relations in agriculture. Chapter 2, by Behrman, Meinzen-Dick, and Quisumbing, presents a suite of quantitative and qualitative methods for collecting and analyzing data on gender relations in agriculture. Recognizing that gender relations are complex and context-specific, the authors do not present definitive recommendations for what kind of data to collect and how to analyze it, but rather recommend more nuanced, context-specific data collection and analytical methods. The strongest recommendation to emerge from this chapter is the suggestion to use both qualitative and quantitative methods for understanding gender relations in agriculture. Three examples illustrate how using integrated, mixed-methods work enabled researchers to understand more about the processes underlying the adoption of agricultural technologies.

The first chapter by Doss (Chap. 3) provides more detailed guidance on the types of quantitative data needed to address key questions on gender roles in agriculture, focusing on farmer surveys, household surveys, labor market surveys, and agricultural censuses. The chapter begins with fundamentals: who should be interviewed and how to structure the interview, linking these issues not only to best practice methods but also to the purpose of the study. It then identifies key types of data needed to examine the gendered roles, contributions, and constraints in agriculture and some of the complexities of collecting this data, from the household roster to the control of land and other assets, labor, production, income, and welfare outcomes. The guidance provided in this chapter would help ensure that women are seen and their voices are heard in agriculture, and to address key questions about women in agriculture.

The second chapter by Doss (Chap. 4) explores the basis for the oft-quoted stylized fact that women produce 60–80 % of food in the developing world. Doss uses three approaches to shed light on this issue: (1) analyzing labor inputs to agriculture, using both employment data and time-use data; (2) analyzing different ways of assigning agricultural output by gender, using four nationally representative household survey datasets; and (3) estimating women's labor productivity relative to men at the macro level, using national-level agricultural productivity data across time and countries. Doss concludes that while it is not possible to substantiate the claim that women produce 60–80 % of the food in developing countries—or even that they provide 60–80 % of the labor in agriculture, this should not be interpreted as evidence that women are insignificant in the agricultural sector. Women contribute a large portion of the measured contributions to agriculture labor and the women's share of the measured agricultural labor force has a positive impact on national-level agricultural productivity. While the "60–80 %" statistic is popular among advocates for women's issues, Doss cautions the reader that the statistical claim obscures the complex underlying reality, that it is difficult to separate women's labor from other uses as well as from men's labor, and that it cannot be understood properly without considering the gender gap in access to land, capital, assets, human capital, and other productive resources.

1.5.2 Part III: Gender, Assets, and Inputs: Issues at the Farm and Household Levels

The chapters in this section delve more deeply into gender gaps in different types of resources, such as land, livestock, or access to markets that may affect women's and men's productivity in agriculture. Control over and ownership of assets is a critical component to well-being. Like income, assets can be converted to cash, but they are also multidimensional. Assets both store wealth and can increase in value. Assets such as homes or buildings may both provide services and may generate rent. Assets can act as collateral and facilitate access to credit and financial services. In short, owning land and livestock, homes and equipment, other resources and wealth enable people to create stable and productive lives. Who controls these assets within the household is critical to household and individual well-being and the intrahousehold allocation of assets has important implications for a range of outcomes.

The chapter by Meinzen-Dick and coauthors (Chap. 5) proposes a conceptual framework to explore the potential linkages between gender, assets, and agricultural development projects in order to gain a better understanding of how agricultural development interventions are likely to (positively or negatively) impact the gendered distribution of assets. It uses a broad definition of tangible and intangible assets—natural capital, physical capital, human capital, social capital, and political capital. The conceptual framework identifies linkages between the gendered distribution of assets and various livelihood strategies, shocks, and well-being, and discusses how agricultural development strategies may affect the gender asset gap—a theme that is then explored in more depth in the remaining chapters of this section.

Land is a starting point for the discussion surrounding the role of assets, because it plays a fundamental role for both agricultural production and security. Lastarrhia-Cornhiel and coauthors provide an overview of the land tenure situation from a gendered perspective and detailed information on a wide range of policy mechanisms put in practice in selected countries to improve gender equality in land access (Chap. 6). These policy mechanisms include joint-titling modalities, land market operations, land leasing, land use certificate issuances, community and territorial land delimitation programs, land allocation through state-managed land reform programs, land law reforms, and other types of land interventions. The chapter also highlights the social and cultural factors that are important to consider when choosing different workable policy options to overcome women's disadvantaged position in accessing land, issues such as existing social protection mechanisms, bargaining power, and other forms of social control and gendered social stereotypes.

A common finding in many empirical studies of agricultural productivity differences is that men and women are equally efficient farmers, once gender differences in use of agricultural inputs are accounted for. Peterman and coauthors review recent evidence for a range of nonland inputs, focusing on three key areas: (1) technological resources, (2) natural resources, and (3) human resources (Chap. 7). While there is comparatively more empirical research on topics such as inorganic fertilizer, seed varieties, and extension services in comparison to that on tools and mechanization, Peterman and coauthors consistently find that across different types of inputs, men generally have higher input measures than women. However, this finding is often sensitive to the use of models that control for other background factors, as well as the type of gender indicator implemented in the analysis. Moreover, most of these studies are based on data from Sub-Saharan Africa, because gender-disaggregated production data are seldom collected in other regions, pointing to a gap in current research.

Financial constraints are often cited as one reason that women use lower levels of many inputs in agriculture. The chapter by Fletschner and Kenney (Chap. 8) reviews existing evidence on rural women's access to financial services, arguing that appropriate financial products for women to be able to save, borrow, and insure are important for strengthening their role as producers and for widening the economic opportunities available to them. The authors discuss how context-specific legal rights, social norms, family responsibilities, and women's access to and control over other resources shape their need for capital and their ability to obtain it. Citing empirical evidence that women have lower access to credit and financial services than men, that women and men have different degrees of risk aversion, and that resources are not always shared within the household, the authors argue that attempts to enhance women's access to financial services must do so *directly*, i.e., not mediated through their husbands. The chapter reviews promising new products and service delivery models introduced to address some of the constraints faced by women, including technical innovations that improve access to existing financial services, changes in product design to better tailor products to women's preferences and constraints, and the development of new products, such as microinsurance.

In societies where legal systems and cultural norms make it difficult for women to accumulate valuable assets such as land, livestock has often emerged as an alternative form for women to hold wealth. Kristjanson and coauthors review the role of livestock in the livelihoods of poor women in Sub-Saharan Africa and South Asia and identify factors that enhance or constrain livestock-related opportunities for women (Chap. 9). They apply a gender lens to three livestock-related pathways out of poverty—securing, building, and safeguarding livestock assets; increasing and sustaining livestock productivity; and enhancing participation in and benefits from livestock markets. For each pathway, the chapter summarizes what is known from an extensive review of evidence and what this knowledge implies for programmatic and policy interventions. Although there are gaps in the evidence base, the authors use the review to identify what kinds of research and development interventions, made in relation to which species and value chains, appear most likely to benefit poor women and their families.

Unlike livestock, which has long been recognized as important in women's asset portfolios, the development community has only recently—but very enthusiastically—paid attention to social capital as an important asset in agriculture. The chapter by Meinzen-Dick and coauthors documents gender differences in social capital, defined in terms of group membership and social networks (Chap. 10). The authors go beyond simple dichotomies of men's and women's groups and networks to investigate whether, and under what circumstances, mixed-sex groups may be more effective than single-sex groups in achieving their development objectives, and how social capital may contribute to women's empowerment in agriculture. The chapter concludes by summarizing the evidence whether women are disadvantaged to men in the accumulation of social capital, and assessing the extent to which programs are helping to overcome this perceived gap.

One type of capital for which there are clear differences between men and women is human capital. There are many studies of gender differences in education and the implications for economic production, including a number of such studies in agriculture. However, there has been relatively less attention to the links between other types of human capital and agricultural production. Although good health and nutrition enhance both men's and women's well-being and productivity in agriculture, life-cycle and reproductive considerations make the implications of health and nutrition different between men and women. The chapter by Harris (Chap. 11) aims to summarize the evidence on gender differences in vulnerabilities to poor nutrition and health, and their potential effects on the agricultural productivity of men and women in farming households. Adopting a life-cycle perspective, Harris examines the implications of four key health and nutritional disorders—undernutrition, iron-deficiency anemia, HIV, and malaria—for the productivity and well-being of men and women in agriculture. She finds that women are generally disadvantaged across disorders, due to both biological and social vulnerabilities. Harris concludes that while investment in nutrition and health does appear to have the potential to increase the productivity of women in agriculture, the social barriers to that investment, such as women's time allocation and traditional roles, must be recognized and addressed. Without attention to broader gender issues, Harris argues that any gains from nutrition and health interventions for women in agriculture are unlikely to achieve their potential.

1.5.3 Part IV: Gender and Markets: Moving Beyond the Farm

Although a large proportion of women are involved directly in smallholder agriculture as farm managers and workers on their own families' farms, ranging from 53.5 % of the rural adult female population in Sub-Saharan Africa to 6.9 % in Europe and South Asia, the residual, which is a significant proportion, consists either of wage earners in agriculture, self-employed persons in nonagricultural rural enterprises, nonagricultural wage earners, and "non-active or not reported." The large number of rural women classified as either non-active or not reported (up to 64 % of the female population in South Asia, and above 50 % both in Latin America and the Middle East and North Africa region) reflects the fact that much of women's work in rural areas is informal or unpaid and still goes unrecorded (Fontana and Paciello 2010). The large number of women participating in labor markets as wage workers or as processors and traders along the agricultural value chain is the subject of this section. The three chapters in this section examine gender in value chains, the specific barriers that women face in accessing high-value markets, and conditions underlying women's employment in labor markets.

There is currently considerable enthusiasm in agricultural development policy for enhancing value chains as a means of increasing farmers' incomes through "pro-poor market development." The chapter by Rubin and Manfre (Chap. 12) examines the growing body of work on reducing gender-based barriers to value-chain development. It highlights key questions that are emerging within the gender and value-chain community related to methodologies for promoting both greater gender equity and efficiency and provides evidence and examples of different gender and value-chain approaches. However, this closer look at the evidence also raises some new questions and challenges facing researchers and practitioners on chain selection, targeting of women, and achieving food security and improved nutrition in value-chain development.

Following the big-picture analysis of value chains, Vigneri and Vargas-Hill (Chap. 13) look in more detail at the barriers that women face in accessing high-value markets for cocoa and coffee in Ghana and Uganda, based on microlevel household data analysis. They examine whether the constraints faced by women arise due to discrimination in input and output markets for cash crops themselves, or arise owing to constraints in assets and other resources. While female farmers in both countries are as productive as male farmers and receive comparable prices to those received by men when they farm with *equal* resources and sell their crops in the *same way*, they rarely have similar access to assets and markets as men, which has consequences for the choice of production technology and marketing channel. They find that women cocoa farmers in Ghana have limited access to liquidity, which induces them to adopt suboptimal production technologies. In Uganda, the low quantities marketed, and the lack of access to bicycles, limit female coffee farmers to marketing channels that have very low transaction costs, but which receive lower prices. The authors conclude with recommendations for improving women's access to high-value markets.

Gender differences in agriculture are not limited to direct production, but also to agricultural labor. Focusing on the rural poor, whose main productive resource is their labor, Dey De Pryck and Termine examine the evidence for gender inequalities in the rural labor market and propose ways of addressing these (Chap. 14). The analysis addresses a number of key questions, including the gender differences in agricultural and nonfarm rural employment, with particular reference to employment in traditional agricultural systems, in modern high-value agro-industries, in emerging rural nonfarm enterprises, and rural public works programs. It examines the barriers to female employment in the rural labor market and the underlying reasons for these barriers (e.g., entry barriers, occupational segregation, wage gaps), and how they differ in traditional agricultural wage employment compared with work in modern agro-industries and rural nonfarm enterprises. Finally, they discuss examples that could be replicated or scaled-up of effective gender-aware or gender-transformative policies, laws and other instruments to remove or at least reduce these barriers in a gender equitable way, and to help women circumvent or tackle these barriers themselves, for example, through their participation in mixed or women-only institutions, such as self-help groups, cooperatives, and workers' organizations.

1.5.4 Part V: Gender, Institutions, and Policy

The agricultural research, development, and extension (RD&E) system at the global, regional, and national level is set up to increase agricultural productivity, profitability, and sustainability. But to what extent does this system recognize the needs of women in agriculture? The chapter by Meinzen-Dick and coauthors examines what this recent evidence on gender differences in access to resources and participation in markets implies for RD&E systems (Chap. 15). They suggest that what is needed is a paradigm shift, from a focus on production toward a broader view of agriculture and food systems, one in which women's distinct role in ensuring the food security of their households is better recognized, and women have greater voice in setting priorities for research. This involves recognizing women's role throughout the value chain for both food and nonfood crops and for both marketed and nonmarketed commodities.

While both men and women can (and should) conduct gender-responsive research, including women in the agricultural research system plays an important role in ensuring that women's needs are addressed. Yet in developing countries, less than one out of four researchers is a woman, although large differences exist across countries. Beintema's chapter (Chap. 16) reviews the evidence on the trends in women's participation in agricultural research in developing countries with more detailed analysis of Sub-Saharan Africa, where more detailed information is available. Although the share of women employed in agricultural research and development has been increasing in most countries, their share disproportionately declines on the higher rungs of the career ladder. The chapter also summarizes the various general human resource challenges in agricultural R&D that developing countries face, specifically in Sub-Saharan Africa, focusing on the challenges women face prior to and during their science careers.

Agricultural research and development alone does not benefit anyone, unless it is adopted. However, lack of information has often constrained women's adoption of improved technologies and agricultural practices. The chapter by Ragasa (Chap. 17) makes the case for improving the gender-responsiveness of agricultural extension systems through consideration of: (1) whether the sex of the extension agent affects the effectiveness of extension services; (2) whether both men and women receive extension advice; and (3) how extension services are delivered. The chapter reviews the evidence on gender differences in access to formal extension agent visits and to other sources of extension information, and the factors that lead to women having lower access to extension services. It then examines the experience of programs and projects that aim to increase women's access to extension, with more detailed analysis of extension system reforms in India, Uganda, Venezuela, and Ethiopia, and the use of Information Communication Technologies (ICTs) in extension.

1.6 The Way Forward: Closing the Gender (Knowledge) Gap in Agriculture

The research and reviews commissioned for *The State of Food and Agriculture 2010–11* and compiled in this volume highlight the extensive body of social and economic research confirming that women make vital contributions to the agricultural sector and rural enterprises, despite the many gender-specific constraints they face in accessing resources and opportunities. If these constraints are lifted, there would be enormous potential benefits for the agricultural sector and society at large. The messages of *The State of Food and Agriculture 2010–11* are clear: (1) gender equality is good for agriculture, food security, and society; and (2) governments, civil society, the private sector, and individuals, working together, can support gender equality in agriculture and rural areas (FAO 2011).

There is no single or simple approach, but there are important lessons to be learned from policies, programs, and interventions aimed at closing the gender gap in agriculture. Many of the approaches to meeting women's needs are similar to those to address the needs of other resource-constrained, small-scale farmers and rural people in general. However, there are additional cultural and behavioral factors that need to be taken into account in efforts to increase gender equality.

Gender-aware agricultural policy decisions and development interventions have to be based on up-to-date, reliable, and context-specific information. Despite the wealth of evidence that has emerged from recently commissioned work, the information base on which gender-sensitive policy decisions can be made needs to be continuously built. Recommendations for future work include

- **Improve the collection and analysis of sex-disaggregated data.** Understanding of many gender issues in agriculture—including crop, livestock, fisheries, and forestry sectors—is hindered by the lack of sex-disaggregated data, and inadequate analysis of the data that exist. Agricultural censuses should focus more attention on areas in which women are relatively more active and collect sex-disaggregated data on ownership, access to, and control over productive resources such as land, water, equipment, inputs, information, and credit. They should avoid gender biases in the concepts and definitions used to ensure that the resulting data accurately highlight gender interactions and inequalities in the agricultural sector. More detailed time-use surveys would lead to greater understanding of women's contributions to household production and welfare as well as to their time constraints. The quantity and quality of sex-disaggregated data for policymaking can be increased through the integration of agricultural censuses and surveys and the re-tabulation of existing census data. Gender differences and their implications may be more visible when sex-disaggregated data are collected, analyzed, and presented by age groups and at subnational levels, including linking it with spatial data to help identify patterns, such as links between gender patterns and agroecological zones.

- **Continue to undertake and validate empirical work on gender issues in agriculture.** Changes in the agricultural and social landscapes as a result of structural transformation, increased integration into market economics, urbanization, and migration may mean that many of the underlying relationships that were studied in the first wave of empirical work on gender issues in agriculture in the 1980s may no longer hold. Gender norms do change, even if slowly. While the studies commissioned for this volume have undertaken or reviewed new empirical work, such work needs to be done continuously, lest we fall into the trap of citing outdated work that is no longer relevant to current conditions. Replicating "classic" studies with different data from other countries, or even with newer data from the same country, and validating the robustness of their results, is an important exercise.
- **Explicitly explore alternative design and delivery mechanisms to meet context-specific gender needs.** Owing to the absence of an evaluation culture, programs are modified in an ad hoc manner, without systematic evaluation. Without evaluation, it is difficult to recommend what programs can be scaled-up. Likewise, it is difficult to know what design features can be modified for local conditions without adversely affecting the overall outcome of the intervention. Paying greater attention to mechanism design in program evaluations would facilitate other organizations' and implementing agencies' learning from these efforts.
- **Explicitly evaluate the gendered impacts of agricultural interventions.** Although governments and civil society have undertaken many programs that attempt to address gender disparities, many of these efforts have not been rigorously evaluated. Where such evaluations have been undertaken, little attention has been paid to gender impacts, and oftentimes, an exclusive focus on quantitative indicators has led to neglect of more subtle, contextual factors that influence project success. While evaluations conducted by external entities help ensure impartiality, the lack of a monitoring and evaluation culture within implementing organizations is also problematic. The absence of a monitoring and evaluation culture prevents organizations from learning from both their successes and mistakes, and from other actors learning from each others' experiences. Moreover, most evaluations narrowly focus on specific communities or localities, are often fragmented with little exchange of experience between local efforts, and lack coordination and monitoring to form a more integrated effective response. Overcoming these constraints is essential to create a "community of learning" around the design and implementation of gender-aware agricultural interventions.

The evidence presented in this volume underscores the key messages of *The State of Food and Agriculture 2010–11* regarding the vital role of women in agriculture and the need to address gender-based constraints in order to increase agricultural productivity and reduce poverty. But addressing these constraints requires going beyond simplistic statements to understand and engage with the complexities and variability of gendered roles and resources in agriculture, based on better data and evidence. We hope this volume contributes to this important effort.

References

Adepoju A (2004) Trends in international migration in and from Africa. In: Massey DS, Taylor JE (eds) International migration prospects and policies in a global market. Oxford University Press, Oxford

Alsop R, Heinsohn N (2005) Measuring empowerment in practice: structuring analysis and framing indicators. World Bank policy research working paper 3510. World Bank, Washington, DC

Bandiera O, Goldstein M, Rasul I, Burgess R, Gulesci S, Sulaiman M (2010) Intentions to participate in adolescent training programs: evidence from Uganda. J Eur Econ Assoc 8(2–3):548–560

Becker G (1973) A theory of marriage: Part I. J Polit Econ 81(4):813–846

Bertini C (2011) "Girls grow: a vital force in rural economies". A Girls Count report on adolescent girls. The Chicago Council on Global Affairs, Chicago

Boserup E (1970) Woman's role in economic development. St. Martin's Press, New York

Budlender D (2003) The debate about household headship. Soc Dyn 29(2):48–72

Catino J, Colom A, Ruiz MJ (2011) Equipping Mayan girls to improve their lives. Promoting healthy, safe, and productive transitions to adulthood. Population Council, New York

Chiappori PA (1988) Rational household labor supply. Econometrica 56(1):63–89

Chiappori PA (1992) Collective labor supply and welfare. J Polit Econ 100(3):437–467

Chiappori PA (1997) Introducing household production in collective models of labor supply. J Polit Econ 105(1):191–209

Cloud K (1983) Women's productivity in agricultural systems: considerations for project design. Harvard Institute for International Development, AID/WID Training Project, Cambridge, MA

Curran SR, Shafer S, Donato KM, Garip F (2006) Mapping gender and migration in sociological scholarship: is it segregation or integration? Int Migr Rev 40(1):199–223

Deshinkar P (2009) Migration, remittances, gender, and agricultural development. Unpublished manuscript, Overseas Development Institute, London

Dey J (1985) Women in African rice farming systems. In: Women in rice farming: proceedings of a conference on women in rice farming systems. International Rice Research Institute. Gower Publishers, Brookfield

Dixon R (1982) Women in agriculture: counting the labor force in developing countries. Popul Dev Rev 8(3):558–559

Donato KM, Gabaccia D, Holdaway J, Manalansan M IV, Pessar PR (2006) A glass half full? Gender in migration studies. Int Migr Rev 40(1):3–26

Doss CR (2001) Designing agricultural technology for African women farmers: lessons from 25 years of experience. World Dev 29(12):2075–2092

Dwyer D, Bruce J (eds) (1988) A home divided: women and income in the third world. Stanford University Press, Stanford

Elmhirst R (2007) Tigers and gangsters: masculinities and feminised migration in Indonesia. Popul Space Place 13:225–238

Erulkar AS, Muthengi E (2009) Evaluation of Berhane Hewan: a program to delay child marriage in rural Ethiopia. Int Perspect Sex Reprod Health 35(1):6–14

Fafchamps M, Quisumbing AR (2008) Household formation and marriage markets in rural areas. In: Schultz TP, Strauss JA (eds) Handbook of development economics, vol 4. North-Holland, Amsterdam

FAO (Food and Agriculture Organization of the United Nations) (2005) Agricultural censuses and gender: lessons learned in Africa. FAO, Rome

FAO (Food and Agriculture Organization of the United Nations) (2011) The State of Food and Agriculture 2010–11. Women in agriculture: closing the gender gap for development. FAO, Rome

Fapohunda E (1988) The nonpooling household: a challenge to theory. In: Dwyer D, Bruce J (eds) A home divided: women and income in the third world. Stanford University Press, Stanford

Feldstein HS, Poats SV (eds) (1989) Working together: gender analysis in agriculture. Volume I: cases, volume II: teaching notes and supplementary materials. Kumarian Press, West Hartford

Fontana M, Paciello C (2010) Gender dimensions of rural and agricultural employment: differentiated pathways out of poverty: a global perspective. In: Gender dimensions of rural and agricultural employment: differentiated pathways out of poverty, status, trends, and gaps. Food and Agriculture Organization of the United Nations, the International Fund for Agricultural Development, and the International Labour Office, Rome

Gladwin CH (1975) A model of the supply of smoked fish from Cape Coast to Kumasi. In: Plattner S (ed) Formal methods in economic anthropology. American Anthropological Association, Washington, DC, pp 77–127

Gladwin CH (1982) Off-farm work and its effect on Florida farm wives' contribution to the family farm. In: Rojas M (ed) World development and women, vol 2. Virginia Tech Title XII Women in Development Office, Blacksburg

Gladwin CH, Macmillan D (1989) Is a turnaround in Africa possible without helping African women to farm? Econ Dev Cult Chang 37(2):345–369

Guyer J (1980) Household budgets and women's incomes. African studies center working paper no 28. Boston University, Boston

Haddad L, Hoddinott J, Alderman H (eds) (1997) Intrahousehold resource allocation in developing countries: methods, models, and policy. Johns Hopkins University Press for the International Food Policy Research Institute, Baltimore

Hill P (1963) Migrant cocoa farmers of Southern Ghana. Cambridge University Press, Cambridge

Hill P (1978) Food farming and migration from Fante villages. Africa 48(3):220–230

Ibrahim S, Alkire S (2007) Agency and empowerment: a proposal for internationally comparable indicators. OPHI working paper. Oxford Poverty and Health Institute, Oxford

IOM (International Organization for Migration) (2004) International migration law: glossary on migration. IOM, Geneva

Jones C (1983) The mobilization of women's labor for cash crop production: a game theoretic approach. Am J Agric Econ 65(5):1049–1054

Joshi S (2004) Female household-headship in rural Bangladesh: incidence, determinants, and impact on children's schooling. Discussion paper no 894. Economic Growth Center, Yale University, New Haven

Kabeer N (1999) The conditions and consequences of choice: reflections on the measurement of women's empowerment. UNRISD discussion paper no 108. United Nations Research Institute for Social Development, Geneva

Kabeer N (2001) Reflections on the measurement of women's empowerment: theory and practice. In: Discussing women's empowerment: theory and practice. Novum Grafiska AB, Stockholm

Kennedy E (1994) Health and nutrition effects of commercialization of agriculture. In: von Braun J, Kennedy E (eds) Agricultural commercialization, economic development, and nutrition. Johns Hopkins University Press for the International Food Policy Research Institute, Baltimore, pp 79–102

Kevane M, Wydick B (2001) Microenterprise lending to female entrepreneurs: sacrificing growth for poverty alleviation? World Dev 29(7):1225–1236

Kirk A, Kutzman D, de Janvry A, Sadoulet E (2011) From analysis to action: adolescent girls and agricultural technology adoption. Commissioned paper for the Nike Foundation. University of California, Berkeley

Manser M, Brown M (1980) Marriage and household decisionmaking: a bargaining analysis. Int Econ Rev 21(1):31–44

McElroy MB (1990) The empirical content of Nash-bargained household behavior. J Hum Resour 25(4):559–583

McElroy MB, Horney MJ (1981) Nash-bargained household decisions: toward a generalization of the theory of demand. Int Econ Rev 22(2):333–349

Miller C, Razavi S (1995) Gender mainstreaming: a study of efforts by the UNDP, the World Bank, and the ILO to institutionalize gender issues. Occasional paper 4. United Nations Research Institute for Social Development, Geneva

Moser C (1989) Gender planning in the third world: meeting practical and strategic gender needs. World Dev 17(11):1799–1825

Mukhopadhyay C (1984) Testing a decision process model of the sexual division of labor in the family. Hum Organ 43(Fall):227–242

Pahl J (1983) The allocation of money within marriage. Sociol Rev 32(May):237–264

Peterman A, Quisumbing AR, Behrman J, Nkonya E (2011) Understanding the complexities surrounding gender differences in agricultural productivity in Nigeria and Uganda. J Dev Stud 47(10):1482–1509

Piper N (2008) Feminisation of migration and the social dimensions of development: the Asian case. Third World Q 29(7):1287–1303

Poats S, Schmink M, Spring A (eds) (1988) Gender issues in farming systems research and extension. Westview Press, Boulder/London

Potash B (1981) Female farmers, mothers-in-law, and extension agents: development planning and a rural Luo community in Kenya. In: Gallin RS, Spring A (eds) Women creating wealth: transforming economic development. Association for Women in Development, Washington, DC

Rao A, Kelleher D (2005) Is there life after gender mainstreaming? Gend Dev 13(2):57–69

Rosenzweig M (1993) Women, insurance capital, and economic development in rural India. J Hum Resour 28(4):735–758

Schroeder RA (1993) Shady practice: gender and the political ecology of resource stabilization in Gambian garden/orchards. Econ Geogr 69(4):349–365

Smith LC, Ramakrishnan U, Haddad L, Martorell R, Ndiaye A (2003) The importance of women's status for child nutrition in developing countries. Research report 131. International Food Policy Research Institute, Washington, DC

Strauss JA, Thomas D (1995) Human resources: empirical modeling of household and family decisions. In: Srinivasan TN, Behrman J (eds) Handbook of development economics. North Holland, Amsterdam

Tacoli C (2010) Internal mobility, migration, and changing gender relations: case study perspectives from Mali, Nigeria, Tanzania, and Vietnam. In: Chant S (ed) The international handbook of gender and poverty: concepts, research, and policy. Edward Elgar, Cheltenham/Northampton

Tinker I, Bo Bramsen M (eds) (1976) Women and world development. Overseas Development Council, Washington, DC

Tripp R (1982) Farmers and traders: some economic determinants of nutritional status in northern Ghana. Food Nutr Bull 8(1):3–12

von Braun J, Puetz D, Webb P (1989) Irrigation technology and commercialization of rice in the Gambia: effects on income and nutrition. Research report 75. International Food Policy Research Institute, Washington, DC

World Bank (2008) Gender equality as smart economics: World Bank Group gender action plan, first year progress report (January 2007–January 2008). Washington, DC

World Bank, Food and Agriculture Organization, and International Fund for Agricultural Development (2009) Gender and agriculture sourcebook. World Bank/FAO/IFAD, Washington, DC

Zlotnik H (2003) The global dimensions of female migration. Available at http://www.migrationinformation.org/Feature/display.cfm?ID=109

Part II
Data and Methods for Gender Analysis in Agriculture

Chapter 2
Understanding Gender and Culture in Agriculture: The Role of Qualitative and Quantitative Approaches

Julia A. Behrman, Ruth Meinzen-Dick, and Agnes R. Quisumbing

Abstract Because gender relations are complex and context-specific, nuanced, context-specific data collection and analytical methods are recommended. This chapter presents a suite of quantitative and qualitative methods for collecting and analyzing data on gender relations in agriculture. It begins with a detailed overview of how quantitative and qualitative methodologies can be employed to collect gender and assets data for agricultural research. It reviews the use of mixed-methods approaches in research projects to strengthen research findings and to create a more complete and convincing picture of gender relationships. Three case studies illustrate the ways in which qualitative and quantitative data can be used together in analyzing the gender dimensions of agriculture: adoption of maize varieties in Mexico, adoption of maize varieties in Zimbabwe, and agricultural technology dissemination in Bangladesh. In these three examples, using integrated mixed-methods enabled researchers to understand more about the processes underlying the adoption of agricultural technologies. The chapter concludes with a number of important data needs for gender work in quantitative and qualitative agricultural research.

J.A. Behrman (✉)
Department of Sociology, New York University,
295 Lafayette Street, 4th floor, New York, NY 10012, USA
e-mail: jab965@nyu.edu

R. Meinzen-Dick
Environment and Production Technology Division, International Food Policy Research Institute (IFPRI), 2033 K Street, N.W., Washington, DC 20006, USA
e-mail: r.meinzen-dick@cgiar.org

A.R. Quisumbing
Poverty, Health, and Nutrition Division, International Food Policy Research Institute (IFPRI), 2033 K Street, N.W., Washington, DC 20006, USA
e-mail: a.quisumbing@cgiar.org

A.R. Quisumbing et al. (eds.), *Gender in Agriculture: Closing the Knowledge Gap*, DOI 10.1007/978-94-017-8616-4_2,

Keywords Agriculture • Gender relations • Qualitative methods • Quantitative methods • Mixed methods

2.1 Introduction

Gender relations are complex and context-specific. The distribution of rights, resources, and responsibilities among men and women is not constant across cultures and contexts. Because this distribution is a product of social, not biological, factors, characterizing gender gaps, understanding their consequences, and evaluating how they are affected by interventions need to go beyond simple quantitative indicators. To be able to understand how gender relations affect outcomes related to agriculture and food security, and are themselves affected by the social, institutional, and political context of a particular society, agricultural researchers and policy analysts must be cognizant of how gender and agriculture affect the livelihood and income strategies of men and women. Moreover, because the well-being of men and women in rural areas cannot be measured only using indices of agricultural productivity, nor only with money-metric indicators such as income and consumption, there must be greater attention to other dimensions such as status, self-esteem, power within and outside the household, access to institutions—that cannot easily be measured using standard household surveys. But while detailed qualitative studies are useful in exploring these complexities, they often do not provide a picture of how widespread these patterns may be. Quantitative indicators therefore play an important role in presenting data that can be more readily compared across regions, socioeconomic categories, or over time.

Understanding gender relations in agriculture requires bringing together different sources of information and different methods of analysis. Traditionally, information on biological variables related to yields, disease resistance, and growth of plant and animal species are generated by agricultural research centers. Information on labor force participation in agriculture, overall production by crop, cropping patterns, and agricultural incomes are routinely collected by the agricultural statistics system, which, until recently, has not collected information on a sex-disaggregated basis. Economists aiming to study relationships among farmer characteristics, access to and use of inputs, and agricultural productivity typically use quantitative household surveys. Yet, such instruments and methods rarely shed light on the complexity of gender relations since they are not designed to capture context- and culture-specific information. The fact that livelihood activities are so varied, and often intermittent or non-commoditized, means that surveys are likely to pick up some activities and miss others. This is particularly the case where women's activities may be excluded from the formal sector or not considered as "real agricultural work" by local communities. Providing examples from their work in Zambia, Norton et al. (1994, 93) argue that

> Most aspects of rural livelihoods are not captured in either income or expenditure-based survey data. This is because they are neither commoditized nor evident enough to the

researchers to be allocated 'imputed values'...Energy (fuelwood) and herbal medicines are two examples. A significant element of the 'safety net' for rural people in times of stress consists of 'famine foods' that can be gathered from bush and fallow lands (Norton et al. 1994, 93).

In contrast, gender relations have been a long-standing area of inquiry of the social sciences outside of economics, particularly anthropology. Although the internally differentiated household was described, analyzed, and widely accepted in mainstream anthropology from the mid-1970s, it took at least a decade for mainstream development economists to take notice (Jackson 2005). Part of the reluctance of economists to draw from anthropological methods arose from the limited geographical coverage of most ethnographic studies, and therefore, the inability to come up with statistically representative results that were "generalizable" across wider areas than one's own study villages. But the costs of not paying attention to detailed, context-specific research would be likelihood of missing out on precisely the most important factors affecting gender relations. Eventually, studies in the 1980s that suggested that men and women systematically spend income under their control in different ways motivated economists to challenge the traditional model of household behavior and to propose alternative models that bear closer resemblance to reality.[1] These studies have added to the evidence rejecting the traditional paradigm of the unitary model of household behavior in favor of the collective model, which allows for differences of opinion regarding economic decisions among household members.[2] Because testing such models requires sex-disaggregated data on factors affecting bargaining power as well as on outcomes of household decision-making, and because formulating the appropriate model of household bargaining must be based on a better understanding of culture and context, increased efforts have been taken by quantitative social scientists to collect more sex-disaggregated data and to use both qualitative and quantitative methods of analysis. This is consistent

[1] These studies included Guyer (1980), Tripp (1982), Pahl (1983), and studies from different countries (for example, Fapohunda 1988) in the volume edited by Dwyer and Bruce (1988). A series of studies on agricultural commercialization and nutrition in developing countries also found that income controlled by women is more likely to be spent on food than is income controlled by men (Kennedy 1994). Evidence that men and women may have different production priorities can be found in the work of Dey (1985) and von Braun and Webb (1989) on irrigated rice in The Gambia, Jones's (1983) formal model of intrahousehold conflict and husband's and wife's gains from cooperation based on work in the Cameroon, Mukhopadhyay's (1984) decision model of the sexual division of labor for household tasks, Hill's (1963, 1978) description of Fante women's entrepreneurial behavior, and Gladwin's (1975, 1982) models of women's marketing and farming decisions. Implications of these differences for project design are discussed by Cloud (1983). These and other studies from Sub-Saharan Africa are reviewed in Gladwin and Macmillan (1989). By the mid-1990s, economists appeared to be developing a considerable level of interest in the issue of the unitary household, to which the research by Chris Udry (1996) and others in Burkina Faso was particularly important and continues to be heavily cited (Jackson 2005). IFPRI played an important role in bringing together current research on gender and intrahousehold issues—and in catalyzing future research—through a 1992 conference, the proceedings of which produced a publication comprising 30 policy briefs and, eventually, a book (Haddad et al. 1997).

[2] See Strauss and Thomas (1995), Haddad et al. (1997), and Behrman (1997) for reviews.

with the more general movement toward mixed methods (qualitative-quantitative or Q-squared) research in the social sciences, for example, in poverty appraisals (see Kanbur 2003 and Kanbur and Shaffer 2007) and in evaluation of social programs (Adato 2008; Maluccio et al. 2010).

By using data from a variety of sources and qualitative and quantitative methods, it is possible to cover a wide range of issues and topics relatively efficiently. Rather than seeing this as a second-best solution, such a combined approach can actually provide a more convincing analysis than any single method (Brewer and Hunter 1989; Creswell 1998; Tashakkori and Teddlie 1998, cited in Adato and Meinzen-Dick 2007). Adato and Meinzen-Dick (2002) argue that people respond differently to quantitative and qualitative information; numbers are required to convince some audiences, while others will be unimpressed by numbers, but relate more to in-depth and contextual information gathered using qualitative techniques. Triangulation, where several types of data are used in a single study and used to cross-check and compare results, enables any weaknesses in one method to be offset by the strengths of another (Denzin 1978; Jick 1979). An assessment of 57 mixed method studies identified five purposes for mixing methods (Greene et al. 1989, cited in Adato and Meinzen-Dick 2007): (1) triangulation—seeking convergence of results; (2) complementarities—examining overlapping and different facets of a phenomenon; (3) initiation—discovering paradoxes, contradictions, fresh perspectives; (4) development—using the methods sequentially, such that results from the first method inform the use of the second method; and (5) expansion—adding breadth and scope to a project.

Because the scope for the use of integrated qualitative and quantitative approaches is quite broad, in this chapter we describe each type of approach, and then illustrate the use of mixed methods or Q-squared approaches in three case studies.

2.2 Using Qualitative and Quantitative Methods: An Overview

One of the traditional divisions between economists, sociologists and anthropologists has been the approach to data collection and analysis. While quantitative analysts (including, but not exclusively restricted to, economists) have traditionally viewed "methods" as equivalent to the analysis of quantitative data, often not questioning the assumptions underlying the collection of that data, sociologists and anthropologists emphasize the importance of methods in the data collection process. Thorbecke (2003) opines that qualitative researchers typically have a direct input in the formulation of questionnaires or protocols used to generate the information required to address the questions they are exploring. These same researchers are often directly involved in the information gathering process in the field so they can modify and clarify questionnaires after pretesting. Within the anthropological and sociological literature, there has been extensive debate about the role, identity and biases of the researcher/interviewer, and the ways in which these factors and the research process itself ultimately influence respondent responses and the broader research product. This is particularly applicable to gender studies, not only because male and female researchers will have different access to respondents and elicit different responses, but also because gender experiences are highly personal, shaping researchers' own experiences. Furthermore, analysis is often part of the qualitative data collection process itself, guiding observations of behavior or follow-up questions in semi-structured interviews. In contrast, there is often a greater separation between data collection and analysis in quantitative research, both in terms of time and personnel. Another way of putting this is to say that qualitative researchers are more inductive than deductive: the "hands on" iterative interviewing technique used by qualitative researchers generates hypotheses that can be formally and quantitatively tested by the more deductive quantitative methodology that relies on econometric and statistical tools (Thorbecke 2003, 164).

Understanding gender issues in agriculture requires drawing on the strengths and complementarities of both methods. Here, we present a summary of methods currently in use, drawing in part on the growing literature on collecting sex-disaggregated data in household surveys (Doss et al. 2008; Doss, Chap. 3), qualitative methods (Hentschel 1999; Moser 2001; Chung 2000), and mixed-methods approaches (Kanbur 2003; Kanbur and Shaffer 2007; Adato 2008).

Qualitative data refer to a broad range of textual or visual information derived from interviews, observations, documents, or records. These data are often associated with methods that require "intensive, often repeated encounters with small numbers of people in their natural environment" (Chung 2000, 337), often called "contextual" methods (Hentschel 1999; Moser 2001) because they attempt to understand human behavior within the social, cultural, economic, and political environment of a locality (Hentschel 1999, 71). Quantitative data may derive from official statistics (such as land records or censuses) and surveys. Survey-based methods involve structured interviews of a representative household sample to obtain information on a range of questions, and preformulated, closed-ended, and codifiable questions are usually asked to one household member (often the head) during one or two visits, although modifications of the standard interview approach can be used to get more detailed information on gender relations.[3]

2.2.1 Quantitative Methods

Household and individual-level data are typically collected using quantitative household surveys with a standardized questionnaire, typically with fixed coded responses, although a few do allow open-ended responses to be coded later. Data for

[3] For a more comprehensive discussion of types of data and methods of data collection, see Hentschel (1999) and Moser (2001). A thorough discussion of qualitative methods in the context of the LSMS is found in Chung (2000).

quantitative analyses may include panel data for the same households over a number of years, which allow for analysis of changes over time. Some of the surveys collect data at the level of the individual household member, which allows for comparison between men and women, and also helps to capture the full range of livelihood strategies within the household. Sampling to cover the range of wealth/poverty categories is critical for these types of surveys. Although some qualitative data are included in the surveys, quantitative researchers analyze most survey data—including qualitative responses—using statistical techniques. Of importance is the issue of who is interviewed within the household—is it the household head? Spouse? The interview subject has implications for the kind of information collected as different members of the household have different stores of knowledge and different perceptions and will therefore report different things.

Ideally, household surveys should collect sex-disaggregated data on a number of topics, including household members, education, asset ownership, agricultural production, income, and other topics of relevance to the project in question. In particular, looking at gender patterns of control over and ownership of assets is important because asset ownership and control is a critical component to well-being.[4] Increasing control/ownership of assets helps create pathways out of poverty more than measures that aim to increase incomes or consumption alone. A body of research indicates that households do not pool resources nor share the same preferences (Alderman et al. 1995; Haddad et al. 1997) and evidence from many countries suggests that increasing resources controlled by women improves child health and nutrition, agricultural productivity, and income growth (Quisumbing 2003; Smith et al. 2003). However, in many contexts, a gender-based asset gap continues to exist. Surveys need to look at both men's assets and women's assets because looking at only one or the other may provide a false impression of what is going on. For example, an agricultural intervention that substantially increases women's control of a particular asset may be deemed a success; however, if researchers do not also look at what is happening to men's assets at the same time (i.e., are they increasing at a faster rate than women's, staying the same, decreasing), false assumptions about project impact may be made.

Analytical methods used to examine gender issues in agriculture will depend on the topic of interest. Often, researchers seeking to analyze gender dimensions of a particular topic spend more time in questionnaire development and pretesting than is typical of most economic analyses. In the case of measuring the gender gap in assets in large-scale household surveys, for example, modules to obtain sex-disaggregated asset data have been developed (see Doss et al. 2008) and have been implemented in three countries (Ecuador, Ghana, and India). Studies that examine how men's and women's assets grow through time and respond to positive and negative events have relied on baseline sex-disaggregated data and resurveys that collect detailed information on shocks and re-administer the sex-disaggregated assets module (Dillon and Quiñones 2010; Quisumbing 2011; Quisumbing et al. 2010). And studies that attempt to evaluate the impact of agricultural development programs on gender-disaggregated

[4] We accept a wide ranging conception of assets that spans both tangible and intangible assets, including physical assets, financial assets, social capital, human capital, and so on.

outcomes typically put a lot of thought into the sampling design, so that an appropriate counterfactual or comparison group can be established in order to evaluate the impact of the intervention (Hallman et al. 2007; Kumar and Quisumbing 2010a, b; de Brauw et al. 2010).[5]

In addition to primary quantitative data from surveys, it is useful to consider secondary data sources from government and other researchers' studies. Secondary data can be used to provide the basis for sampling frames, cross-check the information from more localized primary data collection with other regions or nationally representative samples, and provide direct information for a study. Nationally representative secondary data, especially official statistics, also provide the basis for cross-national comparative studies. The Organisation for Economic Co-operation and Development (OECD) provides a website with a range of sex-disaggregated data, including a series of gender inequality indices.

Unfortunately, sex-disaggregated data related to agriculture are often not collected in national statistical systems, as seen in the many gaps in the statistical annex of sex-disaggregated data on agriculture in the 2011 FAO *State of Food and Agriculture* (FAO 2011). But the very fact that the available sex-disaggregated secondary data are collected in this official publication is an important step forward not only as a resource for research, but also to create pressure on countries to collect such data. Even what is collected may not be very reliable. Basic data on the gender distribution of labor force participation in agriculture, for example, may not be reliable because women are only seen as "helping" their husbands (see Doss, Chap. 4). Recent efforts to collect sex-disaggregated information in the agricultural censuses in Africa by collecting data on enterprises within farms could help to redress this problem (e.g., FAO 2005). Work by Doss et al. (2008) to collect sex-disaggregated assets data in the context of the World Bank's Living Standards Measurement Studies could similarly contribute to future comparative analyses of gender and assets.

Secondary data from government records are also important because they provide more comprehensive coverage than most qualitative studies or even special-purpose studies. Where such data can be linked to administrative units or geo-referenced, it can be put into geographic information systems, which are increasingly being used for planning development interventions and even for prioritizing agricultural research. Unfortunately, because of the lack of sex-disaggregated secondary data on agriculture, gender is not included in such spatial prioritization exercises. Building up reliable secondary databases can therefore help "put gender on the map."

Fortunately, sex disaggregation has become increasingly integrated into national and subnational surveys on agriculture and socioeconomics in the past decade (Wielgosz, personal communication). These large pools of gendered data have not been fully leveraged or distributed for public-access. Geographic Information Systems and geo-referencing of gender statistics from subnational surveys offer an

[5] While randomized placement of the intervention has been viewed as the gold standard for impact evaluation, other approaches may be more feasible, depending on context. This includes matching methods, regression discontinuity designs, and instrumental variables approaches.

avenue for generating maps of gender systems across districts, countries, and regions. Similar approaches to aggregating subnational survey data are being used by the Harvest Choice project to map $2.00 and $1.25 poverty headcounts as well as average farm size. The Spatial Crop Allocation Model is using similar subnational statistics to estimate the distribution of individual crops across districts. Many national and subnational survey datasets now routinely include GPS geo-referencing, which allows for an even finer-grain resolution when mapping such spatial patterns. Livelihood data such as the WOCAT.org SLM Knowledgebase and FAO's Dixon-Farming System data can also offer examples that could be followed for creating useful gender mappings. Once such data are geographically referenced, researchers and policymakers can easily link this information to other datasets that may not have been able or designed to include gender information. Although generating such databases requires dedication and time, the value for integrating gender information into policy choices and program development is significant and justifies the allocation of human resources required. With appropriate funding, a spatial database of gender information could be created to cover Africa, Latin America, and Asia in a period of 1–2 years. Such a database could provide district-level mapping of variables such as the percentage of male- and female-headed households, female and male literacy rates, average maternal age at first birth, maternal mortality rates, and other statistics useful for characterizing gender-related patterns.

2.2.2 *Qualitative Methods*[6]

Many outcomes of interest are not amenable to measurement using standard quantitative survey techniques, particularly when one is interested in processes in addition to outcomes. Understanding gender relations encompasses additional aspects of well-being, status, self-esteem, empowerment (or disempowerment), vulnerability, issues of social differentiation, social norms, and, most important, self-perceptions by individuals and communities of what it means to be "male" or "female" in a given society. Survey-based data collection techniques will frequently prove to be inadequate in capturing the many dimensions of these issues. Risks that are faced by men and women, for example, may be culture-specific and difficult to get at using standard survey questionnaires without prior qualitative work, such as collecting life histories or focus group interviews about major risks. When collecting asset data, there are often important gender differences in the spectrum of asset ownership that may not be accurately captured in household surveys with predetermined answers. For example, what it means to "use" or "control" a given asset may be entirely different from what it means to "own" said asset and differences in categories of asset ownership may fall along gender lines in important dimensions (see Meinzen-Dick et al., Chap. 5). There may be additional qualitative differences in the kinds or

[6] This section draws from some of the background work reported in Hoddinott and Quisumbing (2003) as well as Adato and Meinzen-Dick (2002).

types of assets that male and females own that only emerge from in-depth discussions with the respondents themselves. Qualitative research also allows respondents to express their own opinions freely, thus allowing researchers to better understand why men and women may prefer to collect different types of assets in the first place. Ethnographic methods, such as participant observation, can provide key insights into gender roles in agriculture (and nonagricultural) activities, and prolonged residence in villages may reveal aspects of intrahousehold negotiations, hiding of assets, or sensitive topics that respondents may not reveal in surveys.

Qualitative methods are often conflated with participatory research, but although qualitative and participatory research share many techniques, they are conceptually distinct. The defining characteristic of participatory research is that the respondents are active participants in setting the research agenda and conducting the analysis (Chambers 2004). One benefit of participatory methodologies is that they allow for analysis of the issue at hand by a range of actors, including the respondents themselves, the field staff, and the lead researchers. Thus participatory methodologies can be an interactive learning process for all those involved in the research process, and are often undertaken not only to acquire information, but with an empowerment objective. Participatory research has been championed, in particular, by Robert Chambers and colleagues at the Institute for Development Studies, Sussex, (Chambers 1992) (http://www.ids.ac.uk/). "Participatory rural appraisal" (PRA) often includes a package of methods—including focus group discussions, ranking exercises, venn diagrams, community meetings, transect walks, seasonality calendars, and identification and ranking of livelihood activities and sources of vulnerability—that are all tools used to incite discussion on topics of interest. But many of these methods can be used for conventional research as well, and participatory methods can also be quantitative, for example participatory ranking exercises. The Consultative Group on International Agricultural Research (CGIAR) Participatory Research and Gender Analysis (PRGA) program applied participatory approaches to gender analysis and especially to agricultural research, developing methods for involving farmers in plant breeding and developing new approaches to natural resource management. While not all participatory research is gender-sensitive, there are certainly advantages in involving women in identifying the constraints that they face and exploring ways to overcome these constraints. Lilja and Ashby (1999) provide an overview of how integrating gender analysis, as a form of stakeholder analysis, into participatory agricultural research for plant breeding or natural resource management can lead to better identification of gender-based constraints and adoption of agricultural technologies by women as well as men.

Along with a growing number of researchers on poverty (e.g., Kanbur 2003), we believe that quantitative and qualitative methods should not be considered substitutes for each other, but can be used effectively together to study gender issues in agriculture. Chung (2000) points to three ways in which qualitative methods can be used to improve household surveys. First, qualitative methods can be used to produce hypotheses and to shape a survey's conceptual framework. Second, qualitative methods can be used to clarify the questions and terms that are used in a survey.

Third, qualitative methods can be used to explain inconclusive survey findings or investigate reasons behind unexpected results.

Despite common perceptions that qualitative research is "quick and dirty," in fact, good qualitative data collection is both time-consuming and expensive, because of the need for skilled people to go to the field, and the fact that the time spent in different data collection activities is likely to be far greater than standard household surveys. As a result, qualitative data are almost always collected for a smaller sample size than most surveys. This raises important questions on how representative the data are (Dercon 2001). Particular care therefore needs to be given to the sample selection: even if a site and respondents are purposively selected, the criteria and rationale need to be clear. The same applies for the composition of focus groups: interviewing whoever shows up at a meeting may lead to respondent bias, and the presence of certain people (e.g., high-ranking men or women, or government agents) may repress the voices of others. The reliability of data collected will also be subject to the limits of respondents' memory and knowledge. Rather than assuming away any biases or trying to control for them statistically, qualitative researchers should be aware of the likely biases of respondents, and even make that part of the analysis (Meinzen-Dick et al. 2004).

Combining survey-based and qualitative approaches would enable us to tackle the issue of representativeness head on, while benefiting from the richness of contextual data. Moser (2001), for example, demonstrates that it is possible for participatory research to be quantified and representative; this approach has been taken in an integrated qualitative and quantitative study of chronic poverty in Bangladesh (Baulch and Davis 2008). This involves careful choice of communities and efforts in post-coding of answers in patterns. For example, the sampling frame used in the household survey could be used to generate the subsamples for further study using qualitative methods.

A wide ranging variety of different tools and methods are employed in qualitative research. The following is a brief overview of a number of qualitative methods that may be particularly amenable to research on gender in agriculture.

Focus groups: The most widely-cited type of qualitative methodology is the focus group discussion, in which a trained facilitator leads a discussion among a small group of roughly 5–15 respondents. The facilitator often uses a focus group guide to ensure that topics of interest are covered, but the flow of the discussion will vary based on the participants and their interaction with the facilitator. This method allows groups to elicit collective experience and opinions, while also permitting differing views, experiences, or perceptions of group members to be expressed, discussed, and understood in a group context. Often when doing analysis with a gender dimension, it is useful to hold separate groups with men and women so that both men and women are able to share their opinions freely without external pressure. Additional separation of focus group members by wealth/poverty categories, age, marital status, and/or a variety of other factors of importance allows the researcher to compare how perceptions and experiences differ between categories of groups. Preexisting survey data helps in the disaggregation of wealth groupings for the focus groups, particularly in communities where a wealth ranking exercise may be divisive or difficult to carry out (e.g., because of large community size or time limitations that prevent researchers from getting sufficiently acquainted with a community to comfortably carry out such an exercise). Where possible, households that are selected for the surveys should be included in the focus groups to improve the comparability of the information obtained through different methods.

Focus groups have the advantage of including larger numbers of participants than do other types of qualitative research, and they generate a synergy of ideas when people speak collectively. In addition, they allow for the identification and discussion of unanticipated areas of interest for research based on the information provided by participants. Focus groups can be used following a series of household case studies to further investigate issues raised (including the experiences of households not included in these studies), check whether the findings resonate or contradict, and receive feedback on the research findings. In other studies, focus groups may be the primary means of qualitative data collection, but are followed up with in-depth interviews or case studies of individuals who participated in those groups. However, focus groups should be focused on a few issues. If all issues of interest in the study are included, the interviews will go on too long and energy will wane, or responses will become perfunctory. A disadvantage of focus group discussions is that the views of more dominant participants may be given more weight than they should, and minority or even majority opinions from more timid or less powerful participants may not be heard. Certain issues that are controversial may not be raised at all. These problems can be reduced with good facilitators and careful disaggregation of group participants.

Mapping exercises: A number of mapping tools—including local resource maps, participatory impact diagrams, before and after diagrams, and social network maps—are tools used to gain information from community members or groups about social change, program impact, information diffusion, and other topics of interest. The advantage of these mapping exercises is that they can be used with populations with low literacy levels and can provide large amounts of information in a visual and engaging fashion. Comparing the maps developed by men and women can provide a striking illustration of their different values and experiences, such as what natural or social resources they identify and value, or how they perceive the outcome of programs. The disadvantages of these methods are that they can be time consuming and require attentive supervision.

Interviews: Interviews are another qualitative tool that is commonly used to gain additional in-depth information, a range of insights of a particular topic, and/or general information relevant to specific issues. Interviews may be unstructured and free-flowing or semi-structured, meaning the researcher starts off with guided questions of queries and follows up on relevant topics that emerge during the course of the discussion.

Key informant interviews allow the research team to follow up in more detail with individuals that have specialized knowledge about the subject or topic of interest. Key informants may be identified in advance or over the course of fieldwork, based on information needs of the project or suggestions from other key informants or knowledgeable personnel. Information gleaned from key informants is especially important to address the policies, institutions, and processes affecting the research.

Life history interviews are often conducted with individuals or households, often to find out how particular events or programs have made an impact on them. An important aspect of life histories is that they provide a more complete picture of how agriculture and other livelihood strategies, major shocks, and development interventions interact with each other and with intrahousehold relations. For example, Doss et al. (2011) use life histories to reconstruct patterns of asset accumulation and loss in Uganda, uncovering critical factors shaping women's inheritance of land, and the effect of inherited assets on their livelihood trajectories. In Bangladesh, Baulch and Davis (2008) found that dowry and illness expenses—and particularly the combination of dowry and illness expenses within the same household—were associated with downward trajectories in individual and household well-being.

As with the participatory methods described above, analysis can be an interactive process that is shared between respondents and lead researchers. The advantage of interviews is that they often provide additional analytic power in addition to contextual information and important detail. Furthermore, one-on-one interviews do not run the risk of peer pressure. However, such detailed interviews also tend to be more time consuming and costly.

Ethnographic tools: Ethnographic tools—including participant observation and direct observation,[7] case studies, and diaries/journals—can provide more detail on the complexity of gender relations as they play out in people's own lives. Researchers conducting case studies or ethnographies often live in sample villages for extended time periods, spending time in the homes of a subsample of the survey households, conducting informal interviews, observing and participating in their daily activities, such as farming, extension field days, and social interactions and activities. Long exposure of this type in communities increases trust between researchers and respondents and increases the chances of receiving candid responses and cross-checking responses with observations. Interviews can be conducted—often privately—with household members of different ages, sexes, and roles. Participant observation provides insights that are not available from other methods and informs and refines the questions asked in other, more structured, data collection. The main disadvantage to detailed ethnographic methods is that the number of communities and respondents will often be smaller than in group-based methods, depending on the research budget and the number of qualified researchers that can be found to live in the village. However, the depth of insight gained from this type of research may compensate for other shortcomings.

[7] The difference between participant and direct observation lies in whether the researcher participates in activities along with the household or community. Watching people harvest a crop is direct observation; helping out with the harvest is participant observation.

2.2.2.1 Qualitative Analysis

Qualitative analysis takes a variety of forms, depending on the method employed and desired research output. Implicitly or explicitly, qualitative analysis usually begins in the field. Most data collection methods require the researcher to sift through what is said or observed, identify the key points related to the subject of the study, and follow up accordingly. Writing field notes and reflecting on them as soon as possible (such as by having field teams discuss their observations and emerging findings each night) also helps to ensure that important contextual information is not lost, and increases the possibility of going back to verify or go deeper on certain issues. But the analysis is not all done in the field. Focus group discussions and key informant interviews are often audio recorded and then translated (if necessary) and transcribed so that a final transcription of the entire interchange is produced. In ethnographies or case studies, researchers may write up extensive field notes based on observation, experience, and interviews. Once a written record is produced, qualitative information can be coded according to issues that emerge throughout the fieldwork, as well as issues of interest identified in advance by the research team. The advantage of this approach is that it allows for the issues and themes identified by respondents—rather than researchers—to drive the research agenda rather than the other way around. Qualitative software—such as Nvivo and Atlas TI—are increasingly used to aid in the coding process and allow for disaggregation of respondents by key characteristics or categories. Coded material can then be analyzed by the researcher to identify dominant themes, norms, and trends. Increasingly, non-audio qualitative material—such as photographs, participatory maps, or social network maps—are also being coded and analyzed in a similar fashion.

All of this points to the need for highly skilled qualitative researchers to both collect and analyze the data. This is one of the constraints to sound gender analysis in agriculture: while most national agricultural research systems (NARS) institutes have agricultural economics departments, they are less likely to have anthropology departments or others that teach ethnographic or other qualitative methods. In recent years, while many nongovernmental organizations (NGOs) and some research institutes have been using PRA techniques, there is more need for training on the full range of qualitative methods, and how to integrate them with quantitative approaches. Involving more students in agricultural fieldwork studies could have many advantages: reducing fieldwork costs, training more people in these techniques, and creating a greater understanding of both agricultural and gender realities in the field. If NARS do not have adequate social science departments, collaboration with other universities, both in developing and OECD countries, can provide this bridge.

2.3 Integrating Methods, Data, and Disciplines

Sequenced and integrated qualitative and quantitative analysis can be done in a number of ways, although linking the different sources of data requires explicit attention. Qualitative methods can be complementary to quantitative methods in a number of ways. For example, qualitative formative work can allow for the identification of issues and questions for surveys and hypotheses for testing. Open-ended questions from qualitative work can be used to identify types of indicators or impacts that may feed into survey questions. In a similar manner, qualitative work can be used to identify response options for survey questions, clarify terms/language for use in surveys, and confirm the validity of constructs and proxies. Qualitative work can also provide depth, texture, and context to quantitative findings by providing explanation/interpretation of survey findings. Qualitative approaches will also allow researchers to get at issues of "why" as well as "what" and permit the exploration of topics less amenable to survey questions, including the expression of local voices and interpretations.

Likewise, quantitative methods can also be used to complement qualitative ones in a variety of dimensions. Quantitative data on community and household characteristics can provide a sampling frame and stratification strategy. In addition, quantitative data can allow for identification of issues for qualitative investigation and determine prevalence of qualitative findings in the wider population.

Using both quantitative and qualitative approaches can confirm the validity of data findings or point out contradictions in findings that warrant further exploration. Of importance, a mixed-methods approach can also help establish counterfactuals in impact evaluations. It is well known that in order to fully assess the impact of an intervention, it is important to establish a "counterfactual," i.e., what would have happened in the absence of the interventions. Using qualitative approaches in addition to quantitative methods allows the researcher to take advantage of the strength of in-depth interviewing to probe causality and establish plausible linkages between outcomes, including people's perceptions and experiences of changes. Ultimately, combinations of "before/after" and "with/without," as well as insiders' and outsiders' perspectives, provide the most convincing case of what changes can be attributed to the intervention being evaluated. However, it should be noted that there is no one-fix-all recipe for integrating qualitative and quantitative approaches. In each instance of data collection, quantitative and qualitative data will take a somewhat different relationship to each other and provide different types of interpretative power.

Three case studies illustrate the ways in which qualitative and quantitative data can be used together in analyzing the gender dimensions of agriculture.

2.3.1 Case Study 1: Maize Varieties in Mexico

While improved maize varieties have been available in Mexico for more than 40 years, their diffusion has been limited. Many small-scale, subsistence-oriented farmers have taken up improved varieties, planted them alongside local varieties, and, whether by accident or by design, promoted their hybridization with landraces, producing what they called "creolized" varieties (Bellon et al. 2007). Creolization allows farmers to adapt improved local varieties to local conditions, and to modify improved technologies generated by the formal research system to suit local circumstances and needs. To study the process of adoption of improved germplasm, social and biological scientists from the International Maize and Wheat Improvement Center (CIMMYT) and the International Food Policy Research Institute (IFPRI) used a mixed methods study that involved qualitative and quantitative methods. The qualitative research began with two sets of focus group discussions, followed by household case studies conducted in 4 of the 12 study communities. The quantitative research then involved a representative sample survey of households in the 12 communities. Finally, the project included a collection of all maize types grown in the communities and an agronomic evaluation of maize samples.

Qualitative work explored reasons for men's and women's preferences for different maize varieties in Mexico, and the main risk factors that they faced. The importance of these risk (or vulnerability) factors and perceived advantages of maize characteristics identified in this qualitative work was tested quantitatively through a survey of a wider sample of farmers. Household studies deepened the understanding of how different maize characteristics responded to this vulnerability context, as well as issues such as people's perceptions and trust of the pathways through which seeds enter communities (whether by government channels or informal social networks) and how this influences people's choices. The study found that consumption characteristics—such as yield and quality of dough to make tortillas and quality of *atole* (a traditional beverage made of maize)—were more relevant for female rather than male farmers, because women are in charge of maize processing and preparation. Traits related to vulnerability—tolerance to drought, resistance to rot, and resistance to pests—were also more important for women than men. While both male and female farmers recognized that both types of maize have their advantages and disadvantages, women in Chiapas had a more positive perception of creolized varieties than of hybrids.

A later study by one of the authors (Bellón and Hellin 2011), also using mixed methods and data (secondary data, surveys, and qualitative data), found that commercialization and hybrid adoption have been promoted by government programs. However, cultural preferences, and possibly an antipoverty program coupled with women's empowerment, have fostered landrace retention. A nationwide conditional cash transfer program, *Oportunidades*, may have contributed to the retention of landraces by providing women additional income and empowering them to "purchase" landraces as a valuable consumption product through self-production. Similar transfers from an agricultural program called PROCAMPO—which are controlled by male farmers—had no effect on landrace retention.

2.3.2 Case Study 2: Hybrid vs. Open-Pollinated Maize Varieties in Zimbabwe

A study that examined the diffusion of hybrid maize varieties in Zimbabwe took advantage of the existence of a unique, longitudinal survey, covering three resettlement schemes in three different agroecological zones (Bourdillon et al. 2007). The initial survey was conducted during 1983–1984, and the sample households were reinterviewed in 1987 and annually between 1992 and 2000. Although these surveys were rich in quantitative data, and contained information on the adoption of hybrid maize, there were substantial information gaps. To redress this, the researchers, in consultation with stakeholders, developed a research design using qualitative approaches. The core method was a series of household-level case studies, supplemented by participant observation in villages found in two resettlement areas. The case study work was followed by focus group discussion in the selected villages, together with some focus group discussions in the third resettlement area to confirm findings of the individual case studies, reconcile divergent findings, and allow a wider range of voices to be heard.

The study found that even technologies that are designed to be adapted to less favored areas and poor farmers may not do so without farmers' access to certain assets. Although the HYV maize was designed to do well in drought-prone areas, farmers in better agroecological zones adopted faster than those in middle and poor zones. Results indicate that this is because in these zones, farmers had more agricultural capital stock and livestock to protect them from risk, and had better marketing channels. The ethnographic work revealed many gender dimensions of control over assets. By spending time in the communities and households, researchers were able to observe how household members hid assets from each other and learn how fear of witchcraft accusations restricted sharing of information from one farm to another. Assets and people's access to them also have many gender dimensions, affecting the value of technologies for men and women. In Zimbabwe, men have been found to prefer the improved varieties, while women seek out the open-pollinated varieties. This is because women have less access to the credit and cash required for certified seed and fertilizer. Moreover, women use their social networks to acquire open-pollinated seeds and find these do well with cow manure. Women also have less access to formal maize markets where improved maize is sold, an example of how policies and institutions influence technology adoption choices and consequences among different social categories of farmers.

2.3.3 Case Study 3: The Long-Term Impact of Agricultural Technology Dissemination in Bangladesh

This case study illustrates the value of longitudinal mixed methods approaches to understanding the gendered impact of agricultural technology dissemination in Bangladesh. In 1996–1997, an evaluation of the impact of improved vegetable and fish technologies used an iterative process of survey data collection and qualitative data collection on intrahousehold dynamics and women's empowerment followed by another round that collected individual-level indicators of empowerment (Bouis et al. 1998). Four surveys of 955 households were conducted at 4-month intervals beginning in June 1996, and covered one complete agricultural cycle. Survey data were supplemented with qualitative research on factors affecting intrahousehold bargaining power, which fed into formulation of questions in the last survey round on dowries, assets brought to marriage, and bargaining power (see Box 2.1).

Box 2.1 Qualitative and Quantitative Methods to Examine Gender Impacts of New Agricultural Technologies in Bangladesh

A long-term study of the impact of agricultural technologies (fishponds and improved vegetable gardens) in Bangladesh has used an iterative approach between qualitative and quantitative data collection and analysis, building on datasets collected for previous studies (see Kumar and Quisumbing 2010a, b).

The original study (Bouis et al. 1998) in 1996/1997 began with a census of program villages and comparable nonprogram villages. After the census, a more detailed intrahousehold survey was conducted over four rounds, interviewing men and women separately, in a sample of program participants and nonparticipants. Villages for the qualitative research were selected from among the villages covered by the quantitative survey, with the village census enabling researchers to combine qualitative and quantitative data for analysis and validation of the results. Focus group discussions were conducted for (mostly female) program participants as well as their spouses between the third and fourth rounds to obtain different perceptions by gender. Results of the qualitative study informed the fourth and last round of survey data, by providing indicators of women's empowerment, that could then be used to develop survey modules. Further quantitative analysis raised questions that were addressed in future rounds of qualitative work, such as through focus group discussions and life histories (Hallman et al. 2007).

Table 2.1 below illustrates the range of qualitative techniques employed in the 1996/1997 study, and their respective respondents.

Focus group and key informant interviews with project staff, conducted in 2001, provided additional qualitative data on vulnerability, social relations, disseminating institutions, how the new technologies fit into household livelihoods, and whether they affected men and women differently (Hallman et al. 2007).

Hallman et al.'s (2007) analysis of the impact of the technologies on poverty, vulnerability, and social relations found the strongest poverty impact in the case of the vegetable technology, which is targeted toward women in households with relatively small amounts of land. It is essentially a "nonlumpy" technology that requires a very low level of investment but which has disproportionately significant returns to the very poor and signs of positive impact on female empowerment and child nutritional status. The noneconomic benefits of this technology in terms of network building and reciprocity among women, and intrahousehold empowerment of women also were apparent. The private fishpond had positive effects on the pond and crop profits of adopting households. However, technology had less impact on poverty and empowerment, since better-off households tended to own ponds, and the technology was adopted more by men than by women. Finally, the operation of the women's group fishpond technology, although a potentially beneficial agricultural program for poor households, was significantly undermined by collective action problems.

Table 2.1 Use of qualitative techniques for collecting different types of information

Type of information collected	Technique	Who participated
Village profile	Transect	Team members
	Social map	Men from all socioeconomic categories
	Resource map	Men from all socioeconomic categories
	Crop calendar	Men from all socioeconomic categories
	Event calendar	Men from all socioeconomic categories
	Venn diagram	Program participants (mostly female)
	Mobility maps	Program participants
	Case study	Program participants and their spouses
	Focus group discussion	Spouses of program participants
Program profile	Key informant interviews	Managerial staff of implementing agency
	Observation of group meeting	Field level staff of implementing agency
	Focus group discussion	Field level staff of implementing agency
Income, control over income, savings and investment	Case study	Program participants and their spouses
	Focus group discussion	Program participants and their spouses
	Trend analysis	Program participants and their spouses
	Impact flow chart	Program participants and their spouses
Distribution of benefits	Case study	Program participants
	Trend analysis	Program participants and their spouses
	Impact flow chart	Program participants and their spouses
	Mobility map	Program participants and their spouses
	Focus group discussion	Program participants and their spouses Spouses of program participants

Relative to women who did not have access to this group-based program, however, female fishpond group members appeared to have more mobility, greater likelihood of working for pay, higher off-farm incomes, and better nutritional status.

Ten years later, a follow-up study in 2006–2007 to analyze the determinants and consequences of chronic poverty provided the opportunity to revisit the study sites and evaluate the long-term impacts of the technologies. Using difference-in-difference analysis and a statistical comparison group of early adopters and late adopters of the technology, Kumar and Quisumbing (2010a, b) found that the private fishpond program that had targeted information regarding the technology to the households and by default, husbands, increased husbands' holdings (relative to their wives') of land, livestock, and total value of assets, whereas in programs that targeted technologies to women's groups, women's assets increased faster than their husbands', even though husbands still owned the majority of household assets. Moreover, while the individual fishpond program had the largest gains in terms of consumption expenditures and household assets, improvements in nutritional status of women and children were less than those in the programs targeted to women's groups. Kumar and Quisumbing (2010b) conclude that women's assets increase more relative to men's when technologies are disseminated through women's groups, suggesting that implementation modalities are important in determining the gendered impact of new technologies.

Collaborating for successful mixed methods work. One of the most challenging aspects of such multi-method research involves assembling a research team with the proper mix of skills. Working with interdisciplinary groups from the international to the local level provides a valuable learning process in mixed method research and in integrating economics, sociology, and anthropology, and can provide a model for strengthening the capacity of agricultural research institutions to address poverty in the future. Often one of the greatest obstacles to successful mixed methods work is reluctance on the part of both quantitative and qualitative researchers to understand or acknowledge the methodological validity of research approaches outside of their disciplines.

The challenges of developing such mixed teams lie in ensuring that members can communicate across disciplinary lines, respect each other's contributions, and find the time to integrate the findings or insights from the other members into their own work. Using a common framework is one way to increase coherence and understanding across disciplines. Funding for both quantitative and qualitative aspects of fieldwork is essential, so that both sets of researchers see the other as collaborators, not competitors for resources. Time is also needed for the researchers to get to know and appreciate the other, as well as to allow for iteration between qualitative and quantitative work, as discussed above.

2.4 Conclusions

Both qualitative and quantitative methods have much to contribute to enhancing our understandings of the complex and dynamic role that gender and culture play in agriculture and rural development. Bringing them together in mixed methods provides a more complete, and convincing, picture. For example, qualitative studies have illuminated the different priorities of men and women and allowed researchers to better understand the dynamics behind separate men's and women's activities as opposed to joint domains of activities. Quantitative analyses of gender relations in agriculture have played an important part in convincing broader audiences about the statistical significance of relationships between gender equality and productivity, poverty reduction, and other development outcomes. Mixed methods work allows the qualitative to inform the quantitative and vice versa, thus expanding the depth and breadth of research and providing a more complete picture of gender relationships.

The benefits of mixed methods work have increasingly been documented (Adato and Meinzen-Dick 2007); however, there are still a number of important data needs for gender work in agricultural research. Examples of these include

- Ensuring that quantitative sex-disaggregated information captured in agricultural surveys is included in national statistics, so that there is good secondary data available for interested researchers.
- Geo-referencing of sex disaggregated data on agriculture, so that gender can be considered in GIS and spatial models. Increasingly, household surveys are

collecting geo-referenced data particularly for nationally-representative household surveys. If collected with sex-disaggregated variables, this would be a way of building up the gendered spatially-referenced data system.

- Developing qualitative research capacity in National Agricultural Research Systems (NARS) and other research organizations in developing countries. This could be accomplished by training students in qualitative methods to increase both the number of qualitative studies conducted and the available capacity in developing countries. Likewise, it would be useful if a wide range of qualitative methods were taught to students and researchers, and applied within agriculture research in developing countries. It seems there has been quite a bit of training on PRA in agricultural research, but not necessarily on other qualitative methods, such as ethnography.
- Measuring the gender gap in rights, resources, and responsibilities in agriculture and rural areas and analyzing the consequences of the gap on a variety of outcomes (both monetary and nonmonetary).

Gender-sensitive agricultural research has enormous potential to contribute to productivity growth and poverty reduction. Successful gender-sensitive agricultural research requires the collection of rigorous quantitative and qualitative information that builds on and informs each other.

References

Adato M (2008) Combining survey and ethnographic methods to improve evaluation of conditional cash transfer programs. Int J Mult Res Approach 2(2):222–236

Adato M, Meinzen-Dick R (2002) Assessing the impact of agricultural research on poverty using the sustainable livelihoods framework. Food consumption and nutrition division discussion paper 128. International Food Policy Research Institute, Washington, DC

Adato M, Meinzen-Dick R (eds) (2007) Agricultural research, livelihoods, and poverty: studies of economic and social impacts in six countries. Johns Hopkins University Press for the International Food Policy Research Institute, Baltimore

Alderman H, Chiappori PA, Haddad L, Hoddinott J, Kanbur R (1995) Unitary versus collective models of the household: is it time to shift the burden of proof? World Bank Res Obs 10(1):1–19

Baulch B, Davis P (2008) Poverty dynamics and life trajectories in rural Bangladesh. Int J Mult Res Approach 2(2):176–190

Behrman JR (1997) Intrahousehold distribution and the family. In: Rosenzweig MR, Stark O (eds) Handbook of population and family economics. North-Holland Publishing Company, Amsterdam, pp 126–188

Bellón MR, Hellin J (2011) Planting hybrids, keeping landraces: agricultural modernization and tradition among small-scale maize farmers in Chiapas, Mexico. World Dev 39:1434–1443

Bellón M, Adato M, Becerril J, Mindek D (2007) Improved maize germplasm, creolization, and poverty: the case of Tuxpeño-derived material in Mexico. In: Adato M, Meinzen-Dick R (eds) Agricultural research, livelihoods, and poverty: studies of economic and social impacts in six countries. Johns Hopkins University Press for the International Food Policy Research Institute, Baltimore, pp 239–284

Bouis H, de la Brière B, Guitierrez L, Hallman K, Hassan N, Hels O, Quabili W et al (1998) Commercial vegetable and polyculture fish production in Bangladesh: their impacts on income, household resource allocation, and nutrition. International Food Policy Research

Institute/Bangladesh Institute of Development Studies, Institute of Nutrition and Food Science, Washington, DC/Dhaka

Bourdillon MFC, Hebinck P, Hoddinott J, Kinsey B, Marondo J, Mudege N, Owens T (2007) Assessing the impact of high-yield varieties of maize in resettlement areas of Zimbabwe. In: Adato M, Meinzen-Dick R (eds) Agricultural research, livelihoods, and poverty: studies of economic and social impacts in six countries. Johns Hopkins University Press for the International Food Policy Research Institute, Baltimore, pp 198–237

Brewer J, Hunter A (1989) Multimethod research: a synthesis of styles. Sage, Newbury Park

Chambers R (1992) Rural appraisal: rapid, relaxed and participatory. Institute of Development Studies Discussion Paper 311. Institute of Development Studies, Brighton

Chambers R (2004) Participatory rural appraisal (PRA): analysis of experience. World Dev 22(9):1253–1268

Chung K (2000) Using qualitative methods to improve the collection and analysis of data from LSMS household surveys. In: Grosh M, Glewwe P (eds) Designing household survey questionnaires for developing countries: lessons from 10 years of LSMS experience. World Bank, Washington, DC, pp 337–363

Cloud K (1983) Women's productivity in agricultural systems; considerations for project design. Harvard Institute for International Development, AID/WID Training Project, Cambridge, MA

Creswell JW (1998) Qualitative inquiry and research design: choosing among five traditions. Sage, Thousand Oaks

de Brauw A, Eozenou P, Gilligan DO, Hotz C, Kumar N, Loechl C, McNiven S, Meenakshi JV, Moursi M (2010) The impact of the HarvestPlus Reaching End Users Orange-Fleshed Sweet Potato Project in Mozambique and Uganda. International Food Policy Research Institute, Washington, DC

Denzin NK (1978) Triangulation. In: Denzin NK (ed) The research act: an introduction to sociological methods. McGraw-Hill, New York

Dercon S (2001) Assessing vulnerability to poverty. Mimeo, Centre for the Study of African Economies, University of Oxford, Oxford

Dey J (1985) Women in African rice farming systems. In: International Rice Research Institute (ed) Women in rice farming: proceedings of a conference on women in rice farming systems. Gower Publishers, Brookfield, pp 419–444

Dillon, A, Quiñones E (2010)Asset dynamics in northern Nigeria. IFPRI discussion paper 1049. International Food Policy Research Institute, Washington, DC

Doss C, Grown C, Deere CD (2008) Gender and asset ownership: a guide to collecting individual-level data. Policy research working paper 4704. World Bank, Washington, DC

Doss C, Truong M, Nabanonga G, Namaalwa J (2011) Women, marriage, and asset inheritance in Uganda. Chronic poverty research centre working paper 184. Chronic Poverty Research Centre, Manchester

Dwyer D, Bruce J (eds) (1988) A home divided: women and income in the Third World. Stanford University Press, Stanford

FAO (Food and Agriculture Organization of the United Nations) (2005) Agricultural censuses and gender: lessons learned in Africa. FAO, Rome

FAO (Food and Agriculture Organization of the United Nations) (2011) The state of food and agriculture 2011. Women in agriculture: closing the gender gap for development. FAO, Rome

Fapohunda E (1988) The nonpooling household: a challenge to theory. In: Dwyer D, Bruce J (eds) A home divided: women and income in the Third World. Stanford University Press, Stanford, pp 143–154

Gladwin CH (1975) A model of the supply of smoked fish from Cape Coast to Kumasi. In: Plattner S (ed) Formal methods in economic anthropology. American Anthropological Association, Washington, DC, pp 77–127

Gladwin CH (1982) Off-farm work and its effect on Florida farm wives' contribution to the family farm. In: Rojas M (ed) World development and women, vol 2. Virginia Tech Title XII Women in Development Office, Blacksburg

Gladwin CH, Macmillan D (1989) Is a turnaround in Africa possible without helping African women to farm? Econ Dev Cult Chang 37(2):345–369

Greene JC, Caracelli VJ, Graham WD (1989) Toward a conceptual framework for mixed-method evaluation designs. Educ Eval Policy Anal 11(3):255–274

Guyer J (1980) Household budgets and women's incomes. African studies center working paper no 28. Boston University, Boston

Haddad L, Hoddinott J, Alderman H (eds) (1997) Intrahousehold resource allocation in developing countries: methods, models, and policy. Johns Hopkins University Press for the International Food Policy Research Institute, Baltimore

Hallman K, Lewis D, Begum S (2007) Assessing the impact of vegetable and fishpond technologies on poverty in rural Bangladesh. In: Adato M, Meinzen-Dick R (eds) Agricultural research, livelihoods, and poverty: studies of economic and social impacts in six countries. Johns Hopkins University Press for the International Food Policy Research Institute, Baltimore, pp 103–148

Hentschel J (1999) Contextuality and data collection methods: a framework and application to health service utilization. J Dev Stud 35(4):64–94

Hill P (1963) Migrant cocoa farmers of southern Ghana. Cambridge University Press, Cambridge

Hill P (1978) Food farming and migration from Fante villages. Africa 48(3):220–230

Hoddinott J, Quisumbing A (2003) Data sources for microeconometric risk and vulnerability assessments. Mimeo, International Food Policy Research Institute, Washington, DC

Jackson C (2005) Strengthening food policy through gender and intrahousehold analysis: impact assessment of IFPRI multicountry research. IFPRI impact assessment discussion paper 23. International Food Policy Research Institute, Washington, DC

Jick TD (1979) Mixing qualitative and quantitative methods: triangulation in action. Adm Sci Q 24(4):602–611

Jones C (1983) The mobilization of women's labor for cash crop production: a game theoretic approach. Am J Agric Econ 65(5):1049–1054

Kanbur R (ed) (2003) Q-squared: qualitative and quantitative poverty appraisal. Permanent Black, Delhi

Kanbur R, Shaffer P (2007) Epistemology, normative theory, and poverty analysis: implications for Q-squared in practice. World Dev 35(2):183–196

Kennedy E (1994) Health and nutrition effects of commercialization of agriculture. In: von Braun J, Kennedy E (eds) Agricultural commercialization, economic development, and nutrition. Johns Hopkins University Press for the International Food Policy Research Institute, Baltimore, pp 79–102

Kumar N, Quisumbing A (2010a) Access, adoption, and diffusion: understanding the long-term impacts of improved vegetable and fish technologies in Bangladesh. IFPRI discussion paper no 995. International Food Policy Research Institute, Washington, DC

Kumar N, Quisumbing A (2010b) Does social capital build women's assets? The long-term impacts of group-based and individual dissemination of agricultural technology in Bangladesh. CAPRi working paper 97. International Food Policy Research Institute, Washington, DC. Retrieved online September 27 2010, from http://www.ifpri.org/sites/default/files/publications/capriwp97.pdf

Lilja N, Ashby J (1999) Types of gender analysis in natural resource management and plant breeding. Consultative Group on International Agricultural Research, Cali, Participatory Research and Gender Analysis

Maluccio JA, Adato M, Skoufias E (2010) Combining quantitative and qualitative research methods for the evaluation of conditional cash transfer programs in Latin America. In: Adato M, Hoddinott J (eds) Conditional cash transfers in Latin America. Johns Hopkins University Press for the International Food Policy Research Institute, Baltimore, pp 28–52

Meinzen-Dick RS, Di Gregorio M, McCarthy N (2004) Methods for studying collective action in rural development. Agric Syst 82(3):197–214

Moser C (2001) 'Apt illustration' or 'Anecdotal information'? Can qualitative data be representative or robust? In: Kanbur R (ed) Qualitative and quantitative poverty appraisal: complementarities, tensions, and the way forward, pp 52–55. Draft contributions to a workshop at Cornell University, March 15–16, Ithaca, NY, US

Mukhopadhyay C (1984) Testing a decision process model of the sexual division of labor in the family. Hum Organ 43(Fall):227–242

Norton A, Owen D, Milimo J (1994) Zambia participatory poverty assessment, vol 5: participatory poverty assessment. Southern Africa Department, World Bank, Washington, DC

Pahl J (1983) The allocation of money within marriage. Sociol Rev 32(May):237–264

Quisumbing AR (ed) (2003) Household decisions, gender, and development: a synthesis of recent research. International Food Policy Research Institute, Washington, DC

Quisumbing AR (2011) Do men and women accumulate assets in different ways? Evidence from rural Bangladesh. IFPRI discussion paper 1096. International Food Policy Research Institute, Washington, DC

Quisumbing AR, Kumar N, Behrman J (2010) Do shocks affect men's and women's assets differently? A review of literature and new evidence from Bangladesh and Uganda. International Food Policy Research Institute, Washington, DC (Background paper for the World Development Report)

Smith LC, Haddad L (2000) Explaining child malnutrition in developing countries: a cross-country analysis. Research report 111. International Food Policy Research Institute, Washington, DC

Smith LC, Ramakrishnan U, Ndiaye A, Haddad L, Martorell R (2003) The importance of women's status for child nutrition in developing countries. Research report 131. International Food Policy Research Institute, Washington, DC

Strauss JA, Thomas D (1995) Human resources: empirical modeling of household and family decisions. In: Srinivasan TN, Behrman J (eds) Handbook of development economics. North Holland, Amsterdam, pp 1883–2023

Tashakkori A, Teddlie C (1998) Mixed methodology: combining qualitative and quantitative approaches. Applied social research methods no 46. Sage, Thousand Oaks

Thorbecke E (2003) Tensions, complementarities, and possible convergence between the qualitative and quantitative approaches to poverty assessment. In: Kanbur R (ed) Q-squared: combining qualitative and quantitative methods in poverty appraisal. Permanent Black, Delhi

Tripp R (1982) Farmers and traders: some economic determinants of nutritional status in Northern Ghana. Food Nutr Bull 8(1):3–12

Udry C (1996) Gender, agricultural production, and the theory of the household. J Polit Econ 104(5):1010–1046

von Braun J, Webb P (1989) The impact of new crop technology on the agricultural division of labor in a West African setting. Econ Dev Cult Chang 37(3):513–534

Chapter 3
Data Needs for Gender Analysis in Agriculture

Cheryl Doss

Abstract To support gender analysis in agriculture, household surveys should be better designed to capture gender-specific control and ownership of agricultural resources such as male-, female- and jointly-owned assets. This chapter offers guidelines on how to improve data collection efforts to ensure that women farmers are interviewed and that their voices are heard. Researchers need to clarify who should be interviewed, how to structure the interview, and how to identify which people are involved in various activities, as owners, managers, workers, and decision makers. It is important not simply to assume that one particular individual does these activities based on social norms, but instead to ask the questions to allow for a range of answers that can demonstrate how the gender patterns in agriculture are changing. To assist in these efforts, the chapter provides an overview of relevant questions to include, emphasizing that whenever questions are asked about ownership and access to resources, answers should be associated with individuals. Finally, collecting data on the institutions that are related to agricultural production and marketing allows analysis of the gender-based constraints and opportunities that they present.

Keywords Agriculture • Gender • Survey design • Household decisionmaking

3.1 Introduction

Understanding the role that gender plays in agricultural production is critical for designing agricultural policies to increase productivity and enhance economic growth and to reduce poverty. While there is increasing awareness of the importance of

C. Doss (✉)
African Studies and Economics, Yale Center for International and Area Studies,
Yale University, P.O. Box 208206, New Haven, CT 06520-8206, USA
e-mail: Cheryl.doss@yale.edu

A.R. Quisumbing et al. (eds.), *Gender in Agriculture: Closing the Knowledge Gap*, DOI 10.1007/978-94-017-8616-4_3,

including women in agricultural policies, either through explicit programs for women or through mainstreaming approaches, there are still key data gaps that inhibit the development of appropriate policies and monitoring their progress.

Gender analysis examines how the social roles of men and women are determined and how these roles affect the outcomes being studied. Although the term "gender analysis" is often used to refer to studies that look at women, it is not possible to study women's behavior without considering the broader contexts facing both women and men. Gender analysis examines how the roles, rights, and responsibilities of men and women interact and how this affects outcomes. In agriculture, gender analysis provides insights into how socially constructed roles and responsibilities shape the myriad decisions around agricultural production and processing.

It is critical to understand the constraints and opportunities that people face to develop appropriate agricultural policies. These constraints and opportunities are often influenced and shaped by gender and thus incorporating gender into the analyses is critical. For example, labor markets differ for men and women (see Dey de Pryck and Termine, Chap. 14). The decision regarding the type of work to seek, whether in the formal or informal sector, depends on social norms about household responsibilities the availability of appropriate work and the potential earnings. These all vary by gender. Similarly, both the supply of and the demand for credit varies for men and women. Thus, to understand credit markets, it is necessary to understand how gender influences the decision to seek credit, the sources available, and the probability of obtaining credit.

Two broad sets of questions about agriculture are frequently asked. The first is how to increase agricultural productivity. What are the constraints to improved productivity and what policies are needed to ease them? These may include constraints to the adoption of new technologies or inefficiencies in markets. These questions lead into macroeconomic ones about how to use agriculture to promote economic growth. Gender is often left out of these analyses or included only in a very cursory way, such as including a measure of the sex of the household head in the analysis.

A second set of questions is specifically about women and agriculture, asking whether women are being left behind or made worse off with the introduction of new technologies and new marketing opportunities. These analyses may look at the gender gaps, demonstrating the disadvantages that women face as farmers. *The State of Food and Agriculture 2010–2011, Women in Agriculture: Closing the Gender Gap for Development* 2011 has identified many of these gaps, but the analysis often points to the deficiencies in the data available (FAO 2011). The statistical appendix for *The State of Food and Agriculture* 2011 is impressive for its coverage of countries, but also for its heroic attempt to identify the many gaps. There are many countries and variables for which the information is not available. The data that they analyze clearly demonstrate that there are gender differences with significant consequences. But, in many contexts, the data are not available and local policymakers need national-level data on the gender gaps in their country.

Work on impact assessment is related to both of these sets of questions. Some of it examines how changes in agricultural policy or agricultural programs affect poverty. There is growing awareness that these impacts may be different for men and for women. Yet, the data are not always available or collected to support this analysis. Other work specifically considers the impact on women, often suggesting that the technologies may actually harm women or some groups of women.

To analyze these various questions about agriculture, data are needed at two different levels. The first is that much more of the microlevel data needs to be sex-disaggregated, which will require that the data be collected at the level of the individual, rather than just at the household or farm level or that data are collected both on the agricultural holdings and on the holder. Collecting additional data at the individual level will not only facilitate gender analyses, but also will facilitate a broader range of analyses across individuals based on age, status within the household, and other individual characteristics. Thus, while the benefits of collecting disaggregated data are critical for gender analysis, they will serve a much broader purpose as well. Second, data are needed for researchers to analyze how institutions and structures, such as markets for inputs and outputs, credit markets, and labor markets, are experienced differently by men and women and how this has an impact on the well-being of individuals and communities and the processes of agricultural development and economic growth. This may require information collected at the regional, community, household, and individual levels on control over resources, decisionmaking, contributions of labor, and so on.

Some of the changes in data collection that are needed are quite simple and would require little or no additional resources. These changes should be implemented in all agricultural surveys to considerably improve the data that are available. In other instances, more detailed questions would need to be added and trade-offs will be faced about the breadth and depth of the questionnaires.

Explicitly incorporating gender analysis into discussions of agricultural productivity should also expand the definitions of agricultural production to include a greater level of processing and preparation, much of which is done by women. Incorporating the full range of agricultural production, from farm to table, would provide better insights into some of the constraints facing both male and female farmers.

The remainder of this chapter focuses on the quantitative data needed for gender analysis that can be collected in surveys and censuses. A discussion of combined qualitative and quantitative approaches is provided in Behrman et al. (Chap. 2). This chapter begins with a discussion of the types of agricultural surveys and asks who should be interviewed. It then moves to a discussion of specific types of questions that should be asked.

3.2 Structure and Approach to Data Collection for Gender Analysis

In considering the data needed, it is important to consider who should be interviewed and how to structure the interview. These will depend on the research or policy questions that need to be answered, and the appropriate unit of analysis. Are we interested in the farmer, the household, the plot of land, a particular crop, the farm enterprise? These different units of analysis will lend themselves to different types of surveys.

Broadly, the types of surveys that are used for analyses of agriculture include farmer surveys, household surveys, labor market surveys, and agricultural censuses. Each provides different information and helps to answer different questions. Because they use different sampling frames, the data on similar issues may differ across these types of surveys. Each type of survey has its strengths and weaknesses and should be considered in the context of the broader research questions.

Farmer surveys typically interview individual farmers.[1] They ask detailed questions about the production process and sales and marketing of output. Frequently, farmer surveys focus on a particular crop or set of crops. A farmer survey may interview maize farmers or rice farmers in order to learn about their production challenges and constraints. The strength of farmer surveys is the ability to collect very detailed data about these crops. A disadvantage is that the crops may not be considered in the context of the broader choices that the farmer and his or her household is making. By focusing a survey on a particular crop, especially a cash crop or major staple crop, many of the other agricultural activities may be ignored. These other crops and activities may be small but important sources of income for individuals within the household or contribute to household food security. Involvement with these crops may limit the amount of labor available to the major crops. This focus on the major crops may provide good approximations of the total output from a holding, but miss components, especially those done and controlled by women.

Since agricultural activities, especially among smallholder farmers, are embedded in a range of household activities, household surveys are often used to analyze agricultural decisions. Household surveys usually treat the household as the production and consumption unit.[2] They facilitate analysis that can encompass the range of decisions that households are making, such as which crops to grow, whether or not to engage in off-farm labor, whether to send their children to school, and how much of the farm output to sell in order to purchase other items. Thus, they provide a better means for understanding livelihood decisions, but may have less detailed information on specific crop production practices. When done well, household surveys can provide the data needed for researchers to analyze the trade-offs that people are making across different activities, both agricultural and nonagricultural.

The line between farmer surveys and household surveys is somewhat blurry, but it is useful to conceptualize them differently. A farmer survey focuses on a particular crop or crops and collects detailed data on the production and marketing. Household surveys seek to understand the range of household activities and the interactions among them.

An agricultural census usually involves complete enumeration of all agricultural holdings in a country. This is different from surveys that sample a relatively small number of farmers or households. In a census, the agricultural holder for each

---

[1] Many small-sample farmer surveys are collected through the CGIAR Centers. The analysis by Doss and Morris (2001), discussed later in this chapter, is based on a survey where the unit of analysis was maize farmers in Ghana.

[2] The World Bank Living Standards Measurement Surveys are examples of multipurpose household surveys that usually include a substantial agricultural module or modules.

holding is interviewed and a limited amount of data is obtained about the holding. It is recommended by the FAO that a country conduct an agricultural census once every 10 years in order to provide information on all holdings (FAO 2005a). The amount of information on each holding is relatively small compared to farmer or household surveys, but together they provide the picture of the entire agricultural sector.

Finally, labor force surveys may be relevant for agricultural analyses. They are household-based surveys that collect data on the economically active population, employment, and unemployment. Because they ask about employment in the agricultural sector, they provide aggregate statistics on the agricultural workforce. Data are typically collected on all household members. These data are used to report the share of the population working in agriculture and to examine trends in the agriculture workforce.[3]

Each of these different types of surveys has advantages and disadvantages when considering how they can be used for gender analysis. The structure of the survey influences who is interviewed.

A key challenge for all agricultural surveys is to ensure that women farmers are represented and counted. This challenge is at three levels. First, for surveys that are interviewing "farmers," it is important not to simply assume that the farmer is a man. The farmer is usually defined as the person who makes the major agricultural decisions or the person who knows the most about the agricultural production. It is important that questions be asked to identify the appropriate respondent, rather than assuming that this person is a man. In addition, for agricultural censuses, the FAO is now recommending that the data collection allow for joint holders as well as individual holders of parcels, where the holder is defined as the person or persons making the major decisions.

Second, even when a man makes the major agricultural decisions, a woman may make the decisions for specific crops or activities. The FAO guidelines suggest that agricultural censuses ask about subholdings and subholders, specifically because women often manage small plots within agricultural holdings (FAO 2005a).[4] In many places, if men are culturally defined as the decisionmakers, asking about who makes the agricultural decisions overall will usually result in a male being identified as the farmer. Yet, asking about the decisions over other specific activities, such as specific crops or animals, may result in women being identified.

Finally, even when women are not interviewed as the major decisionmakers for holdings or subholdings, they may make a substantial contribution to agricultural production. It is important to count their contributions and to identify the opportunities and constraints that they face.

---

[3] The labor force results may be affected by how the questions are asked and to whom they are addressed (Bardasi et al. 2011).

[4] The publication, *Agricultural Censuses and Gender: Lessons Learned in Africa* (FAO 2005b) raised many of the issues about collecting agricultural census data that can be used for gender analysis. Some of these concerns have been incorporated into the FAO recommendations.

These latter two concerns suggest that it may be important to interview more than one individual within the household if the survey is interested in obtaining data on the full range of agricultural production done within the household. One recent study suggests that it is important to interview both the husband and wife to obtain complete information on household income from farm households in Malawi (Fisher et al. 2010). Husbands did not report full information on their wives' incomes. For a smallholder farm, it may not be the case that one individual owns the land and makes all of the agricultural decisions from what to plant to how and where to sell the output. For example, the owner of the land may not be the person who makes the key decisions about what crops to plant.[5] This issue of identifying domains of decisionmaking is particularly relevant when a husband migrates, leaving his wife responsible for the day-to-day operations of the farm. Depending on whether or not he is on the farm at the time of the survey, either one of them may be interviewed.[6] Thus, it is useful to ask about the various decisionmaking domains of the respondent identified as the farmer.

Many studies use household-level data and consider gender just at the level of the household head. Thus, comparisons are made of the responses across male-headed and female-headed households. However, since women frequently live in male-headed households, but adult men rarely live in households that are defined as female-headed, this approach conflates measures of household structure and composition with the gender of the head. And considering only the sex of the head of household renders women living in male-headed households invisible.

A number of empirical studies demonstrate the range of problems with simply using the sex of the household head in gender analysis. Doss and Morris (2001) find that in Ghana, after taking into consideration the age and education of the farmer, access to land and labor, contact with extension and market access, there is no difference between male and female farmers as to the probability of planting improved varieties of maize or using fertilizer.[7] Living in a female-headed household, however, does reduce the probability of adopting these technologies. If they had only considered the heads of household, they would have missed the female farmers living in male-headed households who are adopting the improved technologies. Similarly, Peterman et al. (2011) find that using measures at the plot level, rather than indicators of the sex of the household head, results in different conclusions about the extent of gender differences in agricultural productivity. Finally, Deere et al. (2010) demonstrate how using the sex of the household head underestimates women's ownership of assets, including land and housing.

---

[5] See Doss (2001) for a discussion of how the categorization into men's and women's crops varies depending on whether the definitions are based on the sex of the head, landholder, decisionmaker, or person who keeps the revenue.

[6] As cell-phone availability increases rapidly in rural areas, it is not at all clear how this will have an impact on decisionmaking on farms, since a person would not have to be physically present in order to be involved in the process.

[7] Women are less likely overall to adopt the improved technologies because of their lower levels of education, access to land and labor, and contact with extension; see the review in Peterman et al., Chap. 7.

A broader critique of the notion of household headship and how it is defined is articulated by Budlender (2003). Household heads may be defined by age, sex, custom, or economic contributions, among other factors, and differ widely across contexts and, if not clearly defined in surveys, may differ across enumerators. Many OECD countries are moving away from the notion of a household "head" within their data collection exercises and Budlender argues that particularly in the context of households that may be extended family units, polygamous, and/or multigenerational, defining a single household head is not a useful means of understanding the relationships among the household members.

In addition, a multiplicity of definitions of "the household" is used in surveys. Typically, the household is defined as individuals who sleep under the same roof, eat out of the same pot, and share in production and consumption activities. But these different components may result in different people being included within the household. Beaman and Dillon (2012) report on the results of an experiment in survey design where different definitions of the household were used. They find that there are statistically significant differences across reported household size and composition, depending on the definitions that are used. Their results suggest that it is important to consider the appropriate definition of the household for the research questions being analyzed.

Thus, it is not appropriate to simply automatically interview the individual considered to be the male head of household. If focusing on a particular crop or activity, it is important to interview the person who makes decisions or is most knowledgeable; this may be a man or a woman. If the goal is to understand rural livelihoods and to identify and evaluate potential strategies for poverty reduction, it is critical to interview multiple people within the household and ask about multiple activities. And finally, to understand intrahousehold dynamics, detailed data are needed on the key actors (usually the principal couple, but may instead focus on intergenerational dynamics within the household) and needs to include information on the various factors that may affect their bargaining power as well as on their roles and responsibilities in agricultural production.

3.3 Questions to Include

For gender analysis, it is important to attribute the roles and responsibilities of agricultural production to the individual. Thus, whenever questions are asked about ownership and access to resources, answers should be associated with individuals. In addition, to the extent possible, it is useful to attribute the outputs to individuals, acknowledging that there are conceptual issues with this enterprise (see Doss, Chap. 4). Finally, collecting data on the institutions that are related to agricultural production and marketing allows analysis of the gender-based constraints and opportunities that they present.

The discussion below indicates key areas in which individuals need to be identified so that gender can be included in the analysis. Basic demographic data are needed on all of the people involved in the agricultural production process. While this certainly includes the holder and subholders, it also includes those who provide paid and unpaid labor on the holding and those who are involved in the decisionmaking. The minimum demographic data needed are sex, age, education level, marital status, and relationship to household head or respondent for each of these people. All of these may affect the roles that the individual plays in agricultural production. Marital status may affect one's status within the community and one's access to both land and social networks. Widows and women who are divorced, in particular, may have less access to a variety of resources, especially land. Marital status may also frame how decisions are made within the household; even when women are defined as the farmer, their role as decisionmakers may vary, depending on whether or not they are married.

In this section, the assumption is that the gender analysis is going to be carried out within the broader context of a data collection and analysis exercise that is already collecting much of the standard agricultural information. Thus, this does not go into detail on all of the information that should be collected, but simply identifies areas that need specific attention in order to facilitate gender analysis in agriculture.

3.3.1 Productive Resources

One issue is to identify the owners and the individuals who have access to key resources and inputs. The concept of an individual owner who has the entire bundle of rights over the asset may not hold in many rural areas. For example, for land, an individual may have the right to farm the land, but may or may not have the right to sell or mortgage it. Or the individual may have the right to grow annual crops, but may not have the right to plant trees. To understand the agricultural production decisions, especially those involving long-term investment and decisions, it is important to understand people's rights and tenure security over the land. The same may be true of livestock; one individual may have the right to sell the animal while another member of the household may have the milking rights.[8]

Questions about land tenure and the various characteristics of the land are usually asked in agricultural surveys. Yet two key elements of the land tenure systems that are critical for gender analysis are often overlooked. Many surveys ask about the tenure system under which the land is accessed and ask whether or not the respondent (or someone in the household) owns the land. If the land is owned, then the follow-up question is whether there is a title or other document for the land. The additional questions that are needed are which household member(s) own the land and whose names are on the documents. Because multiple household members may be owners, knowing whether the farmer owns it individually or jointly with a spouse or other family member is important. The rights and decisionmaking may vary depending on who owns the land. In addition, asking which names are on the

[8] See Meinzen-Dick et al. (1997) for a discussion of property rights and gender in the context of natural resources.

ownership documents allows for gender analysis because it identifies the gender of the legal owner. It may also provide insights into intergenerational transfers of land, since it is not uncommon for an adult child to farm land that is still legally in a parent's name.

Data on the mode of acquisition of land and other assets may also be important (see Lastarrhia-Cornhiel et al., Chap. 6). Land that is purchased may have a more complete set of rights, including the rights to sell and mortgage, than land that was inherited or allocated under customary tenure. There may be differences among men and women regarding how assets are acquired. (See Quisumbing et al. 2001 for Ghana and Doss et al. 2012 for Uganda.)

Especially in Africa and other places where much land is not formally titled, the issue of access to land and tenure security may be as relevant as the formal ownership of land. The security of tenure and rights of access may influence investment and production decisions. Since women who claim ownership may have fewer rights than men who claim ownership, it is useful to have data on both the reported ownership and the specific rights over the asset. In a recent study in rural Uganda, although many women were reported as jointly owning land with their husband, very few of those households had some form of ownership documents (typically not titles, but sales receipts), very few of the women's names were on the documents, and women claimed fewer rights over the land than men (Bomuhangi et al. 2011).

Similarly, surveys should include individual-level ownership questions on livestock. A frequent generalization is that men own and manage large stock, such as cattle, while women own and manage small stock, such as sheep, goats and poultry. Yet it is important to interrogate these assumptions. Even among pastoralists in Northern Kenya and Southern Ethiopia, groups who are usually considered to be very patriarchal, McPeak et al. (2011) found that women did report owning large animals, including both cattle and camels. Women did own fewer animals than men overall, but assuming that women owned no large animals could lead to wrong policy prescriptions.

Finally, ownership of other key productive resources, such as agricultural equipment, should be included. To the extent that women farmers live in households with some of the agricultural equipment, but are not the owners of the equipment themselves, they may have last claims to use the equipment, making timely planting or harvesting difficult.

In addition to the ownership of these productive assets, it is also useful to know who manages or has control over them. For livestock, in particular, the owner may not be the one actually managing the animals, so information on which individuals are making the daily decisions is also important. The gender patterns in livestock ownership and management are changing, as women and girls may be becoming more involved with livestock as men migrate or have wage jobs (also, see Kristjanson et al., Chap. 9).

3.3.2 Labor

Labor is the other key input into agricultural production. Great strides have been made in collecting individual-level data on formal-sector employment and wages and benefits. The data that are now often routinely available allow for the gendered analysis of patterns of formal employment and wages. Data on formal-sector work are easier to collect than data on nonformal-sector work or work within the household, including agricultural production.

Deere (2005) identifies four reasons why women's labor may be undercounted in population and agricultural censuses, based on her work in Latin America. First, she argues that rural women are likely to report their home as their principal occupation, even when they are actively engaged in agricultural production. Second, the censuses ask about income-generating activities, thus undercounting subsistence production. Third, the definition of agriculture is often narrow, focusing on fieldwork, leaving out a range of predominantly women's activities, including raising livestock, kitchen gardens, and agricultural processing. Finally, Deere claims that the practice of censuses of defining economic activity as involving a minimum number of hours of work in the week prior to the survey for certain categories, such as unpaid family workers, but not for the primary farmer, results in the undercounting of women due to the seasonality of agricultural work.

At a minimum, if agricultural surveys ask about the various tasks being done, they should disaggregate by age and sex. Men and women's labor has often not been interchangeable in agricultural production, but the patterns of activity are changing and data are needed to document this. Women are involved in many tasks that were traditionally male tasks; and men are increasingly involved in activities that have a higher return, regardless of whether the crops were traditionally women's crops.

In addition to details on employment and agricultural work, collecting information on time use provides information about how people are allocating their time across productive and reproductive activities, and leisure.[9] These provide a means to analyze the trade-offs across these various sets of activities.

A number of small sample time-use studies were conducted for rural households in various places in Africa during the 1980s.[10] These studies not only emphasized women's important contributions to agricultural production, but also quantitatively demonstrated women's roles in agricultural processing and food preparation. Fewer time-use studies have focused on the rural sectors in recent years. Yet, as the agricultural sector transforms, time-use data are critical to understand how individuals are allocating their time within and outside the sector. Recent work has emphasized the issue of "time poverty" among the poor and note that it is especially severe for poor women (Blackden and Wodon 2006).

3.3.3 Yield and Output

To the extent possible, it is also useful to attribute the output of agricultural production to individuals. Rather than asking about total output at the household or holding

[9] For a discussion of time use surveys in developing countries, see Hiway (2009).

[10] These are discussed in more detail in Doss, Chap. 4.

level, collecting this data at the plot or subholding level provides one way to attribute output to the individual or joint holders of the land. Similarly, knowing about which individuals are managing and controlling the outputs from livestock, including milk, eggs, and meat, allows the attribution of this output to men and women within households.

Full income, which includes both cash income and the value of the goods and services produced and consumed within the household, is a useful measure of well-being. Income measures are typically used to categorize households—and sometimes individuals—as poor or nonpoor. Income from agricultural production is often not collected in such a way to allow the attribution of any of the income to individuals. But by attributing the output to individuals, based on either management or labor inputs, individual-level income can be estimated.

There is now a substantial literature that suggests that expenditure patterns and other household outcomes vary, depending on whether income is earned or received by men or women.[11] The differences in outcomes may be due to relative changes in the prices of purchased and home-produced goods when individual household members are employed or because individual income confers bargaining power within the household. This suggests that it is important to have data on both household income and on income at the individual level.

In addition, if we are interested in the dynamics within households, the various sources of incomes, or women's status and income-generating opportunities, then collecting information on all sources of income by individual may be important. Without attention to this detail, it is easy to ignore many of the small sources of income that women have. These smaller sums of money may have important implications for the well-being of the women themselves, since they often control this income. And their control of this money may determine how it is spent. Thus, policies that have an impact on these sources of income might have a relatively large impact on household well-being or expenditures on particular types of goods, but these could not be identified if data on only the major sources of income were collected or if the data were only collected at the household level.

To understand the differential constraints that men and women face, it is useful to have data on their participation in a range of agricultural institutions. Women typically have had less access to extension services and agricultural credit (see Ragasa, Chap. 17). As extension is moving increasingly from the government sector to the private sector, it is important to know how these changes are affecting women. And while micro-credit has often targeted women, larger scale agricultural credit is still often out of women's reach. Data on which individuals access these resources are needed to understand who is being left out and to track programs to provide these resources. This would include whether they have contact with extension workers, participate in farmer's groups, obtain credit, and sell produce in the market. Simply knowing who participates is not sufficient to prescribe how to ease the constraints,

---

[11] See, for example, Hoddinott and Haddad (1995) on agricultural income, Schultz (1990) on unearned income, and Doss (2001) and Duflo and Udry (2004) on shocks to income, and the effects on household outcomes, including expenditures and labor-force participation.

but provides useful baseline information. This information may be collected at the individual level or at the institutional level, by examining who uses the services provided.

3.4 Data for Intrahousehold Analyses

Analysis at the household level is often insufficient to understand either the welfare effects of policies or the responses to policies implemented at community or national levels. Household-level responses to policies will often depend on who is living within the household and who within the household is making the decisions. An extensive literature suggests that not all members of the household share equally in the benefits of the household (e.g., Haddad and Kanbur 1990). Evidence suggests that women frequently have less access to household resources than do men. Numerous microlevel studies have demonstrated that households do not respond as expected to incentives, in part because there is not a single household decisionmaker who is optimizing based on incentives (e.g., Udry 1996). Instead, both cooperation and conflict occurs among household members and these shape the responses that are observed as being household responses (Sen 1990).

Many of the proposed changes to data collection proposed in this chapter will provide greater opportunities for intrahousehold analyses. Intrahousehold analyses and gender analyses often overlap, but they are not synonymous. Frequently, intrahousehold analyses assume that the key differences are based on gender and thus test the hypothesis, for example, that giving income to women or giving women increased bargaining power within the household will have an impact on the outcome of household decisions. This approach implicitly assumes that gender defines the key differences in preferences, whereas other considerations, such as age and relationship to the head of household, also come into play. Thus, demonstrating that income in the hands of women is more likely to be spent on children implicitly assumes women, as a group, have a greater preference for spending money on children. It does not allow us to explicitly test the hypothesis that the individual who receives the money can choose how to spend it. More data at the individual level would allow some of these assumptions to be tested, benefiting both intrahousehold analyses and gender analyses. Yet gender analysis is broader than intrahousehold analysis. It examines how gender shapes behavior not only within the household, but also at the community, institutional, and national levels.

Although this paper argues that individual-level data and analysis are critical, household-level analysis of poverty and development issues will continue to be important as well. Many decisions are made within the contexts and constraints of households. Analyses typically take the household structure as given. But household structure can often be affected by the very policies that are being implemented and analyzed. Thus, understanding how households are formed is important. While panel data would be useful for a full analysis of household formation and dissolution, with a few additional questions in multipurpose surveys, we can improve the data available to analyze household structure and to begin to examine the impact of policies on household structure.

One final set of questions is about how decisions are made within the household. These include both questions about production, including agricultural production, and consumption. An extensive literature now concludes that it is inappropriate to assume that the household acts as a unitary decisionmaker (Alderman et al. 1995). Instead, it is important to consider the characteristics of the individuals within the household and the dynamics among them to understand how decisions are made. Thus, the structure and composition of the household and the age and education levels of household members will all influence the outcomes of decisions. In addition, the bargaining power of individual household members may also influence the outcomes. Typically, researchers have analyzed how the bargaining power of the principle couple within the household, often thought of as the head and spouse, affects outcomes.

3.5 Conclusion

To support gender analyses in agriculture, data collection efforts need to do two things. First, they need to ensure that women farmers are interviewed and that their voices are counted. Second, the information should identify which people are involved in various activities, as owners, managers, workers, and decisionmakers. It is important not simply to assume that one particular individual does these activities based on social norms, but instead to ask the questions to allow for a range of answers that can demonstrate how the gender patterns in agriculture are changing.

There are numerous other questions that agricultural researchers and policymakers continue to struggle with. In particular, they are broader questions about how to improve agricultural productivity and how increased agricultural productivity is related to economic growth. Often these questions, especially those at the macro level about agricultural productivity and economic growth, do not include any gender analysis. Yet to effectively answer these questions—and obtain the correct answers—it is important to incorporate gender into the analysis. And thus, this requires appropriate sex-disaggregated data.

References

Alderman H, Chiappori PA, Haddad L, Hoddinott J, Kanbur R (1995) Unitary versus collective models of the household: is it time to shift the burden of proof? World Bank Res Obs 10(1):1–19

Bardasi E, Beegle K, Dillon A, Serneels P (2011) Do labor statistics depend on how and to whom the questions are asked? Results from a survey experiment in Tanzania. World Bank Econ Rev 25(3):418–447

Beaman L, Dillon A (2012) Do household definitions matter in survey design? Results from a randomized survey experiment in Mali. J Dev Econ 98(1):124–135

Blackden CM, Wodon Q (eds) (2006) Gender, time use, and poverty in Sub-Saharan Africa. World Bank working paper no 73. World Bank, Washington, DC

Bomuhangi A, Doss C, Meinzen-Dick R (2011) Who owns the land? Perspectives from rural Ugandans and implications for land acquisitions. IFPRI discussion paper 01136. International Food Policy Research Institute, Washington, DC

Budlender D (2003) The debate about household headship. Soc Dyn 29(2):48–72

Deere CD (2005) The feminization of agriculture? Economic restructuring in rural Latin America. Occasional paper 1. UN Research Institute for Social Development, Geneva

Deere CD, Alvarado GE, Twyman J (2010) Poverty, headship, and gender inequality in asset ownership in Latin America. Working paper 296. Center for Gender in Global Context, Michigan State University, East Lansing

Doss C (2001) Is risk fully pooled within the household? Evidence from Ghana. Econ Dev Cult Chang 50(1):101–130

Doss C, Morris M (2001) How does gender affect the adoption of agricultural innovations? The case of improved maize technology in Ghana. Agric Econ 25(1):27–39

Doss C, Truong M, Nabanoga G, Namaalwa J (2012) Women, marriage, and asset inheritance in Uganda. Dev Policy Rev 30(5):597–616

Duflo E, Udry C (2004) Intrahousehold resource allocation in Côte d'Ivoire: social norms, separate accounts and consumption choices. Working paper 10498. National Bureau of Economic Research, Inc., Washington, DC

FAO (Food and Agriculture Organization of the United Nations) (2005a) A system of integrated agricultural censuses and surveys, vol 1. World programme for the census of agriculture 2010. FAO, Rome

FAO (Food and Agriculture Organization of the United Nations) (2005b) Agricultural censuses and gender: lessons learned in Africa. FAO, Rome

FAO (Food and Agriculture Organization of the United Nations) (2011) The state of food and agriculture 2011. Women in agriculture: closing the gender gap for development. FAO, Rome

Fisher M, Reimer J, Carr E (2010) Who should be interviewed in surveys of household income? World Dev 38(7):966–973

Haddad L, Kanbur R (1990) How serious is the neglect of intrahousehold inequality? Econ J 100(402):866–881

Hiway I (2009) Time use surveys in developing countries: an assessment. In: Antonopoulos R, Hiway I (eds) Unpaid work and the economy: gender, time use, and poverty in developing countries. Palgrave MacMillan, Basingstoke, pp 252–324

Hoddinott J, Haddad L (1995) Does female income share influence household expenditure patterns? Evidence from Côte d'Ivoire. Oxf Bull Econ Stat 57(1):77–96

McPeak J, Little P, Doss C (2011) Risk and social change in an African rural economy: livelihoods in pastoralist communities. Routledge, New York

Meinzen-Dick R, Brown LR, Sims Feldstein H, Quisumbing AR (1997) Gender, property rights, and natural resources. World Dev 25(8):1303–1316

Peterman A, Quisumbing AR, Behrman J, Nkonya E (2011) Understanding the complexities surrounding gender differences in agricultural productivity in Nigeria and Uganda. J Dev Stud 47(10):1482–1509

Quisumbing AR, Payongayong E, Aidoo JB, Otsuka K (2001) Women's land rights in the transition to individualized ownership: implications for the management of tree resources in Western Ghana. Econ Dev Cult Chang 50(1):157–182

Schultz TP (1990) Testing the neoclassical model of family labor supply and fertility. J Hum Resour 25(4):599–634

Sen A (1990) Gender and cooperative conflicts. In: Tinker I (ed) Persistent inequalities: women and world development. Oxford University Press, New York, pp 123–149

Udry C (1996) Gender, agricultural production, and the theory of the household. J Polit Econ 104(5):1010–1046

Chapter 4
If Women Hold Up Half the Sky, How Much of the World's Food Do They Produce?

Cheryl Doss

Abstract This chapter explores—and explodes—the oft-quoted stylized fact that women produce 60–80 % of food in the developing world. It uses three approaches to shed light on this issue: (1) analyzing labor inputs to agriculture, using both employment data and time-use data; (2) analyzing different ways of assigning agricultural output to men or women, based on four nationally representative household survey datasets; and (3) estimating women's labor productivity relative to men at the macro level, using national-level agricultural productivity data across time and countries. While it is not possible to substantiate the claim that women produce 60–80 % of the food in developing countries—or even that they provide 60–80 % of the labor in agriculture, women contribute a large portion of the measured contributions to agricultural labor and women's share of the measured agricultural labor force has a positive impact on national-level agricultural productivity. While women are not the majority of agricultural workers, the agricultural sector is important for women: 48 % of the economically active women in the world—and 79 % in developing countries—report that their primary activity is agriculture. The "60–80 %" statistical claim obscures the complex underlying reality, that it is difficult to separate women's labor from other uses and from men's labor, and that it cannot be understood properly without considering the gender gap in access to land, capital, assets, human capital, and other productive resources.

Keywords Women • Gender • Food production • Labor inputs • Agricultural output

---

C. Doss (✉)
African Studies and Economics, Yale Center for International and Area Studies, Yale University, P.O. Box 208206, New Haven, CT 06520-8206, USA
e-mail: Cheryl.doss@yale.edu

A.R. Quisumbing et al. (eds.), *Gender in Agriculture: Closing the Knowledge Gap*, DOI 10.1007/978-94-017-8616-4_4,

4.1 Introduction

> Women produce between 60 and 80 % of the food in most developing countries and are responsible for half of the world's food production (FAO no date-a).

This statement is widely quoted and used—in varying phrasings and rephrasings—as justification for focusing attention on women farmers throughout the world. Referring to the same FAO document cited above, a similar claim is that "Rural women produce half of the world's food and, in developing countries, between 60 % and 80 % of food crops" (Mehra and Rojas 2008). It has also been adapted to claim, "Women produce more than half of the locally-grown food in developing countries and as much as 80 % in Africa" (Momsen 1991, 2). Possibly the earliest related claim was published in 1972, "Few persons would argue against the estimate that women are responsible for 60–80 % of the agricultural labor supplied on the continent of Africa…" (UNECA 1972). And a recent newspaper column claims, "In reality, half of the world's food and, in developing countries, between 60 and 80 % of food crops, grow from seeds that are planted by a woman's hand" (Gupta 2009).

While all of these claims represent a powerful truth—women are important food producers—the empirical content behind them is less clear.[1] Instead, they should probably be viewed as a metaphor, much like the claim, "Women hold up half the sky."

In addition, each of these claims about women's contribution to food production makes a very different empirical claim. Food production would certainly include both crop and animal products and could include processing and preparation. The claim about women producing locally grown food suggests that in developing countries women are the primary producers of food that is grown and consumed locally.[2] The 1972 claim simply refers to women's labor inputs in agriculture (presumably measured by the number of workers or the number of hours worked), and doesn't address the amount of output that they produce. The metaphor of seeds planted by a woman's hand is also a powerful one, but it leaves out most of the additional steps of food production, such as weeding and harvesting.

In this chapter, I explore various sources of data that can tell us something about the "stylized fact" that women produce 60–80 % of the food in developing countries. One central argument of the chapter is that a claim of this kind does not have any literal meaning, and it does not lend itself to direct empirical tests. Women do not in general produce food separately from men. Quantifying the share of food produced by women involves making many arbitrary assumptions about gender roles in the production process. Since most food is produced with labor contributions of both men and women, to assign the output separately to men and women would be very complex. To take a stylized example, if men provided the labor to clear the field, women planted and weeded the crops, and both men and women

[1] The one reference that indicates that these numbers are not accurate is Jackson (2005), who says, "It is interesting that this statement is so enduring, so effective—*and so wrong*…".

[2] It is not clear what it would mean for food *not* to be locally grown. Usually the context implies that it is consumed near where it is produced.

were involved in harvesting, how would we determine how much of the output was produced by women?

Empirically, this becomes even more of a challenge. Even if the concept were clearly defined, there are not adequate global data to measure the share of food produced by women. Considering only the labor contributions of men and women in crop production—and setting aside gendered patterns of land ownership and control over other resources—we would need very detailed agricultural production data that included labor inputs for different tasks for men and women and agricultural outputs across a wide range of production systems. If we were interested in the contributions to the final food products that people eat, rather than the crops that are harvested, we would need data on the labor inputs in food processing and preparation. As more of this moves out of the home into the commercial sector, we would have to estimate production functions that included sex- and task-disaggregated data in the commercial food processing and production sectors. The level of detail needed for this type of analysis is far beyond what is currently available. And we would still have to match the labor inputs to outputs. This would be a formidable task and one that would not necessarily help to illuminate important policy questions.

In spite of these difficulties, there is a range of data available on labor inputs and on agricultural output that can help to shed light on women's contributions to food and agricultural production. In this chapter, three different bodies of empirical literature and data are used to explore this contribution. First, the chapter analyzes the labor inputs to agricultural production, using both employment data and time-use data. Second, using four nationally representative household survey data sets, it examines several different ways to assign agricultural output by sex, including the sex of the household head, the owner of the land and the decisionmaker or farm manager. Third, national level agricultural productivity data across time and countries is used to estimate women's labor productivity relative to that of men at the macro level. These macro-level estimates are then compared to microlevel evidence on men's and women's agricultural productivity. Thus, unlike the claim that "women hold up half the sky," there is considerable information on women's contributions to food production.

Because of the available data, a variety of different measures of both inputs and outputs are used in this chapter. Each of the measures has its own limitations and strengths. Outputs include both crop production and agricultural production. Crop production excludes animal products—notably milk, eggs, and meat. Agricultural production includes not only food production, both crop and livestock, but also the production of goods such as fiber and flowers. Thus, neither one is a specific measure of food.

None of the analyses presented here differentiates between subsistence production and commercial or export production. Food is produced in all of these sectors. Many types of food are produced by smallholder farmers both for home subsistence consumption and for sale in the market. Many of the discussions about the importance of women's role in food production in developing countries stress women's important role in growing, processing, and preparing the food to feed their families. This is typically portrayed as different—and perhaps more important—than the work that women provide in the commercial and export sectors.

Another measurement issue faced in this chapter concerns the use of the aggregate term “food.” There are, of course, many types of food, including (literally) apples and oranges. The approach that is used here for aggregation is the standard one of adding the monetary value of each type of food. That lets us come up with a consistent aggregate, but implicitly it assigns high weights to the more expensive items. Other methods of aggregating across different food categories might result in very different totals. For example, if we aggregated by the caloric value, then starchy high-calorie staple crops would have a higher value than crops such as coffee which have a higher monetary value. To the extent that women tend to be more heavily involved in the production of staple crops, comparing the caloric value of the food produced by men and women might indicate a significantly higher share being produced by women. This aggregation becomes more of a challenge if we are considering food on the table, rather than simply agricultural products. Much of the work of processing and preparing food for the table is done outside of the market sector, and thus more difficult to value consistently in monetary terms.

4.2 Men’s and Women’s Labor Inputs to Agricultural Production

Food production requires a number of different factors of production, including labor, land, and capital, as well as inputs and intermediate goods such as animal and mechanical power, seeds, fertilizer and water. The claim about women’s contributions to food production is often made in a context in which it appears to be referring simply to labor inputs and comparing men’s and women’s labor inputs, rather than considering a more complete range of inputs. In this section, I examine the labor inputs of men and women in agricultural production.

The most comprehensive data available are the FAO data on the economically active population in agriculture. In this measure, an individual is reported as being in the agricultural labor force if he or she reports that agriculture is his or her main economic activity. Sex-disaggregated data are available for most countries. Table 4.1 presents these data for 235 countries for the period 1961–2006. The unweighted averages simply use each country and time period as an observation. They can be weighted by the size of the agricultural labor force.

Using these measures, the proportion of the agricultural labor force that is female is less than 50 % for the world as a whole and for all the usual country groups. Overall, using the weighted measure, women are only 42.2 % of the agricultural labor force. Even for Sub-Saharan Africa, which is the region where women are typically most involved in agricultural production, these measures suggest that women make up fewer than half of the number of people who are reported as economically active in agriculture.

Many researchers have questioned the gender patterns that emerge from this data. In particular, the reported 16 % share for women in Latin America is strikingly low. Deere (2005) identifies a number of potential sources of under-enumeration of

Table 4.1 Share of women in the agricultural labor force, 1961–2006

	Weighted	Unweighted
World	42.2	34.9
Developed countries	42.1	34
Developing countries	42.3	35.2
Low-income countries	42	46.3
High-income countries	36.4	31.3
Asia	42.9	42
Latin America and the Caribbean	16.2	16
Near East and North Africa	42.1	31.8
Sub-Saharan Africa	46.5	47.1

Source: FAOSTAT
Note: Weighted results adjust for differences in agricultural labor force

women in agriculture. In particular, she notes that rural women in Latin America are likely to reply that "their home" is their primary responsibility, even if they are heavily engaged in agriculture. Other difficulties arise because the censuses tend to emphasize income generating activities, underestimating subsistence production; and because agricultural production is often defined as fieldwork, so that activities such as rearing small livestock, kitchen gardening, and postharvest processing are undercounted. While Deere focuses on critiquing the numbers for Latin America, similar criticisms are relevant for other regions.

Using household survey data collected by the *Comisión Económica para América Latina y el Caribe* (CEPAL), Deere recalculates the female fraction of the population occupied in agriculture in 1999 for 11 Latin American and Caribbean countries. She finds percentages ranging from 3.7 % in Panama to 44.7 % in Brazil. Overall 26.2 % of those reporting themselves as occupied in agriculture are women (Deere 2005, 21).[3] While this is substantially more than the 16 % in the FAO data, the numbers still fall far short of the 60–80 % benchmark.

Using the FAO data, only six countries report that 60 % or more of the agricultural workforce is women.[4] Somewhat surprisingly, these are not African countries in which women work on smallholder farms, but instead they are primarily Middle Eastern countries where women work as wage laborers in agriculture.

Thus, although these data are problematic and probably substantially underestimate women's role in agriculture, they do not support the idea that women produce most of the world's food.

4.2.1 *Evidence from Time-Use Studies*

One shortcoming of data on women's share of the agricultural labor force is that these data do not account for differences in hours worked. If men who are identified

[3] Katz (2003) has similar estimates.

[4] These include Congo, Jordan, Libya, Palestinian Territory, Portugal, and Turkey.

Table 4.2 Proportion of total time in agricultural activities contributed by women

Definition of activity and location	Women's share	Source
Agricultural activities—Burkina Faso	0.54	Saito et al. (1994)
Agricultural activities—Kenya	0.59	
Agricultural activities—Nigeria	0.56	
Agricultural activities—Zambia	0.54	
Crop production—Cameroon	0.58	Leplaideur[a]
Ag production—SW Cameroon (small dry season)	0.73	Charmes (2006)[b]
Ag production—SW Cameroon (rainy season)	0.65	
Ag production—SW Cameroon (dry season)	0.66	
Agriculture—Rural households in Malawi	0.49	
Total food production, incl. processing, marketing, animal husbandry, and water and fuel supply—Upper Volta	0.64	McSweeney (1979)
Food and cash crop production—Upper Volta	0.49	
Tanzania 2006, Tanzania Time-use Survey		
Crop farming (not related to establishments)—Tanzania	0.47	Fontana and Natali (2008)
Agriculture—Rajasthan Ages 14–19	0.60	Jain (1996)
Agriculture—Rajasthan Ages 19–34	0.47	
Agriculture—Rajasthan Ages 34–44	0.36	
Agriculture—Rajasthan Ages 44–70	0.43	
Agriculture—West Bengal Ages 14–19	0.34	
Agriculture—West Bengal Ages 19–34	0.10	
Agriculture—West Bengal Ages 34–44	0.10	
Agriculture—West Bengal Ages 44–70	0.02	
Farm hours worked—China	0.49	de Brauw et al. (2008)
Agriculture—Nepal	0.44	Acharya and Bennett (1982)

[a]Leplaideur, cited from Charmes (2006, 50)
[b]These calculations were made by Charmes (2006, 62), based on data from Pasquet and Koppert

as part of the "economically active population in agriculture" provide fewer hours of agricultural labor than women in the same sector, we could underestimate the importance of women. For example, some literature suggests that men in Africa work fewer hours than women, across all activities (Blackden and Wodon 2006).

To address this possibility, we can look at detailed time-use studies that document the time that men and women spend in agriculture and in other activities. The detailed time-use studies, especially those that include activities for more than one agricultural season, are usually for very small samples. Many studies are now quite old. While some nationally representative time-use studies have been recently collected, the reports do not provide summaries for time spent in agricultural activities.[5]

Table 4.2 summarizes a number of time-use studies for which it is possible to calculate the share of total time spent in agricultural activities that is provided

[5]For example, a discussion of these studies in Africa by Charmes (2006) provides an analysis of the nonmarket time spent in agricultural activities, but does not include the measure of agricultural activities in the market sector.

Table 4.3 Proportion of total time in food production contributed by women

Definition of activity and location	Women's share	Source
Food production, supply and distribution, incl. processing and fuel and water collection, but not cooking—Upper Volta	0.65	McSweeney (1979)
Activities related to food preparation—Tanzania	0.87	Fontana and Natali (2008)
Ag processing and cooking—Rajasthan Age 14–19	1.00	Jain (1996)
Ag processing and cooking—Rajasthan Age 19–34	0.98	
Ag processing and cooking—Rajasthan Age 34–44	0.91	
Ag processing and cooking—Rajasthan Age 44–70	0.90	
Food processing and Cooking—Nepal	0.84	Acharya and Bennett (1982)

by women. Some of the reported numbers are just for crop production, while others include animal agriculture. These studies all support the claim that women are heavily involved in agricultural production, with the exception of the study of West Bengal. In some regions within Africa, women do contribute over 60 % of the total time spent in agricultural activities. Yet there is wide variation across sites and across seasons within sites. And overall, there is no support for the 60–80 % figure at a global scale.

Only a few of the studies report the time spent in other food production activities, such as food processing and preparation (See Table 4.3). Time spent cooking and cleaning is often reported as a single entry, so we cannot separate out the portion that is allocated to food preparation. In one study, fuel and water collection are treated as food preparation activities, but cooking is not included. In others, cooking is combined with other household chores, such as cleaning, so we cannot separate out the time spent in food preparation. Nevertheless, the limited evidence suggests that women are the ones who overwhelmingly provide the greatest proportion of time to food processing and preparation.

The time-use studies are able to give us a sense of how men and women allocate their time across various activities. They are not able to tell us anything about the productivity of that labor, nor do these data provide a means to map the labor inputs into specific outputs. In other words, we are still faced with the question of whether women providing 54 % of the total household hours devoted to agriculture can be interpreted to mean that they have produced 54 % of the agricultural output. An additional concern with the time-use studies is that they are not necessarily drawn from representative samples, so that it is difficult to make meaningful inference to national or international populations. Finally, many of the studies are quite old and the employment patterns in rural areas are changing (see Dey de Pryck and Termine, Chap. 14). They probably still reflect women's time allocation in the subsistence farming sector, but may be less reflective of areas where commercialization is occurring.

To summarize however, the time-use studies do suggest (but do not actually demonstrate in any statistical sense) that if "food production" includes food processing and preparation, then women probably provide 60–80 % of the total labor used in bringing food to the table in developing countries. But if "food production" refers instead to the cultivation and harvest of food crops, or to the labor used in producing unprocessed agricultural outputs, women probably do not account for the majority of hours worked.

The time-use studies also leave unanswered the question of whether men's and women's labor is complementary. Are both required for agricultural production? Or are they entirely substitutable? Social norms, customs, and technologies all impact the extent to which they are complements or substitutes. These differ across and within countries, and these social relationships are not fixed over time.

One final approach to examining women's labor contributions would be to look at the share of agricultural wages that women receive. If men and women were each paid the marginal value of their labor, then the wages would represent their contributions to agricultural output. Unfortunately, this approach would have two problems. First, it would only count agricultural labor in the wage sector, ignoring the unpaid family labor that characterizes much of the agricultural sector worldwide. Second, it would require the assumption that men and women are actually paid their marginal value and that labor markets are competitive. Particularly based on evidence coming out of Latin America,[6] it is clear that the agricultural labor markets are segmented, with women concentrated in jobs that are low paid and seasonal. Similar wage differentials are found in India.[7] There are sufficient claims of gender discrimination in agricultural labor markets to indicate that this would significantly undercount women's contributions.

4.3 Measuring Men and Women's Agricultural Output

If we focus our attention on agricultural output instead of labor inputs, we might try to quantify how much output is produced by men and how much by women. This is complicated, however: as noted above, in many agricultural households, both men and women are involved in crop production, so it is a challenge to determine which individuals are producing it. To compare outputs across men and women, we need a way to assign agricultural outputs by gender. In order to examine this question, household survey data are used from four countries: China, Bosnia-Herzegovenia, Ghana, and Nicaragua. These countries were chosen because available household survey data allow for analysis using multiple measures of assigning output to men and women. There are relatively few nationally-representative household surveys that meet this criterion.

For China, the data come from the China National Rural Survey—a nearly nationally representative sample of 60 villages in six provinces of rural China surveyed during November and December 2000. The numbers presented here are based on the definitions and estimations presented in de Brauw et al. (2008).

[6] See Deere (2005) and Lastarria-Cornhiel (2009) for reviews of this literature and Jarvis and Vera-Toscano (2004) for an econometric analysis of agricultural wages in Chile for men and women.

[7] See Ganguly (2003) and Jose (1989).

Table 4.4 Share of food produced by female-headed household

	Percent of households headed by women	Percent of total value of food produced by female-headed households
China	3.1	2.1
Bosnia-Herzegovenia	25	13.2
Ghana	33	12
Nicaragua	38	17

Source: Calculated by author from data reported in de Brauw et al. (2008), and from the BiH LSMS survey, the Nicaragua MECOVI survey, and Ghana Living Standards Survey

The Bosnia-Herzegovenia data come from the Living Standards Measurement Survey (LSMS) conducted in 2001 (World Bank 2001). The value of the crops (including vineyards and orchards, but not pastures) is based on the amount harvested during the 2000–01 season.

Similarly, the Ghana Living Standards Survey (1998–99) was used to calculate the value of crops produced on land farmed by the household and to calculate the value of crops grown on land held by different individuals. Finally, the Nicaragua data come from the MECOVI survey, a nationally representative survey conducted in 2001 (Gobierno de Nicaragua 2001). The value of the crops produced includes that produced on the agricultural production unit (UPA) and on the *patio* or backyard garden.

None of these surveys provides an adequate means to calculate the value of food processing and preparation. While they do include measures of food processing and preparation for items produced for sale; they do not include a value of that produced for home consumption. As a result, the discussion that follows focuses on the value of crop output.[8]

As a lower bound on women's production, the output could be assigned to women if the crop was produced in a household headed by a woman. While women frequently live in male-headed households, the definition of a female-headed household usually precludes the possibility than men are members of the household. While these female heads may use some male labor on their farms, either from a male relative or through hired labor, it would be reasonable to treat this as output produced by women.

Table 4.4 shows the share of the value of crops produced by female-headed households in these four countries. The proportion of households that are reported as female-headed varies from 3.1 % in China to 38 % in Nicaragua. Yet, in each country, the value of crops produced by female-headed households is less than would be expected if male- and female-headed households produced the same amount of crops. This is, at least in part, because female-headed households are typically smaller (in numbers of people) and have less access to resources (including land and labor) than male-headed households.

[8] Livestock products are also not included in this section of the analysis.

This measure ignores the food produced by women living in male-headed households, but given that female-headed households produce proportionately low amounts of crop output, these numbers cast some doubt on the stylized fact that women produce 60–80 % of agricultural output. Even in the country with the highest level of female-headed households, Nicaragua, to reach the 60 % threshold, women would have to produce 57 % of the food in male-headed households. While this is not outside the realm of possibility, it would be a surprisingly high number.

Another way to assign output to men and women is to allocate it to the owner of the plot of land. In Bosnia-Herzegovenia, 16 % of plots were reported as owned by women. Of the total value of crops produced, 16 % were grown on plots owned by women.

A final approach is to look at the person who is the manager, decisionmaker or person responsible for the agricultural production in a household or on a particular plot.

For Ghana, we have data on the person reported as the holder of each plot and what is grown. In 16 % of male headed households and 56 % of female headed households, at least one plot of land is held by a woman. Of the total value of crops produced, 15 % is grown on plots that are held by women.[9]

In Nicaragua, we can look at several different measures. For each agricultural production unit, we have data on which household member is the primary decisionmaker and for each *patio*, we have data on the person responsible for this production. For each household, there may be more than one agricultural production unit, each with a different decisionmaker. Assigning the production to the individual who is the decisionmaker for the agricultural production unit or the person responsible for *patio* production, 17.6 % of the total value of crops is produced by women.

For China, two measures of female management of farms are used, based on definitions from de Brauw et al. (2008). The first considers women as the managers of the farm based on the employment history of the husband and wife. Using this definition, 16 % of the plots were managed by women, and 13 % of the crop revenues were produced on the plots managed by women.

In addition, de Brauw et al. (2008) create a measure of women's management based on the number of hours that women put into farm labor compared with their husbands. The farm is considered to be managed by the woman if farm work is her primary occupation and if the man had primary work off the farm or worked on the farm only during the harvest season. In this case, women manage 9 % of the plots, and generate 8 % of the revenue.

All of these measures considerably underestimate women's contributions to agricultural output, because they do not include the labor of women working in male-headed households, on male-owned land or on male-managed farms. While they should not be used to suggest that women do not contribute substantially to food production, they also do not provide empirical evidence that women are the primary food producers.

[9] While the data are also available on land owned by men and women, it isn't possible to sort out which outputs are produced on land that is owned by an individual, separate from that simply held by them.

A final way to allocate output by gender would be if we could assume that specific crops were grown by women and others were grown by men. Then we could aggregate the value of women's crops and men's crops to see the share grown by women. Researchers have occasionally used this approach, especially in West Africa, where there are patterns of cropping by gender.[10] Yet, a careful analysis of agriculture in Ghana finds while there are gendered patterns of cropping, the distinctions between men's and women's crops do not hold up well enough to use them to make inferences about men's and women's incomes (Doss 2002). In addition, we know that gendered patterns of cropping may change over time as crops move from subsistence foods to commercial foods or vice versa.

4.4 Women's Labor Productivity

The preceding approaches to calculating women's contributions to agricultural production do not speak of the labor productivity of women. Ideally, we would like to assess labor productivity in some way other than simply counting hours of work. As noted above, wages in agriculture may be a poor measure of relative productivity, because labor markets may not accurately value women's productivity and because some (perhaps most) of women's labor inputs in food production takes the form of nonmarket activities. Even if women are less than half of the agricultural labor force (measured as the proportion of the economically active people who report their primary activity as agriculture) or if women do not provide the majority of total hours worked in agricultural production, it may be that their contributions are particularly critical.

One approach that might help to assess the value of women's time would be to estimate agricultural production functions at the farm level, using micro data and treating women's time as a separate input from men's time. However, neither time-use studies nor household survey data typically provide sufficient information to make these estimates on a systematic basis across countries.

4.4.1 Macro Evidence

One approach would be to use cross-country panel data to estimate agricultural production functions that include a measure of women's share of labor in agriculture. The coefficient on this variable could then be used to calculate a measure of women's effective share of agricultural labor. Thus, if men's labor and women's labor had different impacts on agricultural output, we could capture this.

[10] For example, Hoddinott and Haddad (1995) and Duflo and Udry (2001) use gender patterns of cropping in Côte d'Ivoire to separate men's and women's incomes.

This section uses national-level measures of agricultural productivity and women's share of labor in agriculture to measure the productivity impacts of women in agriculture. This work follows in the footsteps of numerous previous studies that use cross-country regressions to estimate aggregate production functions for the agricultural sector. Almost without exception, these papers are subject to the criticism that they suffer from endogeneity and simultaneity biases. Unfortunately, there are no alternative identification strategies that avoid this criticism, and instrumental variables techniques are not useful because it is difficult to find plausible instruments to use to in these estimations.[11]

The basic structure of the econometric specification used in most of the literature on cross-country agricultural productivity differences[12] is given by a Cobb-Douglas production function in log terms:

$$\ln Y_{it} = \ln A_{it} + \sum_{j=1}^{P} \alpha_j \ln X_{ijt} + \phi \ln R_{it} + \beta \ln L_{it},$$

where Y_{it} is agricultural output, X_{ijt} are a set of inputs (livestock, tractors, fertilizer), R_{it} is land, and L_{it} is labor; i indexes countries and t indexes over time. Dividing through by labor, we obtain

$$\ln \frac{Y_{it}}{L_{it}} = \ln A_{it} + \sum_{j=1}^{P} \alpha_j \ln \frac{X_{ijt}}{L_{it}} + \phi \ln \frac{R_{it}}{L_{it}}.$$

For this exercise, we are interested in the potential impact of women's share of labor on agricultural productivity. We might expect that reported women's agricultural labor has a different impact on agricultural output than men's labor. Thus, we treat women's share of labor as an externality in the model. This formulation allows us to empirically test this across time and country.[13]

We can estimate a model of the form,

$$\ln y_{it} = \ln A_{it} + \sum_{j=1}^{P} \alpha_j \ln x_{ijt} + \phi \ln r_{it} + \gamma \ln S_{it}^{F} + \varepsilon_{it},$$

where $S^F{}_{it}$ is the share of women in the agricultural labor force (female economically active population in agriculture divided by the total economically active population in agriculture.)

[11] The econometric estimations all were done both with and without a measure of gross national product (GNP) per/capita. This is occasionally used as a measure of labor quality, but is certainly simultaneously determined with agricultural output per worker in the agricultural labor force. While lagged GNP could be included, the levels are still highly correlated with agricultural output per worker. The coefficients on women's share of labor were robust to the inclusion or exclusion of this variable.

[12] For example, see Craig et al. (1997), Wiebe et al. (2000), and Vollrath (2007).

[13] This specification assumes constant returns to scale in the production technology.

Table 4.5 Aggregate productivity estimates for female share of agricultural labor

	Estimated relative productivity of female agricultural labor	Share of effective labor units provided by women	
		Weighted	Unweighted
World	1.08	45.6	37.7
Developed countries	0.89	37.5	30.3
Developing countries	1.12	47.3	39.4
Low-income countries	1.26	52.9	58.3
High-income countries	0.89	32.3	27.8
Asia	0.96	41.0	40.2
Latin America & Caribbean	1.42	23.0	22.8
Near East and North Africa	1.19	50.3	38.0
Sub-Saharan Africa	1.14	52.9	53.6

Note: Weighted results adjust for differences in agricultural labor force

The output measure is the value of net agricultural production in dollars divided by total economically active population in agriculture. While we are primarily interested in the coefficient on women's share of the agricultural labor force, we need to control for a variety of other factors that impact agricultural productivity. Land is measured as the total agricultural area in hectares. Fertilizer is measured in metric tons. We include a measure of the total number of agricultural tractors in use. Livestock is the total number of livestock, (aggregated with weights used in Hayami and Ruttan (1985)). In addition, to take land quality into consideration, we use the total area equipped for irrigation divided by the total agricultural area, and the total arable land divided by the total agricultural area. Each estimation also includes both indicators for the year, to control for time trends, and for the country to control for country fixed effects.[14] The observations are unweighted, each country and year is a single observation in the estimation. We have data from 154 countries for each year from 1961 to 2002, resulting in 5,305 observations.

The full results are presented in Appendix Tables 4.6 and 4.7. The coefficients on share of female labor are in Table 4.5. For the world as a whole, over the period 1961–2002, there is a positive coefficient of .08 on the proportion of women in the agricultural labor force. This suggests that, at the margin, the average productivity of a woman reported as being in the agricultural labor force is 1.08 that of men. If we calculate this across all time periods and countries, women's labor contributes 37.7–45.6 % of the labor share of agricultural output, depending on whether we use unweighted or weighted averages.

The estimated coefficients are also calculated by region of the world. The coefficient for Latin America is quite high, 1.42, which corresponds with the research discussed earlier that indicates that women are undercounted in the agricultural

[14] Using the indicator variables for the year does not impose the structure of a smooth growth rate that would be implied by including a time trend; this is a more general specification.

labor data. For Africa, effective labor productivity of women is 53–54 %. The effective shares of women's agricultural labor are higher for developing countries than for developed countries; and for low-income countries compared to high income countries. This may primarily reflect the undercounting that is more likely in subsistence and smallholder production than in commercial agricultural production.

These estimated differential contributions to output could be the result of a number of factors. We expect that women are undercounted in the national level data on employment in agriculture. Thus, each woman reported in the data may actually represent additional women providing labor. To the extent that men and women perform different agricultural tasks that have different impacts on productivity, the coefficient on women's labor might be significant. Finally, because of social norms, other responsibilities, and other factors affecting the allocation of men's and women's labor to agriculture, men and women's measured labor could enter the production function differently. Each woman measured in the data could be involved in other tasks that are complementary or competing with the actual time and effort spent in agricultural production.

Of course it is also true that women's participation in agricultural labor is itself endogenously determined. In countries with good institutions, it may be the case that women have more ability to participate in the labor market. These institutions might include good governance, rule of law, well-functioning state institutions and markets, and political, social, and economic freedoms. This might lead to a correlation between women's labor force participation and productivity levels, but both would essentially be driven by institutional quality. Nevertheless, the correlation is itself interesting.

4.4.2 Micro Evidence

There is an extensive microeconomic literature comparing men's and women's productivity in agriculture. Quisumbing reviews the methodology and results of such studies and concludes, "female farmers are equally efficient as male farmers, once individual characteristics and input levels are controlled for" (Quisumbing 1996, 1590). These inputs include quality and quantity of land, fertilizer, improved varieties, extension services and credit. For example, the study cited above on China finds that they cannot reject the null hypothesis that women are as efficient as farm managers as men, using several different definitions of women as farm managers.

Peterman et al. (2010) propose a decomposition of the gender differences in productivity to determine whether the differences in levels of output are due to differences in resource endowments or due to men being able to use a given set of resources more efficiently. They attribute productivity differences in Nigeria and

Uganda to both of these factors. They argue that choices of crop, technologies used, and management techniques are all endogenous to the production decisions and influenced by gender.

The adoption of technologies for agricultural production frequently differs by gender.[15] Much of the literature on agricultural technology adoption simply demonstrates that women are less likely to adopt these technologies, without exploring why these gender differences exist. Doss and Morris (2001) examine the case of improved maize technology in Ghana and find that the gender differences are a result of the gender linked differences in access to complementary inputs. They also find that famers living in female-headed households are less likely to adopt improved technologies than farmers, whether male or female, living in male headed households.

It is important to note here that even if women were exactly as productive as men, given the same access to resources and inputs, we would expect to see women producing lower levels of output precisely because they have much less access to these resources and inputs.[16] Women often produce food on small plots of land of relatively poor quality, with limited access to improved technologies, credit, or information. Thus, their output, whether calculated per unit of land or per unit of labor is typically lower than that for male farmers.

The macro and micro evidence reported here are not necessarily contradictory. The macro evidence—suggesting that aggregate productivity rises with the proportion of women in the labor force—may simply reflect the fact that women's labor is underrepresented in the macro data. Thus, each woman reported in the data may actually reflect additional women whose labor is unreported.

4.5 Conclusion

Overwhelming empirical evidence points to the importance of women as agricultural producers. Although it is not possible to substantiate the claim that women produce 60–80 % of the food in developing countries—or even that they provide 60–80 % of the labor in agriculture, this should not be interpreted as evidence that women are insignificant in the agricultural sector. In fact, given women's continuing insecurity of land tenure and lack of access to cash and credit for inputs, and information about new technologies (see Lastarrhia-Cornhiel et al., Chap. 6, Peterman et al., Chap. 7, and Ragasa, Chap. 17), and given their primary responsibility for many other household tasks, it would be surprising if they were able to produce over half of food crops. Even if women *did* produce over half of food

[15] See Doss (2001) for a review of this literature on Africa.

[16] See Peterman et al. (2011) for a review of the literature on nonland agricultural inputs, and Peterman et al. (Chap. 7).

crops, it would be even more surprising—given the measurement challenges—if the data supported this claim. Instead, we see that women contribute a large portion of the measured contributions to agriculture labor and the women's share of the measured agricultural labor forces has a positive impact on national level agricultural productivity.

And while women are not the majority of those reported to be working in agriculture, the agricultural sector is important for women. Of those women in the least developed countries who report being economically active, 79 % of them report agriculture as their primary economic activity. Overall, 48 % of the economically active women in the world report that their primary activity is agriculture.[17]

Better data are needed, but data should not be collected simply to demonstrate women's contribution to food production. Instead, better data are needed to document the constraints that women face. Data collection has improved substantially since the 1972 United Nations Economic Commission for Africa (UNECA) estimate that "women are responsible for 60–80 % of the agricultural labour supplied on the continent of Africa." Among other advances, we have learned that to understand women's roles in agriculture, we need to ask not only about their primary economic activities, but about all the activities in which women engage. In addition, data are needed to better understand gender roles in agriculture and how they change over time and in response to new opportunities. Simply finding a better way to calculate women's share of food production will not necessarily help us to understand how gender roles and responsibilities shape agricultural production.

Ultimately, the important issue is not whether women produce 60–80 % of the world's food. We know that women are important as food producers, and we know that development efforts that target food and agriculture must recognize the unique roles and constraints that face women. We also know that interventions targeting women are complicated. The stylized fact that women produce 60–80 % of the world's food resonates with many people—researchers, policymakers and activists—who work on rural women's issues. But perhaps the statistical claim obscures the complex underlying reality, which is that women's labor in agriculture cannot be neatly separated from their other time uses; neither can it be separated from men's labor; nor can women's labor in agriculture be understood properly without also understanding their differential access to land, capital, assets, human capital, and other productive resources.

For many reasons, we need to support women as food producers and to ensure that the structural changes now underway in world agriculture benefit women, as well as men, both as producers and consumers of food.

[17] FAOSTAT.

Appendix Tables

Table 4.6 Agricultural productivity and female share of total population active in agriculture–Developed and developing countries, high and low income countries

	(1) World	(2) Developed countries	(3) Developing countries	(4) High income countries	(5) Low income countries
Constant	3.838***	4.317***	3.837***	5.036***	3.068***
	(0.110)	(0.189)	(0.127)	(0.185)	(0.203)
Female share of total Agricultural population	0.0794***	−0.109***	0.118***	−0.113***	0.259**
	(0.0118)	(0.0136)	(0.0170)	(0.0159)	(0.102)
Conventional inputs					
Livestock	0.205***	0.269***	0.200***	0.193***	0.279***
	(0.0118)	(0.0195)	(0.0140)	(0.0223)	(0.0195)
Land	0.647***	0.402***	0.634***	0.511***	0.251***
	(0.0129)	(0.0241)	(0.0193)	(0.0233)	(0.0412)
Tractors	0.0725***	0.0226***	0.0865***	0.0351***	0.0936***
	(0.00416)	(0.00691)	(0.00486)	(0.00549)	(0.00687)
Fertilizer	0.0573***	0.0908***	0.0553***	0.132***	0.0383***
	(0.00318)	(0.00699)	(0.00360)	(0.00719)	(0.00393)
Land quality					
Percent equipped for irrigation	0.0269***	0.0356***	0.0224***	−0.00943	0.0695***
	(0.00602)	(0.00693)	(0.00803)	(0.00845)	(0.00903)
Percent arable land	0.0845***	0.0869**	0.0910***	−0.0122	0.197***
	(0.0222)	(0.0354)	(0.0261)	(0.0378)	(0.0366)
Observations	5,305	1,447	3,858	1,294	1,806
R-squared	0.804	0.952	0.703	0.947	0.441
Number of countries	154	52	102	37	51
Years	1961–2002	1961–2002	1961–2002	1961–2002	1961–2002

Notes: Dependent variable is the log of agricultural output per worker. Standard errors are in parentheses. *** $p<0.01$, ** $p<0.05$, * $p<0.1$. All regressions include year dummies and control for country fixed effects

Table 4.7 Agricultural productivity and female share of total population active in agriculture

	Asia	Latin America	Near East and North Africa	Sub-Saharan Africa
Constant	4.250***	4.649***	2.044***	5.930***
	(0.310)	(0.282)	(0.217)	(0.308)
Female share of total agricultural population	−0.0438	0.422***	0.191	0.137***
	(0.0645)	(0.0565)	(0.122)	(0.0218)
Conventional inputs				
Livestock	0.269***	0.242***	0.231***	0.171***
	(0.0220)	(0.0283)	(0.0213)	(0.0348)
Land	0.277***	0.581***	0.691***	0.251***
	(0.0392)	(0.0399)	(0.0407)	(0.0409)
Tractors	0.0815***	0.173***	0.0510***	0.0289*
	(0.00568)	(0.0218)	(0.00798)	(0.0163)
Fertilizer	0.0658***	0.0234**	0.0164***	0.0883***
	(0.00623)	(0.0102)	(0.00457)	(0.00886)
Land quality				
Percent equipped for irrigation	−0.108***	0.0650*	0.0561***	−0.00392
	(0.0150)	(0.0332)	(0.00978)	(0.0201)
Percent arable land	0.0808	−0.223**	0.406***	−0.0924**
	(0.0683)	(0.0909)	(0.0374)	(0.0433)
Observations	664	499	1,504	1,106
R-squared	0.873	0.947	0.498	0.673
Number of countries	17	12	43	27
Years	1961–2002	1961–2002	1961–2002	1961–2002

Notes: Dependent variable is the log of agricultural output per worker. Standard errors are in parentheses. *** $p<0.01$, ** $p<0.05$, * $p<0.1$. All regressions include year dummies and control for country fixed effects

References

Acharya M, Bennett L (1982) Women and the subsistence sector: economic participation and household decisionmaking in Nepal. World Bank staff working paper number 526. World Bank, Washington, DC

Blackden CM, Wodon Q (eds) (2006) Gender, time use, and poverty in Sub-Saharan Africa. World Bank working paper no 73. World Bank, Washington, DC

Charmes J (2006) A review of empirical evidence on time use in Africa from UN-sponsored surveys. In: Blackden CM, Wodon Q (eds) Gender, time use, and poverty in Sub-Saharan Africa. World Bank working paper no 73. World Bank, Washington, DC, pp 39–73

Craig B, Pardey P, Roseboom J (1997) International productivity patterns: Accounting for input quality, infrastructure, and research. Am J Agric Econ 79(4):1064–1076

de Brauw A, Li Q, Liu C, Rozelle S, Zhang L (2008) Feminization of agriculture in China? Myths surrounding women's participation in farming. China Q 194(June):327–348

Deere CD (2005) The feminization of agriculture? Economic restructuring in rural Latin America. Occasional paper 1. UN Research Institute for Social Development, Geneva

Doss CR (2001) Designing agricultural technology for African women farmers: lessons from 25 years of experience. World Dev 29(12):2075–2092

Doss C (2002) Men's crops? Women's crops? The gender patterns of cropping in Ghana. World Dev 30(11):1987–2000

Doss C, Morris M (2001) How does gender affect the adoption of agricultural innovations? The case of improved maize technology in Ghana. Agric Econ 25(1):27–39

Duflo E, Udry C (2001) Intrahousehold resource allocation in Côte d'Ivoire: social norms, separate accounts, and consumption choices. Mimeo, Department of Economics, Massachusetts Institute of Technology, Cambridge, MA

FAO (Food and Agriculture Organization of the United Nations) (no date-a) FAO focus on women and food security. Rome. Retrieved from http://www.fao.org

FAO (Food and Agriculture Organization of the United Nations) (no date-b) FAOSTAT. Rome. Retrieved from http://faostat.fao.org

Fontana M, Natali L (2008) Gendered patterns of time use in Tanzania: public investment in infrastructure can help? Paper prepared for the IFPRI project on evaluating the long-term impact of gender-focused policy interventions, University of Sussex, Brighton

Ganguly ATS (2003) Feminization of India's agricultural work force. Labor Educ 131–132:29–30

Ghana Statistical Services (1999) Ghana living standards survey, 1998–99. Accra, Ghana

Gobierno de Nicaragua (2001) Encuesta Nacional de Hogares sobre medición de nivel de vida. Instituto de Estadísticas y Censos. Proyecto MECOVI, Managua

Gupta GR (2009) When women farm, crops and economies grow. Des Moines Register

Hayami Y, Ruttan V (1985) Agricultural development: an international perspective. Johns Hopkins University Press, Baltimore

Hoddinott J, Haddad L (1995) Does female income share influence household expenditure patterns? Evidence from Côte d'Ivoire. Oxf Bull Econ Stat 57(1):77–96

Jackson C (2005) Strengthening food policy through gender and intrahousehold analysis: impact assessment of IFPRI multicountry research. IFPRI impact assessment discussion paper 23. International Food Policy Research Institute, Washington, DC

Jain D (1996) Valuing work: time as a measure. Econ Polit Wkly 31(43):48–49

Jarvis L, Vera-Toscano E (2004) Seasonal adjustment in a market for female agricultural workers. Am J Agric Econ 86(1):254–266

Jose AV (1989) Limited options: women workers in rural India. International Labour Organization, Geneva

Katz E (2003) The changing role of women in the rural economies of Latin America. In: Davis B (ed) Food, agriculture, and rural development: current and emerging issues for economic analysis and policy research–II, vol I, Latin America and the Caribbean. Food and Agriculture Organization of the United Nations, Rome, pp 31–66

Lastarria-Cornhiel S (2009) Feminization of agriculture: trends and driving forces. RIMISP. Background paper for the world development report 2008. World Bank, Washington, DC

McSweeney BG (1979) Collection and analysis of data on rural women's time use. Learning about rural women. Stud Fam Plan 10(11/12):379–383

Mehra R, Hill Rojas M (2008) A significant shift: women, food security and agriculture in a global marketplace. International Center for Research on Women (ICRW) citing FAO focus on women and food security. Food and Agriculture Organization of the United Nations, Rome. Retrieved at http://www.icrw.org/publications/women-food-security-and-agriculture-global-marketplace

Momsen JH (1991) Women and development in the Third World. Routledge, New York

Peterman A, Behram J, Quisumbing A (2010) A review of empirical evidence on gender differences in nonland agricultural inputs, technology, and services in developing countries. IFPRI discussion paper 00975. International Food Policy Research Institute, Washington, DC

Peterman A, Quisumbing A, Behrman J, Nkonya E (2011) Understanding the complexities surrounding gender differences in agricultural productivity in Nigeria and Uganda. J Dev Stud 47(10):1482–1509

Quisumbing A (1996) Male–female differences in agricultural productivity: methodological issues and empirical evidence. World Dev 24(10):1579–1595

Saito K, Mekonnen H, Spurling D (1994) Raising the productivity of women farmers in Sub-Saharan Africa. Africa Technical Department series no 230. World Bank, Washington, DC

UNECA (United Nations Economic Commission for Africa) (1972) Women: the neglected human resource for African development. Can J Afr Stud/Revue Canadienne des Études Africaines. Special issue: the roles of African women: past, present, and future 6(2):359–370

Vollrath D (2007) Land distribution and international agricultural productivity. Am J Agric Econ 89(1):202–216

Wiebe K, Soule M, Narrod C, Breneman V (2000) Resource quality and agricultural productivity: a multicountry comparison. Paper presented at the annual meeting of the American Agricultural Economics Association, Tampa. Retrieved at http://ageconsearch.umn.edu/bitstream/21723/1/sp00wi01.pdf

World Bank (2001) Bosnia-Herzogovenia. Living standards study survey. Washington, DC

Part III
Gender, Assets, and Inputs: Issues at the Farm and Household Levels

Chapter 5
The Gender Asset Gap and Its Implications for Agricultural and Rural Development

Ruth Meinzen-Dick, Nancy Johnson, Agnes R. Quisumbing, Jemimah Njuki, Julia A. Behrman, Deborah Rubin, Amber Peterman, and Elizabeth Waithanji

Abstract Because gender differences in access, control, and use of assets are pervasive in the agricultural sector, agricultural development interventions are likely to have gender-differentiated impacts. This chapter proposes a conceptual framework to explore the potential linkages between gender, assets, and agricultural development projects in order to gain a better understanding of how agricultural development interventions may be expected to (positively or negatively) impact the gendered distribution of assets. It uses a broad definition of tangible and intangible assets—natural capital, physical capital, human capital, social capital, and political capital. The conceptual framework identifies linkages between the gendered distribution of assets and various livelihood strategies, shocks, and well-being, and discusses how agricultural development strategies may affect the gender asset gap. In addition, the framework explores the gendered pathways through which asset accumulation

R. Meinzen-Dick (✉)
Environment and Production Technology Division, International Food Policy Research Institute (IFPRI), 2033 K Street, N.W., Washington, DC 20006, USA
e-mail: r.meinzen-dick@cgiar.org

N. Johnson
IFPRI, Agriculture for Nutrition and Health (A4NH), 2033 K Street, N.W., Washington, DC 20006, USA
e-mail: n.johnson@cgiar.org

A.R. Quisumbing
Poverty, Health, and Nutrition Division, International Food Policy Research Institute (IFPRI), 2033 K Street, N.W., Washington, DC 20006, USA
e-mail: a.quisumbing@cgiar.org

J. Njuki
Senior Program Officer, International Development Research Center, PO Box 62084 00200, Nairobi, Kenya
e-mail: jnjuki@idrc.ca

A.R. Quisumbing et al. (eds.), *Gender in Agriculture: Closing the Knowledge Gap*, DOI 10.1007/978-94-017-8616-4_5,

occurs, including attention to not only men's and women's assets but also those they share in joint control and ownership. Unlike previous frameworks, this model depicts the gendered dimensions of each component of the pathway in recognition of the evidence that men and women not only control, own, or dispose of assets in different ways, but also access, control, and own different kinds of assets.

Keywords Gender • Assets • Control and ownership • Inequalities • Conceptual framework • Agricultural development

5.1 Introduction

Access to, control over, and ownership of assets are critical components of well-being (Sherraden 1991; Carter and Barrett 2006). Productive assets can generate products or services that can be consumed or sold to generate income. Assets are also stores of wealth that can increase in value. Assets can act as collateral and facilitate access to credit and financial services as well as increase social status. Flexibility of assets to serve multiple functions provides both security through emergencies and opportunities in periods of growth (Deere and Doss 2006, 1). In her study of "voices of the poor," Narayan (2000, 5) found that "the poor rarely speak of income, but focus instead on managing assets—physical, human, social, and environmental—as a way to cope with their vulnerability." Access to, control over, and ownership of assets including land and livestock, homes and equipment, and other resources enable people to create stable and productive lives. Increasing the nexus of control over assets also enables more permanent pathways out of poverty compared to measures that aim to increase incomes or consumption alone.

Similar to typical measures of income and consumption, not only are assets unequally distributed between rich and poor, but they are also unequally distributed

J.A. Behrman
Department of Sociology, New York University, 295 Lafayette Street, 4th floor, New York, NY 10012, USA
e-mail: jab965@nyu.edu

D. Rubin
Cultural Practice, LLC, 4300 Montgomery Avenue, Suite 305, Bethesda, MD 20814, USA
e-mail: drubin@culturalpractice.com

A. Peterman
Department of Public Policy, University of North Carolina, Chapel Hill, Chapel Hill, NC, USA
e-mail: amberpeterman@gmail.com

E. Waithanji
International Livestock Research Institute, P.O. Box 30709, Nairobi 00100, Kenya
e-mail: e.waithanji@cgiar.org

between men and women, nationally as well as within communities and households (Hausmann et al. 2010). A growing empirical literature from both developed and developing countries has shown that distribution of these assets within the household is critical to household and individual well-being, as measured by outcomes such as food security, nutrition, and education (Deere and Doss 2006; Quisumbing 2003). Thus, an understanding of the gendered nature of asset distribution and how this influences individual and household livelihoods is essential to designing effective development policies and interventions.

Agricultural development programs are increasingly seen as vehicles for poverty alleviation, nutrition, and food security as well as agricultural growth, yet relatively little is known about how they affect or are affected by differential access to and control over assets by men and women or how the interventions lead to differential accumulation of assets by men and women. Sabates-Wheeler's (2006) review of the relationship between ownership and control over tangible assets such as land, livestock, machinery, and the patterns of agricultural growth concluded that the combination of asset inequality and market failure has a negative impact on growth, and that inequalities tend to reproduce inequalities. This suggests that without specific attention to addressing asset inequalities, interventions that promote agricultural growth are likely to reinforce inequalities, which could ultimately undermine their poverty-alleviation objectives.

This chapter explores the potential linkages between gender, assets, and agricultural development and aims at gaining a better understanding of how agricultural development interventions are likely to (positively or negatively) have an impact on the gendered distribution of assets. Section 5.2 briefly summarizes the literature on gender and assets. Section 5.3 presents a conceptual framework for identifying the linkages between the gendered distribution of assets and various livelihood strategies, shocks, and well-being. Section 5.4 examines the implications of the framework for agricultural development interventions to identify issues concerning gender and assets that are relevant to the intervention, how these issues might be addressed, and what kind of information would be needed to be able to fully assess the impact of projects on the gender-asset gap. The final section summarizes the preceding discussion and identifies gaps in knowledge.

5.2 Why Does the Gender-Asset Gap Matter?

5.2.1 Gendered Asset and "Capital" Typologies

Households and individuals hold and invest in different types of assets, including tangible assets such as land, livestock, and machinery, as well as intangible assets such as education and social relationships. Several chapters in this volume explore different dimensions of the gender gap among a range of assets from "standard" agricultural assets such as land, equipment, and livestock, to those not typically regarded as inputs into agriculture (social capital, political capital). These different forms of asset holdings have been categorized as

- ***Natural resource capital:*** land, water, trees, genetic resources, soil fertility (see Chap. 7 by Lastarria-Cornhiel et al.);
- ***Physical capital:*** agricultural and business equipment, houses, consumer durables, vehicles and transportation, water supply and sanitation facilities, and communications infrastructure (see Chap. 7 by Peterman et al.);
- ***Human capital:*** education, skills, knowledge, health, nutrition (see Chap. 11 on health and nutrition by Harris);
- ***Financial capital:*** savings, credit, and inflows (state transfers and remittances) (see Chap. 8 by Fletschner and Kenney);
- ***Social capital:*** membership in organizations and groups, social and professional networks (see Chap. 10 by Meinzen-Dick et al.);
- ***Political capital:*** citizenship, enfranchisement, and effective participation in governance.

As Bebbington (1999) argues, people's livelihoods are based on a range of assets, income sources, and products as well as interactions with labor markets. Assets are, however, not just a means through which people earn a living; they also give meaning to people's lives (Bebbington 1999). They are not only resources that people *use* in building their livelihoods; assets give individuals the *capability* to be and to act. Bebbington's framework of capitals and capabilities treats assets as "vehicles for instrumental action (making a living), hermeneutic action (making living meaningful), and emancipatory action (challenging the structures under which one makes a living)" (Bebbington 1999, 2022).

There is now substantial evidence to contradict the still common assumption made in economics (and in many development projects) that households are groups of individuals who have the same preferences and fully pool their resources. This unitary model has been rejected in both developed and developing countries, with important implications for policy (Strauss and Thomas 1995; Haddad et al. 1997; Behrman 1997). An alternative, the collective model, allows for differences of opinion regarding economic and other decisions among household members. Within households, assets are not always pooled, but rather can be held individually by men, women, and/or children (Haddad et al. 1997). Under the collective model, when there is a disagreement, its resolution may depend on the bargaining power of individuals within the household (Manser and Brown 1980; McElroy and Horney 1981). One of the determinants of the bargaining power of individuals is the ownership and the nexus of control over assets. Who within a household has access to which resources and for what purposes is conditioned both by the broader socio-cultural context as well as by intrahousehold allocation rules.

Figure 5.1 provides a conceptual illustration of what the "gender gap" in asset allocation would look like for a given context. The radar graph illustratively plots the extent of men's and women's control over assets in each of the aforementioned types of "capital" (ignoring, for the moment, the fact that each of these types of assets are multidimensional in themselves, and consequently collapsing any one

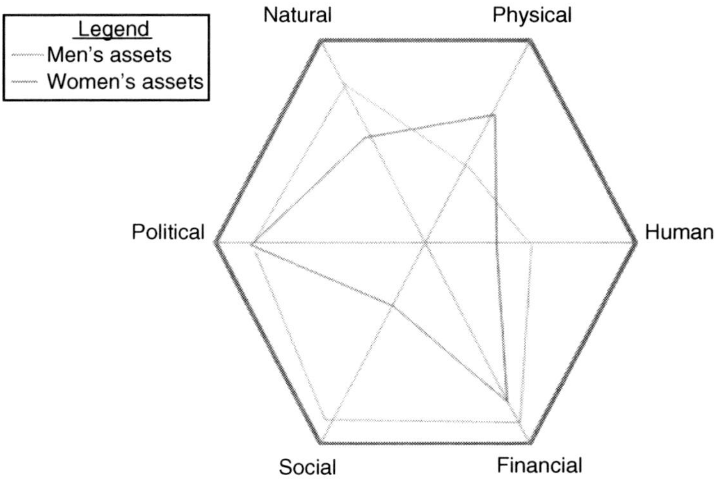

Fig. 5.1 Illustration of gender-asset gap

dimension into a single index would be extremely problematic). A third line could be used to map joint assets, and the sum of all the lines would represent household asset holdings (represented by an outer hexagon). Note that the context used below is hypothetical and is only one example of how the gender asset gap might look. In fact, the gender asset gap will vary across different contexts—for example, in some contexts men will have more human capital than women; in other contexts women's human capital may be on par or even greater than men's—thus any conceptual illustration of the gender asset gap would vary along these lines.

The graphical depiction suggests, and empirical evidence supports, that for a given context, men and women will own different types of assets. For example, in the rural Philippines, women may have, on average, higher education levels, while men, on average, own greater areas of land (Quisumbing et al. 2004). According to Antonopoulos and Floro (2005), Thai women were more likely to own jewelry, while men were more likely to own transport vehicles. Examining patterns of livestock ownership by men and women, Kristjanson et al. (Chap. 9) found that women were more likely to own small livestock such as poultry and goats, while men were more likely to own large livestock such as cattle and buffaloes. Men's and women's different patterns of control and ownership of assets exist both nationally (Hausmann et al. 2010) and within communities and households.

While women may have greater asset ownership of certain types of assets as compared to men, a growing body of empirical evidence shows that in contexts throughout the world, women typically have fewer overall assets than men. For example, Antonopoulos and Floro (2005) found that in Thailand, men's assets were, on average, worth more than those of women. Likewise, Quisumbing and Maluccio (2003) found that husbands brought greater wealth to marriage than wives in

Bangladesh, Indonesia, Ethiopia, and South Africa. Men accounted for a larger proportion of property owners and owned more land in Brazil, Nicaragua, Mexico, and Paraguay (Deere and León 2003), while men owned more assets than women in Ghana, Kenya, Northern Nigeria, Mexico, and urban Guatemala (see the survey of the literature documenting gender-asset gaps in Deere and Doss 2006). In formal education, many countries still have a large gender-schooling gap. In countries such as Ghana, Uganda, Cambodia, India, Guinea, Bolivia, and Iraq, men still have, on average, at least 1 year more of schooling than women, while others, such as Rwanda, Kenya, Palestine, and China, show average gaps in the range of 0.6–0.8 years of schooling (Hausmann et al. 2010).

5.2.2 Consequences of Increasing Women's Control of Assets

Previous research has found that not only do gender disparities in asset control exist, but increasing women's control over assets, mainly land, and physical and financial assets, has positive effects on a number of important development outcomes for the household, including food security, child nutrition, and education, as well as women's own well-being (Quisumbing 2003). For example, the greater a woman's asset holdings at marriage, the larger the share the household spends on children's education (Quisumbing and Maluccio 2003). In Bangladesh, a higher share of women's assets is associated with better health outcomes for girls (Hallman 2000). A study by Smith and Haddad (2000) using cross-country data found that increases in women's education (investment in human capital) have made the greatest contribution to reducing the rate of child malnutrition, accounting for 43 % of the total reduction. Improvements in food availability came in a distant second to women's education, contributing 26 % to the reduction, while improvements in women's status, proxied by the ratio of female to male life expectancy, make up 12 %. These gains to household welfare suggest that strengthening women's control of resources need not result in a zero-sum gain. Indeed, if the existing distribution of resources is not Pareto-optimal—that is, if welfare can be improved by redistributing resources from one household member to another—there may be efficiency gains in increasing women's control of resources.

The importance of strengthening women's control over resources is also reflected in five of the eight United Nations Millennium Development Goals, and none of the goals can be fully achieved without addressing gender disparities in rights to resources. Abu-Ghaida and Klasen (2004) estimate that those countries that are off track of meeting MDG 3 on gender parity in primary and secondary education are likely to lose an average of 0.4 percentage points in annual economic growth between 2005 and 2015. To ensure that Goal 3, Promote Gender Equality and Empower Women, is achieved by 2015, the UN Millennium Project Task Force on Education and Gender Equality has identified seven strategic priorities, including strengthening women's and girls' rights to education and property (Grown et al. 2005).

The rationale for reducing the gender-asset gap extends beyond disparities in education to other productive resources. Based on results of various empirical studies, Alderman et al. (1996) suggest that agricultural productivity could increase by 10–20 % if women had access to the same range of inputs (education, fertilizer, labor) as men. Estimates by the FAO in the *State of Food and Agriculture 2011* suggest that if women had the same access to productive resources as men, they could increase yields on their farms by 20–30 %. The FAO (2011) extrapolates these yields as increasing total agricultural output in developing countries by 2.5–4 %, and reducing the number of hungry people in the world by 12–17 % (FAO 2011, 5), which would imply 100–150 million less hungry people in the world.

Women's control of tangible assets may also affect the outcomes of household decisions; in Ghana, expenditure patterns differed based on the share of assets owned by women in the household (Doss 2006). In the context of a burgeoning HIV/AIDS epidemic, securing women's property and inheritance rights to land can promote women's economic security and thus reduce their vulnerability to domestic violence, unsafe sex, and other AIDS-related risk factors (Bhatla et al. 2006; Gillespie and Kadiyala 2005). However, increasing women's control over key assets, such as land, may not be straightforward, as discussed below.

The gender gap in specific assets has consequences not only because of their direct impact, but also because of the interactive effects of ownership and control of different *types* of assets. Several longitudinal studies using income and/or expenditure capture the welfare enhancing capability of land in different settings in Latin America and Africa (Scott 2000; Gunning et al. 2000; Grootaert et al. 1997). While these accounts were able to isolate the welfare effects of land among a broad range of assets, including human capital (proxied by education and labor), livestock, machinery, transfers, and over a broad range of household-level and environmental shocks, the interactions of land—including both individual and common property—with these assets need further understanding. Early attempts at fleshing out these interactions in Mexico by Finan et al. (2005) show that household characteristics, complementary assets, and contextual circumstances greatly influence the welfare generating potential of land. For example, one hectare of land can be sufficient to escape poverty for households living in villages with access to a paved road (physical capital), in large part because Mexican farm households are engaged in off-farm activities that complement incomes derived from land. Recent studies in South Africa demonstrate that land restitution and redistribution programs have done little to contribute to poverty reduction (Bradstock 2005). Instead, access to the labor market for rich households and access to social grants for poor households remain key to avoiding poverty—highlighting the importance of contextual circumstances in addition to assets. Conversely, Friedemann-Sanchez (2006) shows how women in Colombia use social capital for acquiring land and home ownership. Recent work is also highlighting the potential interaction between natural capital, especially land and water, physical capital, and other assets much more broadly defined to include access to education, information, and institutions in reducing poverty (Moser 2006).

These and other studies make a convincing case that closing the gap between men's and women's ownership of assets is a necessary step toward poverty reduction and achieving global development goals. As explained by Doss et al. (2008, 3):

> The gender-asset gap arguably provides a much firmer basis for understanding gender economic inequality and women's empowerment than just a focus on income or wages and may be a more powerful indicator of progress than others toward Millennium Development Goal 3 (Promoting gender equality and empowering women). Besides being a measure of opportunities (i.e., through the ability to generate income or additional wealth) or outcomes (net wealth), ownership of assets is critically important to women's bargaining power and hence their economic empowerment.

5.2.3 Social and Institutional Bases for Property Rights

Even with recognition of the importance of women's control over natural capital, particularly land and water, there is no straightforward path to strengthening women's rights to these assets (Meinzen-Dick 2006).[1] Statutory law is only one (albeit important) source of rights and it is one that many women fail to use, owing to lack of legal knowledge and poor legal implementation. Thus, it is important to begin with an understanding of existing statutory and customary rights systems, which often involve complex relationships between different uses and users of the resources. It is also important to recognize that the ability to control resources has many components, including use of the resource, modification, sale, consumption, and so on. Rather than outright "ownership" of resources, women often have separate "bundles" of rights in which they are able to use or control select aspects or components of resources. For example, a woman may be allowed to collect water or fallen branches for firewood from a piece of land, but have no rights to plant trees on that land. Land titling programs that fail to acknowledge the different bundles of rights have, in many cases, decreased rather than increased women's tenure security by strengthening the claims of men without recognizing the rights women have had over land under customary systems (Lastarria-Cornhiel 1997; Mwangi 2007).

Claims about who is able to use or control a given resource may be based on a range of customary or religious laws or local norms. In fact, there may be competing statutory or customary rules or norms about the same resource or behavior. For example, a country may have laws specifying that all children are entitled to inherit an equal share of assets from their parents, while Islamic law specifies that daughters

[1] Even in industrialized countries, the strengthening of women's property rights has not been a linear path. An exhaustive review of women's control of assets (Deere and Doss 2006) found that some of the reasons for the large increase in women's wealth between the 1920–1950s in the United Kingdom and the United States were longer life expectancy for women than for men, an increase in joint purchases of homes, and changes in estate laws for tax purposes that favored division of wealth among spouses. The applicability of "lessons learned" from industrialized country experience needs to be tempered by the knowledge that "ownership" is more complex in developing countries and the timing and composition of wealth transfers may differ substantially cross-culturally.

receive one-half the share of sons, and local norms may prescribe that women should not cultivate their land, but rather give it to their brothers. Rights to water or trees may differ depending on whether they are to be used for domestic consumption on production and market-related activities. Even where there is agreement on rights, it may be difficult for people to operationalize their claims, particularly in the face of social pressures. On the other hand, men and women also use their social connections to access land as well as the labor needed to use it, particularly for agroforestry. Thus, both state law and local norms, particularly the interplay of gender and power relations, play a crucial role in shaping women's rights to assets (Meinzen-Dick 2006).

Men and women also acquire assets differently. For example, in Sub-Saharan Africa, men often acquire use and certain management rights over land through inheritance or allocation by their clan or lineage, whereas, for women, marriage is the most common way to gain access to land. In this case, women's rights to land may be either use rights or permanent rights (also see Lastarrhia-Cornhiel et al., Chap. 6). In Latin America, women become landowners mainly through inheritance, while men are much more likely to acquire land through purchases in land markets (Doss et al. 2008). There are also varying levels of knowledge on the gendered patterns of asset ownership with relatively more knowledge surrounding physical and tangible assets and education compared to other non-tangible assets. Therefore, not only have gendered asset inequalities and their impacts been documented, but there is also evidence pointing to nuanced gendered differences in the exclusivity of asset ownership, bundles of rights, and acquisition.

5.3 A Conceptual Framework Linking Gender and Assets

In spite of the recognized importance of assets, few development interventions explicitly consider their impacts on men's and women's assets. This omission points to critical gaps in basic research about the extent and consequences of the gender gap in assets, how assets are accumulated by men and women, and which mechanisms best strengthen women's access to productive assets. Although many programs aim to increase women's asset ownership, very few have documented successful and sustained ability to reduce the asset gap between men and women. In some cases, while women may accumulate assets, men may acquire them at a faster rate, or even take over the control of women's assets, thereby worsening the gender-asset gap.

The conceptual framework presented here (Fig. 5.2) draws from several other efforts to articulate the critical relationships between assets and poverty, notably the Sustainable Livelihoods (SL) framework (Carney et al. 1999; DfID 1997; Scoones 2009), the Capacities and Capabilities Framework (Bebbington 1999), as well as other work on poverty traps (Carter and Barrett 2006; Barrett and Swallow 2006), and pathways from poverty (IFPRI 2003).[2] While these frameworks are all useful

[2] For details on key ideas from each framework that we draw upon, see Meinzen-Dick et al. 2011.

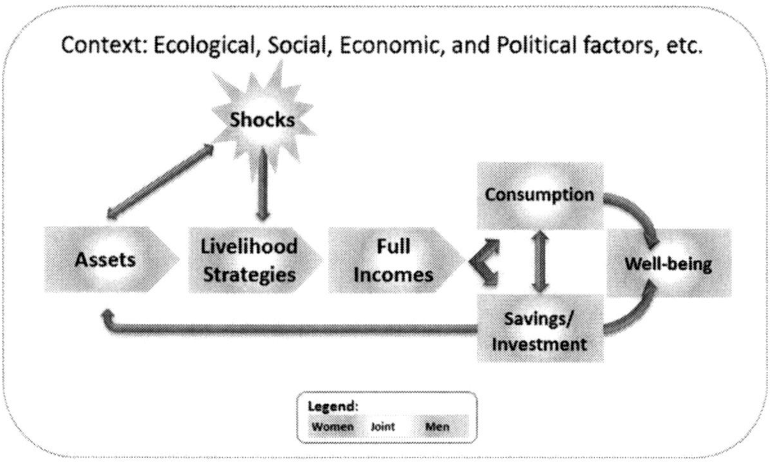

Fig. 5.2 Schematic representation of a gendered livelihood conceptual framework

mechanisms to assess various relationships between assets and poverty, most of these approaches are not explicitly gendered, and applying a "gender lens" to each framework revealed enough shortcomings to warrant a new framework that explicitly examines assets and livelihoods from a gendered perspective.

The aforementioned frameworks may be applied to a household as a whole, or to individuals, but do not capture the complexity of both individual and shared assets, decisionmaking, and outcomes of men and women of different ages, within households. The unitary model of the household does not adequately capture gender dynamics, treating all assets, enterprises, and consumption at the individual level (as in the Pathways from Poverty framework). This is inadequate, as it ignores the instances of sharing that occur within households. Both joint production and consumption are important.

The Gender, Agriculture, and Assets Project (GAAP) framework (see Fig. 5.2) shows the links between assets and well-being while making clear that gender relations influence the constraints and opportunities that occur in each pathway. In our framework, each component is gendered. The shading of each component is a reminder that we need to consider separation and jointness in each box. Women and men have separate assets, activities, and consumption and savings or investment strategies, but households can also have joint assets, activities, and consumption strategies, among others.

The first element of the conceptual framework is the *context*, which may include a broad range of ecological, social, economic, and political factors that affect men's and women's control over and access to assets. Even if individuals are living in the same household, men and women typically experience this context differently based on their roles and responsibilities and other social, economic, and cultural factors. In some cases, the gendered nature of the context is explicit. For example, cultural norms may define roles and responsibilities for men and women, and in some cases, men and women are treated differently by laws or legal provisions.

In other cases, it is necessary to look deeper to fully realize how contextual differences shape men's and women's differential access to and control over assets. For example, while the biophysical context applies to the entire household, its effect may also differ between men and women, depending on other contextual factors and assets. Low rainfall may be less of a constraint to men if their fields have irrigation and women's do not (or conversely), or if women have primary responsibility for collecting water for the household and must devote more time to this activity. Availability of wild plants may be more important for women than for men, if gender roles assign them to be responsible for basic food security and health care of their families. Conversely, availability of wild animals may be more important to men who typically hunt. Geographical location, such as proximity to roads and market centers, might appear to have homogenous effects on all members of a household, but, in fact, men and women may differ in terms of their ability to travel or to engage in certain types of markets. Security conditions may also differentially affect men and women, particularly with respect to women's safety in traveling, working fields, and engaging in gathering activities for firewood or water.

In addition to the visible or concrete contextual factors, it is essential to consider the institutional context. The "institutions of exchange" are especially important, as they condition the ways through which livelihood strategies are translated into incomes. Markets are the most obvious institutions of exchange, but social reciprocity norms (such as mutual help groups for labor or norms of sharing food) also play a critical role. Access to markets is often gendered: for example, women are restricted from participating in markets in parts of South Asia; however, equal participation by men and women is more common in Latin America; and women are active as traders, as well as clients, in much of West Africa.[3] These institutions include not only markets for agricultural production, but also labor markets, in which women's participation tends to be more limited than men's (see Chaps. 12, 13, and 14 by Rubin and Manfre, Hill and Vigneri, and Dey de Pryck and Termine).

It is important to keep in mind that in general, context should not be considered as static: weather patterns, access to markets, and certainly institutions (even those considered "traditional") change over time, and should be regularly re-assessed.

Access to, control over, and ownership of *assets*, as described in the previous section, are key determinants of individual agency. The shading in this and all other components of the diagram represent the fact that within a household there are assets that are held by women, some that are held by men, and others that are jointly owned and/or utilized jointly. The distribution of assets in a particular household will influence how the household and its members use their assets to further their livelihoods and improve well-being.

Livelihood strategies represent decisions that individuals and households make about how to invest their assets in productive and reproductive activities in order to

[3] For an analysis of gender relations and control over capital in Indian and West African marketing, see Harriss-White 1998.

generate expected returns. The livelihood strategies available in a particular area will depend on many of the contextual factors (agro-ecology, market access) and may be heavily influenced by gender roles. Whether men and women will be able to pursue the available strategies will further depend on what assets those livelihood strategies require, and on how "household assets" are allocated across different household members to enable them to engage in specific livelihood strategies. In some cases, men and women pursue different livelihood strategies; in other cases, these may be joint, as with "family farms" or family businesses.

The actual returns to different activities will depend on *shocks* (negative or positive). Weather, disease, illness (particularly of prime-age adults, see Harris, Chap. 11), violent conflicts, theft, and even sudden policy changes represent potential shocks. The majority of shocks we list here have a negative effect, but there are also positive shocks, as well as shocks that have both positive and negative effects for different people in a given household. For example, a drought that reduces crop yields on a broad scale and leads to higher prices can benefit the people who have irrigation and can still produce, or people who produce a particularly drought-resistant crop.

How are shocks gendered? First, men and women experience shocks differently, depending on their different roles and responsibilities. Men who own livestock are more directly affected by cattle rustling or by drought that reduces the availability of good forage; women who keep poultry will be more affected by diseases such as avian influenza. Human diseases are likely to have a disproportionately large effect on women, who are often affected not only by their own illnesses and typically have lower access to healthcare, but are also responsible for taking care of other sick family members (see Harris, Chap. 11).

Second, men and women have differential ability to withstand shocks. For example, do men and women have equal access to irrigation or rain harvesting methods to address the effects of droughts, or have the same access to insurance to deal with extreme weather or pests? Third, assets can play an important role in withstanding or responding to shocks, and men's and women's assets are often used differently to respond to shocks. For example, in Bangladesh, Quisumbing (2011) found that women's assets are disposed of to respond to family illnesses, whereas men's assets are used for marriage expenses and dowry. This has important implications for gendered asset accumulation if the incidence and magnitude of both shocks and asset disposition vary over time.

In addition to general and gendered shocks, there are also shocks that specifically affect women and lead to loss of their assets and a threat to their livelihood strategies. For example, divorce or death of a husband can lead to women losing their assets, especially in cases where marriage is governed under customary laws that do not protect women's rights to property (Peterman 2010).

The livelihoods strategies and shocks result in a household's *full income*, which is defined as the total value of products and services produced by the household members, some of which are consumed directly and others sold for cash or traded for other goods or services. The concept of full income also includes leisure time of household members. Because it is more likely for women's time to be devoted to nonmarket or reproductive activities—including growing food consumed at home, caring for children, and caring for the ill—measures of income that do not take into account the value of time spent in maintaining the household will tend to underestimate women's contribution.

Household members differ in their contributions to household income, and they also differ in their control over how that income is used. Under the unitary model of the household, this distinction is not an issue, but where household members have different preferences, household expenditures will differ depending on how control over income is distributed within the household. A large body of evidence shows that, in many parts of the world, men and women spend money differently: women are more likely to spend the income they control on food, healthcare, and education of their children (Haddad et al. 1997; Lundberg et al. 1997). Evidence from Malawi and Uganda showed that women were more likely to spend more of their income on food compared to men, while men were likely to spend more of their income on assets than women. On average, women spent 23 % of their income on food and 14 % on assets, while men only spent 8 % of their income on food and 25 % on assets (Njuki et al. 2011). Asset ownership, in particular, is among the factors that may influence women's control over income and bargaining power in household negotiations (Doss 1999; Thomas et al. 2002; Quisumbing and Maluccio 2003).

Neither the unitary model of a household pooling all income and allocating for the needs of all, nor the bargaining model of individuals bargaining based on their individual interests, is likely to fit most situations. Rather, when considering *consumption* (of goods, services, and leisure) and *savings and investment*, it is useful to consider which decisions are made individually, and which collectively. Because full income also includes the value of time, it is important to consider the distribution of leisure within the household—or conversely, whose time is most occupied in productive and reproductive activities. When both market and household work are taken into account, time allocation studies show that women work significantly more hours than men (Juster and Stafford 1991; also, see World Bank 2001, 66).

In simplified terms, *savings* are the balance of income that is not consumed. How savings are *invested* will affect asset accumulation (or loss) for the future. If kept in a bank account, savings would increase financial capital; if used to purchase equipment or build a house, savings build physical capital; if used to buy land, plant a tree, or install irrigation (water control), then savings increase natural capital.

Although much economic theory dichotomizes between consumption and savings, in fact the dividing line is not so clear. Certain types of consumption can also increase intangible assets of human and social capital. Consumption of healthy food, clean water, adequate shelter, and a clean environment improves nutrition and health outcomes for adults and children, which is an important aspect of human capital. Ceremonial expenses, hospitality, the ability to wear decent clothing, and even some types of conspicuous consumption of prestige goods, as well as spending time with others (either informally or in group meetings), can all contribute to social capital (Cancian 1972).

Again, the relevant question is, "how are investment patterns gendered?" For example, how are women's, men's, and joint income used for different types of investment by different family members? What affects their respective decisions on investment? What are the common and differential opportunities for men and women to invest? These include both formal opportunities as well as the practical obstacles. For example, even if a country legally allows women to own land, if most land is held under a customary tenure regime where decisions are dominated by men, then women will be effectively excluded from this avenue of asset accumulation. Policies that give husbands and wives joint tenure over land acquired during marriage and implement practical steps to ensure application of the policy can result in an increase in joint asset ownership. In the case that women (or men) are precluded from investing in one type of asset, are there other types of assets that they can accumulate, and how valuable are they for creating good livelihood options, or for strengthening bargaining power?

Changes in savings and assets are not always positive. In the case of a severe shock (such as a major drought, or family illness), a household may need to dip into its savings or liquidate particular assets in order to maintain a certain level of consumption. As described above, men's and women's assets may be used differently to buffer shocks. Children (often girls) may be kept out of school, reducing human capital accumulation. In cases of negative savings (debt) and investment, it is important to ask whose savings or assets are being liquidated to keep the individual or household consumption levels and whether there will be other mechanisms for those who lose assets to replace them. For example, women's jewelry is often used to pay for family emergencies. Where banks or pawn shops are available to provide loans against the jewelry, there is a greater chance that the women can reclaim their asset, compared to having to sell the asset outright.

The shading in the framework is a reminder that all of the key components may be different for men and women, or may be shared by members of a household (or even community). This prompts us to consider how the differences in context, assets, livelihood strategies, risks, and other components for men and women may affect outcomes for individuals and households (with particularly important implications for children and the intergenerational transmission of poverty). Degrees of shading—whether control and ownership of assets tend to be concentrated in men, or be more equitably distributed—represent the extent of the gender gap in assets and consequently in bargaining power.

5.4 Gender, Assets, and Agricultural Development Interventions

By articulating the importance of gender and assets in livelihoods and welfare, our framework can inform the design and implementation of a range of policy and programmatic interventions. In this section, we use the framework to assess the implications of gender and assets for agricultural development interventions. However, the overall framework could be used to examine nonagricultural development projects (such as microfinance) as well.

Despite the lack of knowledge about what works to reduce the asset-gap, we now have substantial experience demonstrating "what works" in gender targeting of agricultural development interventions. This includes knowledge surrounding improving women's participation, and increasing the chances that women will benefit from the project activities, and working with men to change attitudes and behaviors that limit women's economic opportunities (Quisumbing and Pandolfelli 2010; Kristjanson et al., Chap. 9). While these methods are well known, they are still not widely used in development projects. Lack of knowledge and/or capacity—from field implementers through to project managers and donors—has led to a situation where what is common practice is often far from what is known to be "good practice." For example, good practice involves going beyond simple participation to ensure that women capture meaningful benefits and are empowered by the intervention process. Only by conducting rigorous analyses of alternative interventions that include well-designed and implemented strategies for reaching women can we begin to identify which pathways provide the greatest opportunities to build women's assets and offer guidance about policies that help reduce the gender asset gap.

The current framework can assist in the design of development programs, by better conceptualizing how the gendered asset distribution affects the uptake and eventual outcomes of programs, and how the accumulation of assets by men and women is affected by interventions. By specifying the linkages between assets, livelihood strategies, risks, and outcomes, it can also help to design better impact assessments that show which strategies are most effective in different contexts.

5.4.1 How Agricultural Development Programs Influence Assets

Agricultural development interventions tend to influence assets in three major ways. First, some interventions increase the stock of agricultural assets such as land, livestock, water, or machinery, enabling farmers to increase production, or build up the stock of intangible assets (human capital, social capital, political capital) that may be complements to traditional agricultural assets. Second, they can increase the returns to assets such as land or labor that are used in agriculture by increasing productivity, for example, through improved technologies, or ameliorating market failures. Third, they can reduce risk, thereby protecting assets. In reality, many projects affect assets through a combination of these three pathways. Agricultural development programs may also include interventions to strengthen markets and increase income, even if they do not directly involve specific assets as transfers or target outcomes.

5.4.1.1 Programs That Increase the Stock of Agricultural Assets

Many land reform, redistribution, and/or titling programs have the goal of stimulating agricultural productivity by improving access to land, security of tenure, and providing means and incentives, via credit markets, to increase investment in agricultural production (see Lastarria-Cornhiel et al., Chap. 6). Irrigation development programs, fishponds, and livestock distribution schemes similarly seek to increase the asset base so that people can increase their productivity in agriculture (see Kristjanson et al., Chap. 9). Most technology transfer programs can be seen as increasing the physical capital on farms, and even the introduction of new seeds can be seen as improving natural capital. Beyond natural and physical capital, programs also invest in strengthening human capital (via training, including extension services) or social capital by building or strengthening organizations (see Chap. 17 by Ragasa and Chap. 10 by Meinzen-Dick et al.).

In terms of application to the framework, these programs translate into an increase in the size or value of the asset component. Its impact on the asset hexagon and ultimately the shading of the box indicating jointness or relative control by men or women will depend not only on how the program assigns rights, but also on whether these rights defined by the program can be defended against other competing claims in the household and community. Whether the individuals and households are willing and able to maintain the asset will depend on how a particular increase in asset stock contributes to welfare.

The impact of the asset building programs on food production, income, and ultimately on well-being will depend in part on who ultimately uses the asset in what livelihood strategies. This, in turn, will depend on current gender roles, especially in reference to labor and access to complementary inputs (e.g., credit, knowledge). Control over the income generated by the assets (whether in kind or cash) will also be important both in terms of incentives of household members to use the asset and in terms of how the products and services it generates translate into well-being for household members. Finally, the ownership of the asset itself may alter intrahousehold negotiations by strengthening the bargaining position of its owners at the expense of others.

In practice, many agricultural projects provide more than one type of asset, because they recognize that complementary assets may be needed for people to take advantage of the main asset being transferred. For example, Heifer International (http://www.heifer.org) does not only transfer an animal; it organizes recipients into groups that will receive training on how to care for the animal and help each other to raise the animal, thereby strengthening human and social capital as well. In Mali, the Millennium Challenge Program (MCC http://www.mcc.gov/pages/countries/program/mali-compact) not only provided women with irrigation, but also training, seeds, and assistance in forming women's farming associations. Landesa's (http://www.landesa.org/women-and-land/) work to transfer homestead plus garden land titles to poor families in India is an example of a program that seeks to intervene directly in strengthening assets. It is also a good example of one that seeks to strengthen joint assets, with attention to ensuring daughter's inheritance rights. The project organizes community discussions and boys' and girls' groups to address gender discrimination and early marriage, in order to ensure that the provision of land to poor households will also benefit the daughters.

Certainly not all (or even a majority) of agricultural programs that aim to increase the stock of assets, whether through distribution, subsidized purchase, or other means, target women. To the contrary, many assume that men are the farmers, and therefore transfer assets to the (male) "head of household." Such gender-blind programs are likely to increase the gender-asset gap.

5.4.1.2 Programs That Increase Returns to Agricultural Assets

Many types of agricultural programs focus on introducing new or strengthening existing livelihood strategies, through new crops or inputs to make existing crops more productive, or improving market access and value chains so that farmers receive higher prices for their output. The GAAP conceptual framework can be used as a diagnostic tool to make such programs more successful: if a particular type of intensified production requires certain assets or increases returns to certain assets, and if poor households or women farmers (in households headed either by men or by women) do not have those assets, then they will not be able to benefit from the intervention. Ensuring that they benefit may require either selecting interventions that increase returns to assets they do have, or looking for ways to improve their access to the assets they need.

If women do not have control over land or water resources, for example, a project can work to identify land for group gardens or collective irrigation. Alternatively, the intervention may be adapted so that the lack of these assets is not a constraint to adoption. In developing soil fertility replenishment strategies in Kenya, the World Agroforestry Centre recognized that women had limited property rights to plant trees on their land, and often lacked the cash or transport needed to acquire chemical fertilizer. They therefore used plants that grow in hedgerows and "interstitial spaces" where women could harvest the leaves and transfer biomass to improve their soil fertility on land that they cultivate, but do not "own", Adapting the outreach materials so that they were understandable by illiterate women further meant that lack of human capital (education) was not a barrier to adoption; the result was that women adopted on a par with men (Place et al. 2007).

A study in Uganda by Nkedi-Kizza et al. (2002) found that 22 % of women and 52 % of men thought there was a gender bias in allocation of plots to men and women, with men taking up the more fertile plots. The authors found slightly higher levels of soil organic matter (available nitrogen and available phosphorous) in plots managed by husbands compared to those managed by wives. This will influence women's ability to cultivate certain crops successfully. Projects seeking to benefit women need to focus on crops that can be grown on their plots, or on making fertilizer more accessible. Recommendations to make fertilizer adoption more feasible for poor female farmers in Africa include the sale of fertilizer in smaller bags, rather than the 50-kg bags that poor farmers are unable to afford (Gladwin 2002), and that women may especially have difficulty transporting.

Strengthening human or social capital might appear to be an option for targeting asset-poor individuals and households, but programs often inadvertently put in place asset-based barriers. While often well intentioned—it is true that human or social capital may not translate into new livelihood strategies if people cannot access complementary inputs—poorly designed or overly simplistic criteria for program participation exclude people who could benefit. For example, producer associations often require landownership as a prerequisite for membership, thus limiting the participation of women and youth who may have access to household or community land but no claim to ownership (Meinzen-Dick and Zwarteveen 1998).

Diagnosing these types of situations ex ante can encourage agricultural programs to make provisions for more equitable participation. For example, a polyculture fishpond technology program implemented through the government extension system in Bangladesh and targeted to households primarily benefited men from wealthier families, even if women were required by the donor to account for 30 % of project beneficiaries, because adopting the technology required ownership of a pond, or land on which to construct the pond. To reach landless women with this technology, an NGO made provision for groups of women to rent water bodies that they could use collectively to grow fish (Hallman et al. 2007).

Many agricultural development interventions seek to influence not only production and income but also outcomes such as nutrition or health status. For example, the Harvest Plus initiative (http://www.harvestplus.org) promotes varieties of staple crops that are higher in micronutrients. This intervention essentially improves the quality rather than the quantity of food produced by the household—which is another way to improve productivity—however, whether or not that will translate into improved nutrition depends on the willingness and ability of household members not only to plant the new crop, but also to feed it to members of the household who are nutritionally vulnerable. To help ensure that this happens, seed distribution is often accompanied by social marketing and behavior change campaigns to encourage consumption of these nutritious products. Similarly, dairy development projects such as the East African Dairy Development (EADD) (http://www.heifer.org/eadd/index.html) project are testing nutrition awareness messaging to encourage households to dedicate some of their increased milk production to the nutrition needs of target groups such as children and pregnant women. While awareness-raising is important, our framework suggests that these programs also need to pay attention to whether women are actually able to influence not only how food is allocated (which foods and to whom) but also how much of household expenditure will go to food, a function of their control over full income.

5.4.1.3 Innovations to Reduce Risk

A growing number of agricultural development programs seek to address shocks through financial instruments such as insurance; however, typically, most of these products only target men. Whereas participation in crop insurance programs is often restricted to landholders, newer weather-based index insurance can be purchased by landless families or women. Nevertheless, if women are less involved in agricultural production, or if weather shocks do not directly affect their asset holdings, they may be less willing to pay for weather insurance. Conversely, if women's assets are disposed of to cope with illness shocks, a health insurance project might be an important avenue for social protection for women. In general, the GAAP framework can serve as a reminder of the importance of shocks in the lives of the poor, and to assess whether new livelihood strategies being introduced will increase or decrease vulnerability to such shocks. Are the new crop varieties more susceptible to fluctuations in water, temperature, or pests? If aiming for specialized markets, will that introduce price fluctuations and the risk of produce not meeting grading standards? If so, is there a differential ability for men and women farmers to bear these shocks? Use of this framework can draw attention to other types of shocks that may affect particular livelihood strategies. For example, if malaria, HIV, or other diseases are a constraint on labor availability, teaming with health interventions to redress those shocks may be essential to the outcome of the program. As noted above, women often bear a disproportionate burden for illness shocks, so health interventions may be especially important for gender-equitable participation.

The same logic applies to government programs such as public works projects that act as a form of insurance. Provision of childcare facilities (as in India's National Rural Employment Guarantee Scheme) increases women's ability to participate. In a recent study of livestock insurance, 42 % of insurance contracts were purchased by women, although it was not clear whether it was insurance against women-owned livestock. For example, with evidence that most of the camels were owned by men, 37 % of the contracts for insuring camels were sold directly to women (Mude, Andrew, 2011, personal communication). Focus group discussions showed the high number of women purchasing insurance was as a result of the absence of men due to migration with livestock.

5.4.2 *Interventions to Strengthen Markets and Increase Income*

Market expansion, linking smallholders to high-value markets, is the avowed aim of many current agricultural programs by governments and NGOs. Examples of market-oriented interventions include infrastructure—roads, communication systems, collection and storage facilities—as well as investments in better information and better organization on the part of producers and/or other actors in the value chain. Market investments are often accompanied by technology investments based on the logic that increased market opportunities will provide an incentive to invest in improved productivity. As with other elements of the framework, examining these interventions as part of the entire cycle can draw attention to complementary interventions that may be needed, as well as to the other factors that condition returns to program interventions and how they are distributed within households.

It is important to consider how participation in different types of markets is gendered (see Chap. 13 by Hill and Vigneri, and Chap. 12 by Rubin and Manfre). For example, in Bangladesh, where women are restricted from going to markets, a dairy value-chain project hired women to be milk collectors and redesigned the vehicle they were to use to make it easier for them to visit the homes and collect the milk from other women producers. In some cases, special training, for example, in negotiation skills, may be needed for women to participate in markets. Women often participate in more informal markets, accepting buyers' offers rather than negotiating for better prices for their commodities. Training in negotiation skills can enhance women's bargaining power. As markets get more formalized and further away from their homes, women can be disadvantaged if interventions to increase their participation and benefits from these markets are not implemented.

In many cases, access to markets depends on other assets, such as transportation or communication equipment like carts to get produce to markets or cell phones and radios to find out market opportunities and prices. Here again, our framework would draw attention to the question of whether women and men have these necessary complementary assets.

In addition to intrahousehold bargaining power, the frequency and size of income receipts matter. Small incremental payments may not allow for much savings unless there are microfinance institutions available; likewise, lumpy income receipts (such as at an annual harvest) can lead to disproportionately high consumption followed by a hungry season unless there are appropriate ways to save and reinvest. Looking at how access to these channels are gendered will help in ensuring that increased income does, indeed, translate into improved assets and a reduction in the gender gap in assets.

The key questions revolve around the strategies needed to ensure that women both earn income and have control of the income they earn. Many agricultural development programs, even those that ostensibly attempt to increase women's production and income, result in men taking control of the output that women have produced (for classic examples, see von Braun and Webb (1989) and Jones (1983)). In cases wherein men take the produce to market and get paid for it, they may also be taking the decisions on consumption and investment by themselves, leaving women with little influence over these critical decisions that affect their own welfare and those of the children. New options to make payment into women's microfinance accounts, or to pay via cell phone, can help to ensure that women retain control over income and consumption decisions. However, these innovations also depend on women having access to financial capital (savings accounts) or physical capital (cell phones) to equitably implement a given program design. Technological approaches are not the only way to ensure income for women. In Malawi, for example, integrating gender training in a market development program, having multiple crops and livestock enterprises, and focusing on different types of markets led to more income under the control of women (Njuki et al. 2011).

5.4.3 Impacts of Agricultural Programs on the Gender Distribution of Assets

While agricultural development programs may affect the distribution of assets within the household, very few efforts have been made to examine these impacts, partly because sex-disaggregated asset data are scarce. There is suggestive evidence that interventions that attempt to equalize the gender-asset gap may have better impacts on health and nutrition outcomes. A gender-blind Bangladesh fishpond program that targeted information regarding the technology to the households and by default, husbands (Hallman et al. 2007; Kumar and Quisumbing 2010) found that participation in the program increased husbands' holdings (relative to their wives') of land, livestock, and total value of assets; whereas in programs targeted to women's groups, women's assets increased faster than their husbands', even though husbands still owned the majority of household assets. Moreover, while the gender-blind program had the largest gains in terms of consumption expenditures and household assets, improvements in nutritional status of women and children were less than those in the programs targeted to women's groups. One could argue that one way to reduce the gender gap would be to reduce men's assets, and consequently, the overall household holdings; however, this is obviously not desirable. Ideally, men's, women's, and joint assets would increase, but women's would increase more rapidly in situations where they have had less control over assets.

It is important to also note that targeting women alone for market-oriented agricultural interventions may backfire, leading to appropriation by men as women's enterprises become profitable. In this case, working with both men and women and with multiple enterprises may secure women's participation and management of income. The type of product or commodity as well as the type of market can also influence who markets and subsequently who controls the income from the commodity (Njuki et al. 2011). In a study in Malawi and Uganda, women were more likely to participate in local markets for legumes and livestock products such as milk than in cattle markets or markets for cash crops such as tobacco. The setup of marketing arrangements can have an important influence on the degree of separate or joint control of incomes within the household, as well. As previously mentioned, if men take the produce to market and receive the payment, women may receive little or none, even if they grew the crop or cared for the animals. In this case, if payments are made to a woman's account, or using payment means that she has access to such as through mobile phones, she may retain a greater share. In addition, providing price information can improve transparency, both within and outside the household.

5.5 Summary and Implications

Tangible and intangible assets play a multifaceted role in increasing well-being: they are required to pursue certain livelihood strategies, buffer against shocks, and provide for status and bargaining power for those who hold them. There is also increasing evidence that assets are not shared or distributed equally, even within households, with women usually controlling fewer assets than men, as detailed in the following chapters of this book. However, the implications of the gender gap in assets have not been fully examined, nor has the knowledge that does exist been consistently applied to development programs.

The GAAP conceptual framework presented in this chapter offers a starting point for examining how gender and assets influence the well-being of households and individuals. The first step in applying the framework is to identify the relevant contextual factors, then consider how access to and control of assets affects livelihood strategies and ability to withstand shocks, to result in full income (including not only cash, but in-kind products and leisure). Rather than focusing exclusively on income, the framework highlights how income is allocated between consumption and savings or investment, affecting welfare of household members and asset accumulation (or loss). Once again, assets may influence bargaining power over the decisions on how income is used.

A key element of this framework is that each component is gendered, allowing for men and women to have different assets, livelihoods, shocks, income, and consumption, but also for some elements of each of these to be shared within the household.

This framework can be used to generate hypotheses about individual and household decisionmaking and to measure the impacts of agricultural development programs. In addition, beyond the research and impact assessment applications, the framework can also be used by program designers and implementers to examine how their interventions are gendered, and are likely to interact with other elements and play out in terms of ultimate welfare outcomes and long-term asset accumulation. While we still need a stronger evidence base on how programs can reduce the gender gap in assets, understanding the linkages and the impact of these programs on key outcomes of food security, health, nutrition, and even empowerment and agency can contribute to more effective development programs, particularly in the agricultural sector.

Acknowledgments This chapter is a revised version of a paper written for a research project on "Evaluating the Impacts of Agricultural Development Programming on Gender Inequalities, Asset Disparities, and Rural Livelihoods," funded by the Bill and Melinda Gates Foundation. We thank Haven Ley for her support in conceptualizing the Gender, Assets, and Agricultural Program (GAAP) project as well as Jere Behrman, Cheryl Doss, Shelly Feldman, Susan Kaaria, Anirudh Krishna, Yvonne Pinto, and all the participants in the Nairobi inception workshop of the GAAP project for helpful comments and discussions. We thank Nicole Rosenvaigue for assistance with figures. All errors and omissions are ours.

References

Abu-Ghaida D, Klasen S (2004) The costs of missing the millennium development goal on gender equity. World Dev 32(7):1075–1107

Alderman H, Hoddinott J, Haddad L, Udry C (1996) Gender differentials in farm productivity: implications for household efficiency and agricultural policy. Food Consumption and Nutrition Division discussion paper 6. International Food Policy Research Institute, Washington, DC

Antonopoulos R, Floro MS (2005) Asset ownership along gender lines: evidence from Thailand. Gender, equality, and the economy working paper 418. Levy Economics Institute, Annandale-on-Hudson

Barrett CB, Swallow BM (2006) Fractal poverty traps. World Dev 34(1):1–15

Bebbington A (1999) Capitals and capabilities: a framework for analyzing peasant viability, rural livelihoods and poverty. World Dev 27(12):2021–2044

Behrman JR (1997) Intrahousehold distribution and the family. In: Rosenzweig MR, Stark O (eds) Handbook of population and family economics. North-Holland Publishing Company, Amsterdam, pp 126–188

Bhatla N, Chakraborty S, Duvvury N (2006) Property ownership and inheritance rights of women as social protection from domestic violence: cross-site analysis. In: International Center for Research on Women (ed) Property ownership and inheritance rights of women for social protection—the South Asia experience. International Center for Research on Women, Washington, DC, pp 71–101

Bradstock A (2005) Changing livelihoods and land reform: evidence from the Northern Cape province of South Africa. World Dev 33(11):1979–1992

Cancian F (1972) Change and uncertainty in a peasant economy. Stanford University Press, Palo Alto

Carney D, Drinkwater M, Rusinow T, Neefjes K, Wanmali S, Singh N (1999) Livelihoods approaches compared A brief comparison of the livelihoods approaches for the UK Department for International Development (DfID), CARE, Oxfam, and the United Nations Development Programme (UNDP). Department for International Development (DfID), London

Carter MR, Barrett CB (2006) The economics of poverty traps and persistent poverty: an asset-based approach. J Dev Stud 42(2):178–199

Deere CD, Doss C (2006) The gender asset gap: what do we know and why does it matter? Fem Econ 12(1–2):1–50

Deere CD, León M (2003) The gender asset gap: land in Latin America. World Dev 31(6):925–947

DfID (Department for International Development) (1997) The UK White Paper on international development and beyond. Department for International Development (DfID), London

Doss CR (1999) Intrahousehold resource allocation in Ghana: the impact of the distribution of asset ownership within the household. In: Peters GH, von Braun J (eds) Food security, diversification, and resource management: refocusing the role of agriculture? Dartmouth Publishing, Aldershot, pp 309–316

Doss CR (2006) The effects of intrahousehold property ownership on expenditure patterns in Ghana. J Afr Econ 15(1):149–180

Doss C, Grown C, Deere CD (2008) Gender and asset ownership: a guide to collecting individual level data. Policy research working paper 4704. World Bank, Washington, DC

FAO (Food and Agriculture Organization of the United Nations) (2011) The state of food and agriculture 2011—women in agriculture: closing the gender gap for development. Food and Agriculture Organization of the United Nations (FAO), Rome

Finan F, Sadoulet E, de Janvry A (2005) Measuring the poverty reduction potential of land in rural Mexico. J Dev Econ 77(1):27–51

Friedemann-Sanchez G (2006) Assets in intrahousehold bargaining among women workers in Colombia's cut-flower industry. Fem Econ 12(1–2):247–269

Gillespie S, Kadiyala S (2005) HIV/AIDS and food and nutrition security: from evidence to action. Food policy review 7. International Food Policy Research Institute, Washington, DC

Gladwin C (2002) Gender and soil fertility in Africa: introduction. African Studies Quarterly 6(1/2). Available at http://web.africa.ufl.edu/asq/v6/v6i1a1.htm

Grootaert C, Kanbur R, Oh G (1997) The dynamic of welfare gains and losses: an African case study. J Dev Stud 33(5):635–657

Grown C, Gupta GR, Kes A (2005) Taking action: achieving gender equality and empowering women. UN Millennium Project 2005. Task Force on Education and Gender Equality. Earthscan, London

Gunning JW, Hoddinott J, Kinsey B, Owens T (2000) Revisiting forever gained: income dynamics in the resettlement areas of Zimbabwe, 1983–96. J Dev Stud 36(6):131–154

Haddad L, Hoddinott J, Alderman H (eds) (1997) Intrahousehold resource allocation in developing countries: methods, models, and policy. Johns Hopkins University Press for the International Food Policy Research Institute, Baltimore

Hallman K (2000) Mother-father resources, marriage payments, and girl-boy health in rural Bangladesh. Food Consumption and Nutrition Division discussion paper 93. International Food Policy Research Institute, Washington, DC

Hallman K, Lewis D, Begum S (2007) Assessing the impact of vegetable and fishpond technologies on poverty in rural Bangladesh. In: Adato M, Meinzen-Dick R (eds) Agricultural research, livelihoods, and poverty: studies of economic and social impacts in six countries. Johns Hopkins University Press for the International Food Policy Research Institute, Baltimore, pp 103–148

Harriss-White B (1998) Female and male grain marketing systems: analytic and policy issues for West Africa and India. In: Jackson C, Pearson R (eds) Feminist visions of development: gender analysis and policy. Routledge, New York, pp 189–214

Hausmann R, Tyson L, Zahidi S (2010) The global gender gap report. World Economic Forum, Geneva

IFPRI (International Food Policy Research Institute) (2003) Pathways from poverty: a proposal for a global research program (GRP26). Washington, DC

Jones C (1983) The mobilization of women's labor for cash crop production: a game theoretic approach. Am J Agric Econ 65(5):1049–1054

Juster FT, Stafford FP (1991) The allocation of time: empirical findings, behavioral models, and problems of measurement. J Econ Lit 29(4):471–522

Kumar N, Quisumbing A (2010) Does social capital build women's assets? The long-term impacts of group-based and individual dissemination of agricultural technology in Bangladesh. CAPRi working paper 97. International Food Policy Research Institute, Washington, DC. Retrieved online September 27, 2010, from http://www.ifpri.org/sites/default/files/publications/capriwp97.pdf

Lastarria-Cornhiel S (1997) Impact of privatization on gender and property rights in Africa. World Dev 25(8):1317–1333

Lundberg S, Pollak R, Wales TJ (1997) Do husbands and wives pool their resources? Evidence from the United Kingdom child benefit. J Hum Resour 32(3):463–480

Manser M, Brown M (1980) Marriage and household decisionmaking: a bargaining analysis. Int Econ Rev 21(1):31–44

McElroy MB, Horney MJ (1981) Nash-Bargained household decisions: toward a generalization of the theory of demand. Int Econ Rev 22(2):333–349

Meinzen-Dick RS (2006) Women, land, and trees. In: Garrity D, Okono A, Grayson M, Parrott S (eds) World agroforestry into the future. World Agroforestry Centre, Nairobi, pp 173–180. http://www.worldagroforestry.org/downloads/publications/PDFS/b14409.pdf

Meinzen-Dick RS, Zwarteveen M (1998) Gendered participation in water management: issues and illustrations from water users associations in South Asia. Agric Hum Values 15(4):337–345

Meinzen-Dick RS, Johnson N, Quisumbing A, Njuki J, Behrman J, Rubin D, Peterman A, Waithanji E (2011) Gender, assets, and agricultural development programs: a conceptual framework. CAPRi working paper no. 99. International Food Policy Research Institute, Washington, DC

Moser C (2006) Asset-based approaches to poverty reduction in a globalized context. Global economy and development working paper #01. Brookings Institution, Washington, DC

Mwangi EM (2007) Socioeconomic change and landuse in Africa: the transformation of property rights in Kenya's Maasailand. Palgrave MacMillan, New York

Narayan D (2000) Voices of the poor: can anyone hear us? World Bank, Washington, DC

Njuki J, Kaaria S, Chamunorwa A, Chiuri W (2011) Linking smallholder farmers to markets, gender, and intrahousehold dynamics: does the choice of commodity matter? Eur J Dev Res 23(3):426–443

Nkedi-Kizza P, Aniku J, Awuma K, Gladwin C (2002) Gender and soil fertility in Uganda: a comparison of soil fertility indicators for women and men's agricultural plots. Afr Stud Q 6(1): 27–43. Online: http://web.africa.ufl.edu/asq/v6/v6i1a2.htm

Peterman A (2010) Widowhood and asset inheritance in Sub-Saharan Africa: empirical evidence from 15 countries. Presented at the Chronic Poverty Research Centre/Overseas Development Institute Roundtable on Inheritance and the Intergenerational Transmission of Poverty, ODI,

London, October 11, 2010. http://www.chronicpoverty.org/uploads/assets/files/IGTinheritance/Papers_-_IGT_Peterman_2.pdf)

Place F, Adato M, Hebinck P, Omosa M (2007) Impacts of agroforestry-based soil fertility replenishment practices on the poor in Western Kenya. In: Adato M, Meinzen-Dick RS (eds) Agricultural research, livelihoods, and poverty: studies of economic and social impacts in six countries. Johns Hopkins University Press for the International Food Policy Research Institute, Baltimore, pp 149–197

Quisumbing AR (ed) (2003) Household decisions, gender, and development: a synthesis of recent research. International Food Policy Research Institute, Washington, DC

Quisumbing AR (2011) Do men and women accumulate assets in different ways? Evidence from rural Bangladesh. IFPRI discussion paper 1096. International Food Policy Research Institute, Washington, DC

Quisumbing A, Maluccio J (2003) Resources at marriage and intrahousehold allocation: evidence from Bangladesh, Ethiopia, Indonesia, and South Africa. Oxf Bull Econ Stat 65(3):283–328

Quisumbing AR, Pandolfelli L (2010) Promising approaches to address the needs of poor female farmers: resources, constraints, and interventions. World Dev 38(4):581–592

Quisumbing AR, Estudillo JP, Otsuka K (2004) Land and schooling: transferring wealth across generations. Johns Hopkins University Press for the International Food Policy Research Institute, Baltimore

Sabates-Wheeler R (2006) Asset inequality and agricultural growth: how are patterns of asset inequality established and reproduced? WDR background paper on asset inequality and agricultural productivity. World Bank, Washington, DC

Scoones I (2009) Livelihoods perspectives and rural development. J Peasant Stud 36(1):171–196. doi:10.1080/03066150902820503

Scott CD (2000) Mixed fortunes: a study of poverty mobility among small-farm households in Chile, 1968–86. J Dev Stud 36(6):155–180

Sherraden M (1991) Assets and the poor: a new American welfare policy. ME Sharpe Inc, Armonk

Smith LC, Haddad L (2000) Explaining child malnutrition in developing countries: a cross-country analysis. Research report 111. International Food Policy Research Institute, Washington, DC

Strauss JA, Thomas D (1995) Human resources: empirical modeling of household and family decisions. In: Srinivasan TN, Behrman J (eds) Handbook of development economics. North Holland, Amsterdam, pp 1883–2023

Thomas D, Contreras D, Frankenberg E (2002) Distribution of power within the household and child health. Mimeo, RAND, Santa Monica

von Braun J, Webb P (1989) The impact of new crop technology on the agricultural division of labor in a West African setting. Econ Dev Cult Chang 37(3):513–534

World Bank (2001) Engendering development through gender equality in rights, resources, and voice. World Bank policy research report. Management 1: report no. 36546-MW. World Bank, Washington, DC

Chapter 6
Gender Equity and Land: Toward Secure and Effective Access for Rural Women

Susana Lastarria-Cornhiel, Julia A. Behrman, Ruth Meinzen-Dick, and Agnes R. Quisumbing

Abstract Land is one of the most fundamental assets in the agricultural sector because it is the gateway through which people gain access to many other assets and opportunities. This chapter examines gender and land issues, identifying the gender gap in land rights and examining ways to redress this gap. The first section frames the discussion in terms of the four major ways by which women acquire legal and customary rights to land, and the obstacles to women's secure land tenure. The second section explores the nature and extent of the global gender land gap and the importance of going beyond common notions such as management, ownership, and headship, when discussing land tenure security. The third section looks at a number of strategies undertaken by a variety of actors—including governments, aid agencies, and civil society organizations—to lessen the gender land gap, organized broadly around

S. Lastarria-Cornhiel (✉)
Emeritus, Department of Urban and Regional Planning, Affiliate, Land Tenure Center and The Women's Studies Program, University of Wisconsin-Madison, 925 Bascom Mall, Madison, WI 53706, USA
e-mail: slastarr@wisc.edu

J.A. Behrman
Department of Sociology, New York University, 295 Lafayette Street, 4th floor, New York, NY 10012, USA
e-mail: jab965@nyu.edu

R. Meinzen-Dick
Environment and Production Technology Division, International Food Policy Research Institute (IFPRI), 2033 K Street, N.W., Washington, DC 20006, USA
e-mail: r.meinzen-dick@cgiar.org

A.R. Quisumbing
Poverty, Health, and Nutrition Division, International Food Policy Research Institute (IFPRI), 2033 K Street, N.W., Washington, DC 20006, USA
e-mail: a.quisumbing@cgiar.org

A.R. Quisumbing et al. (eds.), *Gender in Agriculture: Closing the Knowledge Gap*, DOI 10.1007/978-94-017-8616-4_6,

three types of interventions: strengthening women's land rights, redistribution of land rights, and improving the implementation of reforms. The chapter concludes that closing the gender land gap must go beyond reforms that affect only landownership, to include those that affect the multiple ways through which women and men acquire land, whether through legal or statutory means, the family, the market, or civil society.

Keywords Gender • Land • Property rights • Inheritance • Communal land • Land reform

6.1 Introduction

Secure land tenure is essential for stable livelihoods and poverty reduction in rural areas. Although land is clearly a key asset in rural areas throughout the developing world and is necessary to access many other services, land policy reform has typically focused on changing household rights to land, and not the rights of individuals *within* the household. This view has changed, partly because of accumulating evidence from South Asia, Africa, and Latin America demonstrating that women are disadvantaged in both statutory and customary land tenure systems (Agarwal 1994; Lastarria-Cornhiel 1997; Kevane 2004; Deere and León 2001; Deere et al. 2011), as well as the increased recognition that men and women within households do not necessarily pool resources (Haddad et al. 1997). Thus, strengthening household rights to land does not necessarily imply that women within those households have equal and secure land rights.

Justifications for paying attention to women's land rights have ranged from the need to give rural women equal access to resources to increase agricultural investment and productivity (e.g., Saito et al. 1994; Udry 1996; World Bank 2003), to rights-based approaches that view women's rights to land as a basic human right (e.g., Ikdahl et al. 2005). Moreover, the fallback options that secure land tenure provides can reduce the vulnerability of women in times of economic hardship, divorce, or widowhood, and can even strengthen their bargaining power within the household.

Thus, legislation has increasingly paid attention to strengthening women's land rights. However, even when such legislation is enacted, women often lack the legal know-how to claim their rights, or effective enforcement mechanisms are missing (Giovarelli 2009). Moreover, statutory rights and regulations operate alongside local customary norms and practices around land allocation, marriage, and inheritance, so that legislation alone does not determine the gendered distribution of property rights. Such legal pluralism (coexistence of statutory, customary, and religious laws and rules) must be taken into account in understanding, as well as seeking to change, women's rights to land. Analysts of land law have increasingly concluded that neither state nor customary law provides the best protection for women's land rights; both customary and statutory regimes have their strengths and weaknesses (Ikdahl et al. 2005).

This chapter looks in-depth at gender and land issues, identifying the gender gap in land rights and examining ways to redress this gap. The first section of this chapter frames the discussion in terms of the four major ways by which women acquire legal and customary rights to land, and the obstacles to women's secure land tenure. The second section explores the nature of the gender land gap, providing evidence on the extent of the global gender land gap and the importance of going beyond common distinctions, such as management, ownership, and headship, when discussing land tenure security. The third section looks at a number of strategies undertaken by a variety of actors—including governments, aid agencies, and civil society organizations—to lessen the gender land gap, organized broadly around three types of interventions: strengthening women's land rights, redistribution of land rights, and improving the implementation of reforms. The chapter concludes that closing the gender land gap must go beyond reforms that affect only landownership, to include those that affect the multiple ways through which women and men acquire land, whether through legal or statutory means, the family, the market, or civil society.

6.2 Understanding the Channels Through Which Women and Men Obtain Access to Land

Joint titling and reform of inheritance laws have been much discussed in debates regarding women's rights to land. However, the *Gender in Agriculture Sourcebook* (World Bank et al. 2009) argues that while these legal issues are critical, understanding the gendered nature of land rights requires going beyond these two forms of land acquisition, to examine the whole range of statutory and customary rights affecting women's access, use, and control of land. Women's (and men's) rights to land mainly come through five channels: (1) family allocations, specifically at marriage and from inheritance; (2) customary or community allocations, including common property; (3) state allocations such as land reform and resettlement programs (and housing and urban upgrading programs in urban areas); (4) civil society or NGO programs; and (5) through the market. All these rights are determined by both formal and customary law as well as the market economy's influence on agricultural production and land market structures.

6.2.1 Land and Marriage

Perhaps the most obvious area of gender disparity in land rights is women's unequal access to and ownership of land held by the household. Women typically have weaker rights to land *within the same household* than their husbands because community land allocations generally go to men and land transfers within families occur among men (Udry 1996; Quisumbing and Maluccio 2003; Deere et al. 2011; Swaminathan et al. 2011). Studies on intrahousehold resource allocation have revealed that land, similar to other resources within the household, is not always shared equally by husband and wife. For example, using nationally representative surveys from Latin America, Deere et al. (2011) point out that women's control of land is much less relative to men's, even if it is not uncommon for women to inherit land.

Changes in household structure, whether through marital dissolution (divorce, separation, or death) or the practice of polygamy, also have implications for women's land rights. Deere and Doss (2006) point out that women's rights to property depend critically on marital (property) regimes. Marital regimes differ according to how property acquired prior to and during the marriage is treated both while married and in case the relationship dissolves for whatever reason (separation, divorce, or death). In general, there are three main types of marital regimes: separation of property, partial community property, and full community property (Deere and León 2001).[1] Under separation of property, both spouses retain individual ownership of the assets they acquire both before and during marriage. If the marriage dissolves, each spouse is automatically entitled only to their own property, since there is no community property to divide. Under full community property, all property acquired before and during marriage is considered the joint property of the couple; if the marriage dissolves, all assets are divided equally between the two spouses. Partial community property combines features of both of these regimes. Property acquired prior to marriage remains the individual property of each spouse both during the marriage and after marital dissolution; however, all assets acquired during marriage (with the major exception of inheritances) are considered joint community property and divided into equal shares should the marriage dissolve. Each of these three types is more common in different regions. For example, the partial community property regime is often practiced in Latin America, while the separation of property regime is found in some non-Hispanic Caribbean countries (e.g., Surinam) and in most of Sub-Saharan Africa. These property regimes are usually based on customary norms but are often legally recognized in most countries.

Assets brought to marriage by husband and wife, as well as their claims to those assets if the marriage were to dissolve, have also been viewed as determining the bargaining power of spouses within a marriage (Thomas et al. 2002; Quisumbing and Maluccio 2003; Fafchamps et al. 2009). The expectation of disposition of assets upon divorce also affects the nutrition differential between spouses (Dercon and Krishnan 2000; Fafchamps et al. 2009). However, women typically bring fewer assets to marriage than men in a wide range of countries (Quisumbing and Maluccio 2003), and evidence on assets at marriage across time suggests that the gender-asset gap at marriage is widening, even if the gender gap in schooling has narrowed over time (Quisumbing and Hallman 2005). Most of the existing evidence also suggests that the bulk of land brought to marriage, even if it is part of the wife's dowry, is controlled by the husband.

[1] In the context of marital regimes, community property refers to the joint property of husband and wife, not of the community where they live or to which they belong.

Practices of polygamy, which exist in several non-Western countries, whether legally or not, have implications for the welfare of women.[2] While first wives are vulnerable when their husbands take second wives—household resources, including land, are divided among two or more families—second wives are often unprotected, particularly where polygamy is illegal. It is possible that allocation of land of differing quality depends on the status of co-wives within the same household. Polygamy also makes titling guidelines tricky and ambiguous: if a man supports two households, should the land be titled in the husband's and both wives' names? While it is outside the scope of land policy reform to change long-standing marital practices, where polygamy is widely practiced (whether legally recognized or not), legislators should consider carefully the effect of any land tenure reforms on the welfare of first and subsequent wives.

Regardless of the marital property regime in place, protection of women's rights upon divorce is weak. Particularly where separation of property is practiced, wives do not have a claim to marital property upon divorce. Even if wives may have a claim on marital property, divorce allocations often depend on who is perceived to be at fault. Prior to amendments in the Ethiopian Family Law in 2000, for example, half of rural monogamous households who were surveyed in 1997 expected the land and house to go to the husband upon a no-fault divorce; another 40 % expected them to be divided equally between husband and wife (Fafchamps and Quisumbing 2002). However, the allocation of assets upon fault-based divorce in Ethiopia varies considerably, depending on who is at fault. If it is the husband, the wife is slightly more likely to be granted land and livestock; if the wife is deemed to be at fault, asset distribution is dramatically changed in favor of the husband. Even her own livestock is likely to go to her husband. Fault-based divorce thus encompasses an element of punishment, which is particularly harsh for wives.

In Tanzania, The Law of Marriage Act (1971), which aimed at providing uniformity in marriage and divorce and recognizing equity between husbands and wives, does not provide specificity in the division of assets upon widowhood or divorce. For example, a woman's rights are retained for any property she may have brought to the marriage; however, there is ambiguity as to whether unpaid labor, including improvements to land, subsistence farming, housekeeping, or childcare, constitutes contributions to marital assets or are simply "wifely duties" (Mbilinyi 1972). As a result, though on paper Tanzanian women have strong rights to land and possessions, it is widely agreed that considerable resistance and interpretation of the law occurs, especially in rural areas where customary law is strong or where populations lack knowledge about their rights concerning land and property (Peterman 2011a).

6.2.2 *Land and Inheritance*

Women inherit land either as daughters from their parents, or as wives from their husbands. In the former, the transfer of land from parent to child need not occur at

[2] This discussion draws from World Bank et al. (2009).

the time of death, as land is often (but not always) transferred at the time of marriage. Women's ability to inherit land from parents is highly influenced by culture and tradition, which, in turn, influences legal rights that women have to land. In many patrilineal and patriarchal societies, women have very limited rights to inherit as daughters, particularly after they marry, while in some matrilineal regimes, women are favored.[3] In bilateral societies, daughters and sons have equal rights to inherit their parents' land. Where inheritance is influenced by religious practice, it is often the case that underlying custom influences the practice of inheritance law. Muslim law, for example, provides that daughters inherit half their brothers' share. But in predominantly Muslim Indonesia, which practices bilateral kinship and inheritance, sons and daughters have equal inheritance rights; and among matrilineal societies in Indonesia, women are favored in inheritance of paddy land (Quisumbing and Otsuka 2001). On the other hand, in Muslim Africa, daughters often do not claim any of their inheritance, leaving all the land to their brothers (Aldashev et al. 2009). Similar cases of legal pluralism, even within statutory law, can be observed in inheritance through marriage: if a person marries as a Muslim in India, for example, the inheritance rules are different than if a person marries as a Hindu (World Bank et al. 2009).

Widows' rights to their deceased husbands' lands are also tenuous, even where the marital property regime includes joint ownership of property acquired during marriage. The *Gender in Agriculture Sourcebook* (World Bank et al. 2009) posits that inheritance of land by spouses is even less likely to occur than inheritance by daughters in patrilineal and patrilocal communities. Since family land tends to be closely guarded in those communities, wives, who have no blood relationship to their husband or their clan or community, are often given use rights to the house and land, but not recognized as owners. Often, these use rights continue to be observed only if the woman has had a child with the deceased. Peterman (2011b) analyzed widow inheritance using cross-country, nationally representative Demographic and Health Survey (DHS) data from 15 Sub-Saharan African countries. Results indicate that across these DHS countries, less than half of widows report inheriting any assets (average inheritance of any assets is 47 %, ranging from 22 % in Sierra Leone to 66 % in Rwanda), and those that report of inheriting the majority of assets is lower (average of 32 % ranging from 13 % in Sierra Leone to 60 % in Rwanda). Across countries, inheritance is positively correlated with higher education and wealth, indicating that women who are better off may be more able to negotiate favorable asset inheritance outcomes. Unfortunately, because the DHS for Sub-Saharan Africa does not ask questions separately about land and other assets, it is possible that land was not part of the asset "package" that a widow inherits. This data gap should be addressed in future surveys on inheritance. Widow's rights to land are particularly important in countries with high mortality rates owing to HIV/AIDS.

[3] In many matrilineal regimes, land is not necessarily vested in women; rather, land is transferred to men via the uterine (mother's) line.

6.2.3 Community Land Allocations and Common Property

In addition to considering gender differences in how people acquire land as individuals or as members of households, it is important to also consider how men and women obtain rights to land through their communities. In much of Sub-Saharan Africa, as well as many other areas under indigenous land tenure in Latin America, land is allocated to a household by a tribe, clan, or lineage.[4] Rights to this land include long-term use and heirship, but neither men nor women usually have full "ownership" rights to such land, because they cannot sell or otherwise alienate the land without approval of the clan or lineage. The security of tenure and the extent of management and control rights enjoyed by women are often much less than those for men, especially in patrilineal systems where they usually enjoy only use rights. Especially if a wife is not considered to join her husband's clan, she can lose her use rights to land under customary tenure if her husband dies, unless she remarries her husband's brother, or has a son for whom she can act as custodian. However, there are indications of some changes in such systems, with allocation of customary land to daughters (UN-HABITAT 2005).

Where common lands are still available, they provide an important source of water, fodder, fuel, medicinal plants, or forest products. Even in areas where private tenure predominates, men and women may derive part of the resources for their livelihoods, as well as for domestic consumption, from the commons. These are often of particular importance to women, who are responsible for providing domestic water and firewood, medicinal plants, and may be responsible for grazing animals or providing fodder. Especially for households with little private property, the commons are a critical source of livelihoods. Rights to use common resources usually derive from being a member of a particular community, and decisionmaking rights (management or exclusion) derive from participation in some form of management entity. In some cases, a government agency such as the irrigation or forest department claims the rights to make these decisions, but often there is at least co-management with local organizations. In either case, women's effective rights to the common property resources will depend on whether they have a voice in the governing institutions, but many of these local institutions are male-dominated and do not take full account of women's differential priorities for use of common property (Zwarteveen and Meinzen-Dick 2001). Furthermore, common property is often itself not as secure as private property—often designated as "wastelands" by government with less legal protection than private property, and subject to reallocation without consent of the users, whether male or female (Alden Wily 2011).

[4] These systems are often referred to as "communal tenure," but in this context, "community" or "communal" refers to the clan, lineage, or residential community, not the households, as in the context of property rights within marital regimes.

6.2.4 Land Allocations from the State

The most common type of land allocation by the state in rural areas is redistributive land reform, which seeks to modify the size of operational holdings by transferring land from those with larger holdings to the landless or to wage laborers. Another state-run land allocation is resettlement, where rural households are resettled in another location where either land is abundant or some agricultural infrastructural program (such as irrigation or drainage) has been built. The two principal objectives of these types of land allocation are to bring about increases in agricultural output and to decrease rural inequality (Byres 2004).

A common characteristic of all land reforms enacted throughout Latin America, Eastern Europe, Africa, and Asia in the twentieth century has been disregard of gender issues. Policymakers, legislators, and peasant leaders assumed that tillers of the land are men and therefore men should be the owners of the land. Little thought has been given to who in the household owned or controlled land and who worked the land, so gender-disaggregated data were not collected. Even where more progressive redistributive land reforms have been implemented, they encounter certain limits on reducing gender bias because custom often prevails or interacts with changing statutory law to influence how land is distributed on the ground.

In all of these land reforms, land was redistributed mainly to men. Deere and León (2001) in their analysis of redistributive land reforms in Latin America found that very few women benefitted directly from land redistribution. Land reform regulations stated that heads of household and full-time or permanent farmworkers were eligible to be land reform beneficiaries. This meant that wives and children who worked as seasonal or part-time workers were usually not eligible beneficiaries.

Countries in most regions have been experimenting with resettlement programs, opening up land for agricultural cultivation. In Zimbabwe, for example, the senior Minister in charge of the land resettlement program in 1998 rejected women's demands that land certificates be automatically registered in both spouses' names. Neither did he permit that land earmarked for redistribution be offered to women heads of households and single unmarried women. The Minister maintained such moves would lead to the break up of families, since they would accord women too much freedom (United Nations-OCHA 2002). In Asia, Indonesia, Malaysia, and Thailand have implemented land redevelopment programs. Nuclear families were recruited for settlement in these redeveloped areas, but only the husbands were seen as the landowner and head of household. Although women in these countries have traditionally been very active and independent agriculturalists, the state and these programs do not recognize them as equal partners with their husbands. In Thailand, a woman in the Land Allocation program cannot have land in her name as long as her husband is alive. In Malaysia, as a consequence of the construction of dams near river ecosystems or forests, a number of ethnic groups, including matrilineal ones, were forced to resettle to other areas. Indigenous women's control over their lands and resources was undermined because they did not have official titles or deeds. Moreover, these lands were often excluded from compensation payment although they were recognized under customary law (Asian Development Bank 2002).

6.2.5 *Market-Based Land Acquisition*

Finally, women can obtain access to land from others outside the family through land sales and rental markets. Although the market might be thought of as a gender-neutral institution, women typically do not have enough financial resources (cash or credit), thus their ability to purchase land may be very limited (Meinzen-Dick et al. 1997; see Fletschner and Kenney, Chap. 8). Landownership may also be viewed as empowering to women, making husbands reluctant for women to purchase land in their own names.

Land rental markets may provide a less controversial way for women to obtain access to land, because leasing land does not create long-term secure property rights in the borrower/lessee. In Burkina Faso, the increased market value of land unexpectedly created avenues for women to lease land anonymously over the long term (Bruce 2006). Male landholders who have excess land are more willing to lease to women because women cannot claim permanent rights to land. Husbands generally support this borrowing of land by their wives, and women are therefore better able to cultivate land independently, even though they do not own it (Giovarelli 2006).

6.3 Evidence on Global Gender Gaps in Land Distribution

Field evidence and statistics available at country and regional levels demonstrate that most land tenure systems are inherently gender biased, preferentially allocating primary rights to land to male members of the community and family (Deere and León 2001; FAO 2002; Razavi 2003; World Bank et al. 2009). While available data on landholdings by sex are not always comparable across region and countries, existing data show that men control most agricultural land, even in regions such as Sub-Saharan Africa and Asia where women are heavily involved in agriculture (World Bank 2008). Despite the importance of the different modes of land acquisition in determining the strength of women's and men's property rights to land, existing cross-national data on the gendered distribution of land typically focus on landholdings rather than ownership, and even when ownership is considered, rarely examines the different ways by which men and women acquire land rights.

6.3.1 *Regional Gender Gaps in Land Distribution*

Most agricultural census data focus on the agricultural landholder, following the FAO definition that the holder is a "civil or juridical person who makes major decisions regarding resource use and exercises management control over the agricultural holding operation." According to this definition, the holder can be an owner or a manager, although being an agricultural holder does not necessarily have a one-to-one relationship with ownership (Deere et al. 2011). This leads us to question some of the assumptions underlying cross-national statistics that are used to document gender gaps in landownership. In Latin America, all the information provided in the censuses makes reference to agricultural producers. In Africa, while cross-national

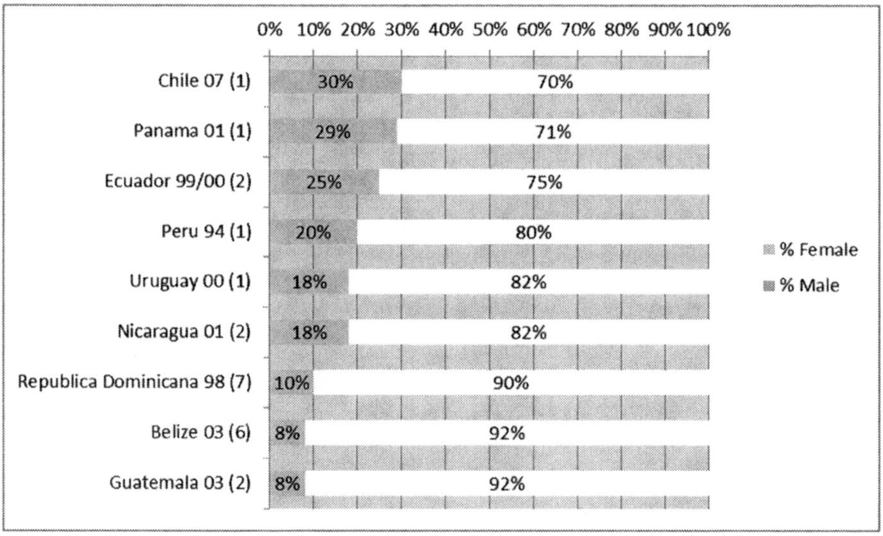

Fig. 6.1 Individual holders of agricultural land, by sex, in selected countries in Latin America (Source: FAO, Gender and Land Rights Database)

statistics assume that the household controls the holding, this does not take into account the cultivation of separate plots by men and women, and the possibility that men and women may have different rights to land, even within the same household. In Asia, the censuses generally identify holders as "operational holders" (that is, person actually cultivating the land), which would be an imperfect proxy for ownership, given active land rental markets in much of land-scarce Asia. To sum up, the data provided in the graphs below include individual civil/private holders as the person who exercises management control and takes major decisions over the agricultural holding, but this person can be an owner, manager, or producer.

In Latin America, the great majority of Civil Codes gives sons and daughters equal rights to inherit their parents' assets (Deere and León 2001). In addition, in the great majority of countries, either full community property or partial community property is the legal default marital property regime. Nevertheless, a review of data gathered by FAO on individual holders of agricultural land in Latin America from a sample of countries (Fig. 6.1) shows that in no country do women hold more than 30 % of private landholdings. The most influential set of land policies since the 1950s has been redistributive land reform. Deere and León (2001) have posited that the great majority of land redistributed in Latin America was given to men. While some countries that underwent land redistribution, such as Chile and Ecuador, have higher levels of women landholders than others, in Peru, a country that had one of the more radical land reforms with regard to the amount of land redistributed, only 20 % of the land is held by women. It would seem that neither legal inheritance norms nor land reform in Latin America have had a significant positive effect on distribution of land between women and men.

Data for Africa, where customary tenure systems based on patriliny are quite strong, show that a small percentage of individual holders of agricultural land are women.

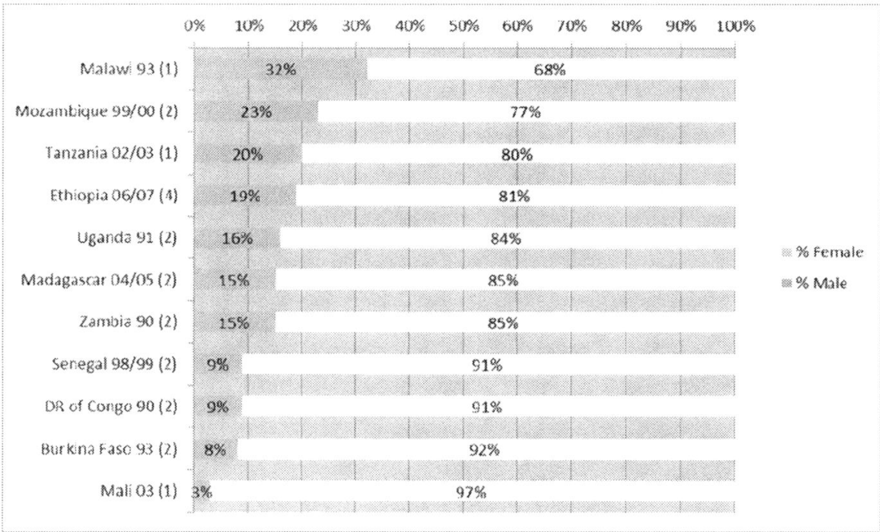

Fig. 6.2 Individual holders of agricultural land, by sex, in selected countries in Africa (Source: FAO, Gender and Land Rights Database)

The data we have for seven countries in east and southern Africa show that between 15 % and 20 % of individual landholders are women in predominantly patrilineal countries, including Uganda and Tanzania (see Fig. 6.2). In countries where there are significant numbers of matrilineal communities, such as Malawi and Mozambique, between one-quarter and one-third of landholders are women. Although men also inherit land in most matrilineal societies, women are able to exert some control over land. In western and central Africa, where Muslim personal laws are more prevalent, the data show that even though Muslim inheritance law allows daughters to inherit (albeit half the amount a son inherits), the actual practice is that very few women (less than 10 %) hold land (Fig. 6.2).

The available data for Asia show that (with the exception of Thailand) a very low percentage of women hold land (Fig. 6.3). This is even the case in Laos and Vietnam, where gender-equal legislation has been in place for several decades, and in India, where a number of states have targeted women in their land reform programs. In Indonesia, where numerous matrilineal societies hand land down from mother to daughter, some groups practice bilateral inheritance, and marital community property is observed, the number of women landholders is still under 10 %. Thailand also has a number of ethnic groups that are matrilineal and that pass land from mother to daughter. This is also the case in Laos and Vietnam, but they are small minority ethnic groups.

6.3.2 *Beyond Land "Ownership"*

Sex-disaggregated data are more likely to be available for individual holdings (and even that is incomplete for many countries), but they provide a very limited picture

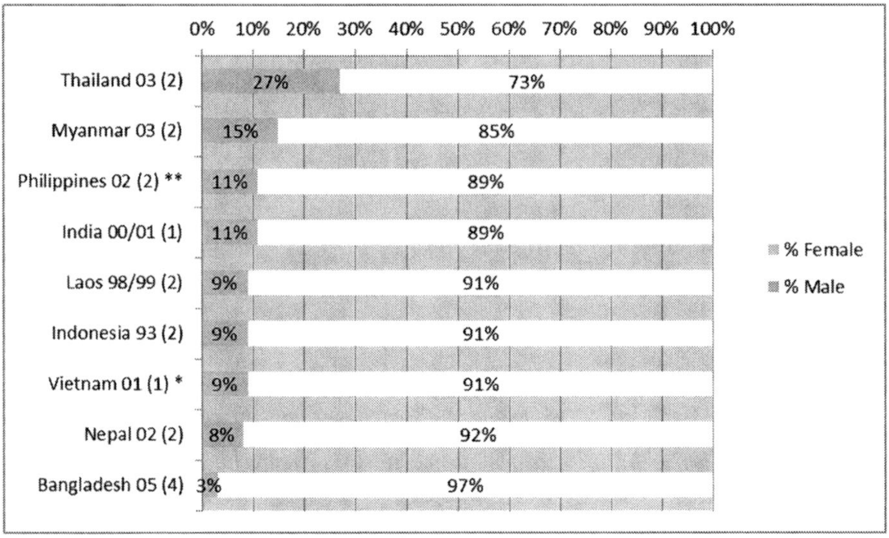

Fig. 6.3 Individual holders of agricultural land, by sex, in selected countries in Asia (Source: FAO, Gender and Land Rights Database)

of the gender distribution of rights over land. First, they do not consider the extent of joint property rights within marriage—whether husband and wife share equally in land acquired during marriage, and even in land acquired before marriage or inherited during marriage. In Ecuador, for example, a recent nationally representative survey of 2,892 households found that 36 % of agricultural land parcels were reported as jointly owned by husband and wife, 30 % as women's individual property, and 25 % as men's individual property—reflecting the effects of gender-equitable customary norms and a legal framework with joint property (Deere and Contreras Diaz 2011).

Second, much rural land is held under customary tenure and is therefore not reflected in the official data on individual holdings. This is especially important in Sub-Saharan Africa, as well as in areas of indigenous landholdings in Asia and Latin America. For example, in the West African region as a whole, only 2–3 % of land is held by written title, this being largely confined to a few major cities and development areas, such as irrigation schemes. In Burundi, it is estimated that less than 1 % of land is registered (Toulmin 2008, 12). Thus, it is essential to look also at the gender distribution of land within customary tenure.

Moreover, it is not enough to solely look at who owns land; it is important to also understand different types of property rights, which can often overlap on the same piece of land (Meinzen-Dick and Mwangi 2008). The literature on property rights defines bundles of rights, which refer to gradients of control over a given resource, usually applied to land and other natural resources. For example, bundles of rights for land can be divided into the right to use the asset (including the right to access and the right to extract resources), the right to appropriate the return from the asset (including earnings and income), the right to change its form, substance, and location (including decisionmaking rights such as management, and the exclusion of other users), and alienation (including transfer of rights to others) (Di Gregorio et al. 2008). These bundles of rights are applied at different levels (individuals, families, groups, the state), and actors often overlap in their levels of rights.

A recent study in three districts in Uganda illustrates how property rights can have different interpretations, depending on how questions are asked (Bomuhangi et al. 2011). When asked who was the "owner" of the land, 68 % of men and 58 % of women reported owning land, with considerable reporting of joint ownership. Documented land rights do not necessarily align with local perceptions: 52 % of men and only 20 % of women said they had any documentation of land rights in their own name, and only 1 % of either men or women had a registered deed in their own name.

Related to land rights are land use patterns. In addition to private "agricultural" land uses, one should also consider uses of common lands for purposes such as collecting firewood, water, and medicinal plants, grazing, and other uses often not counted in official statistics. Such common lands are often very valuable especially to women; they may, however, have the most insecure tenure, even being designated as "wastelands" by governments (Alden Wily 2011; Rossi and Lambrou 2008). For example, the government of India strove to bring 400,000 ha of classified marginal lands, which were de facto common property resources of the villages, under cultivation of nonedible oilseed crops (mostly Jatropha) for biodiesel production (Rajagopal 2007). On the other hand, Fortmann et al. (1997) point out that in Zimbabwe women and men were equally likely to plant trees on community woodlots because rights over those trees derived from community membership and investment, not marital status, and hence there were fewer gender differences in tenure security for trees than on household land.

6.4 Interventions for Reducing the Gender Gap in Land Rights

Interventions to reduce the gender gap in land rights can be roughly classified into three types: (1) those that strengthen women's land rights but do not necessarily involve land redistribution; (2) those that transfer land to women; and (3) those that improve the implementation of reforms to strengthen women's land rights. These can be undertaken by governments (such as redistributive land reform and changes in land legislation), civil society (group-based approaches to acquiring land), or both.

6.4.1 Interventions That Strengthen Women's Land Rights

6.4.1.1 Legal Reform

Both statutory and customary legal traditions lay the foundation for men's and women's property rights, and thus, legal reform can be a powerful tool for strengthening women's land rights. Legal reform can have potentially far-reaching effects, such as by eliminating discriminatory provisions in property rights law, and redefining the basis of property rights to give women the right to own property. Legal reform can also encompass reforms in marriage and inheritance laws. Finally, legal reform can include attempts to change tenure forms and rules.

Many countries have passed statutory laws that aim to reform discriminatory customary practices (Gopal 2001). Ethiopia provides an example where the legal framework has sought to protect the interests of women. The constitution prohibits discrimination on the basis of sex, although it upholds the application of customary laws to personal matters. While women have obtained equal legal rights to hold property, are able to participate freely in economic activities, and have been given inheritance rights equal to those of men, the application of customary law, as well as long-seated traditions, often result in women's having weaker land and property rights compared to men. In 1960, for example, the new Civil Code in Ethiopia imposed a set of norms different from the current customary rules used to govern personal relationships and property. This collision of values and norms meant that the more egalitarian principles of the Civil Code did not take root (Gopal 2001). The passing of the Revised Family Code in 2000 gave equal rights to spouses during the conclusion, duration, and dissolution of marriage, and required equal division of all assets between the husband and wife upon divorce, although adoption of the law has not been uniform across all the regions within Ethiopia, which has a federal system.[5] Nevertheless, compared to baseline perceptions obtained from rural households in 1997, the perception regarding the division of assets upon divorce in 2009 has shifted toward an equal division between the husband and the wife after the passage of the Revised Family Code (Kumar and Quisumbing 2012).

Rwanda provides an example of state institutions and civil society organizations working together to secure women's land rights. Beginning with the reform of inheritance law (Succession Law of 1999, in which all children, regardless of sex, inherit equally and in the majority of marriages both spouses own property together as community property) to the reform of land tenure (Organic Land Law of 2005, in which women and men, married or single, have equal rights to land), Rwanda now has the best legal conditions for gender equity with regard to land rights in Sub-Saharan Africa. During the years this legislation was being prepared and enacted, there were awareness-raising campaigns and extensive field consultations in preparation for implementing the tenure reform program. A recent study reports that

[5] Federal Negarit Gazetta Extra Ordinary Issue No. 1/2000 The Revised Family Code Proclamation No. 213/2000.

daughters are now inheriting land and many wives feel increased security and autonomy within their households (Daley et al. 2010). There remains resistance to these changes, however, by some men and local authorities. Many husbands continue to manage land as if it belongs to them exclusively, without consulting their wives. Few wives actually proceed with divorce proceedings, partly because of the social stigma of divorce and partly out of fear that they will not be able to actually claim their part of marital property. Sisters sometimes hesitate to claim their share of their inheritance in order to maintain good relations with their brothers (Daley et al. 2010).

Reforms in inheritance law that give sons and daughters equal rights to inherit can go a long way to improving gender equality in land rights. India has had a long history of legal activity to overcome a historical legacy of discrimination and high inequality, with varying levels of success. In 1994, the states of Maharashtra and Karnataka took the lead in amending the inheritance law applying to Hindus to grant daughters equal shares in inheritance relative to sons. In 2005, the national inheritance legislation was amended to eliminate gender discrimination that had thus far prevailed in all but a few southern states. Using data on three generations of individuals,[6] Deininger et al. (2010) compare land inherited by males and females in Maharashtra and Karnataka, depending on whether their fathers died before or after the 1994 amendment of the Hindu Succession Act. They find that the amendment of the Act had a significant positive impact on the probability that daughters inherited land, although they did not eliminate inequality in landholdings between males and females. The amendment of the Act also had a significant positive impact on age at marriage of females relative to males, and was associated with increased females' educational attainment.

6.4.1.2 Joint Titling and Land Certification Programs

In the recent past, the discourse around land policy and reform has focused on policies that create or deepen individual and private rights to land and expand land markets. It has often been argued that lack of formal land rights reduces the incentive to invest in productivity-enhancing technologies (e.g., Besley 1995).[7] This argument has been used to justify individualization of land rights, commonly exemplified in land titling programs. However, as with most land tenure approaches, these land policies and reforms have also been gender biased in that men have been the major beneficiaries and women are usually excluded. The process of individualization of

[6] The data come from the nationally representative Rural Economic and Demographic Survey (REDS) administered over 2006–2009. An analysis across all states (using the same REDS database) by Brulé (2011) of the 2005 amendment to the HAS shows that for the great majority of states, the impact has been mixed and insignificant. Brulé argues that an analysis of the 2005 amendment impact is more reliable because the passage of that amendment is more plausibly exogenous to local socio-political dynamics.

[7] See Platteau (2008) for a critical review of this literature with regard to Africa.

land rights in some regions may even result in a weakening of women's customary rights to land (see, for example, Lastarria-Cornhiel 1997 on Africa). Isolated efforts to remedy this gender bias have had negligible effects. However, as women's movements in some regions become strong enough to challenge norms and practices that privilege men with regard to land rights and to lobby for legislative and programmatic changes, more equal land rights are slowly emerging. One intervention has been the joint titling of land to both spouses, and titling programs either offer joint titling as an option (for example, Honduras) or make it mandatory in cases of marital property or land reform parcels.

In Bolivia, the 1996 land law that authorized its ongoing titling program explicitly recognizes equal rights to land by women as well as men, stating that equity criteria will be applied in the distribution, administration, tenure, and use of land in favor of women, independent of their civil status. This last phrase is important because it does not require that a woman be head of household or married in order to be eligible for land rights. Nevertheless, in the first few years of the titling program, no measures were taken to ensure that women received titles and very few women were being titled, either as individuals or as co-owners. Beginning in 2000, explicit efforts were undertaken to include women in the titling process and the percentage of titles issued to women increased significantly; men, however, continue to receive over 50 % of titles. In addition, the amount of land being titled to men far exceeds the amount titled to women as individuals and as joint owners (Lastarria-Cornhiel 2010).

In Vietnam, while the State holds ownership rights to land, landholders have been issued land certificates that grant them most land rights, including inheritance, transfer, and mortgaging. While joint land certificates were an option, most of the land certificates were issued in the husband's name as head of household. Only single women such as widows were issued land certificates and these accounted for 10–12 % of all certificates (Hatcher et al. 2005).

The Ethiopian land registration scheme provides some evidence that land registration does not necessarily have to work against women. In Ethiopia's low-cost, community-based land certification scheme, land administration committees at the *kebele* level (the smallest administrative unit in Ethiopia) were required to have at least one female member. Both the land registration exercise and the issuance of certifications were conducted publicly in village assemblies for transparency (Deininger et al. 2008). In regions where a photo, in addition to a name, was required for certification, females were considerably more likely to have their names on a deed for two notable reasons: (1) photos made it more difficult for husbands to sell or rent out land without a wife's consent and (2) photos are a useful form of identification in a society with low literacy rates.

Rwanda provides an example of how some societies are dealing with problematic situations such as polygamous marriage and common-law marriages, neither of which is legally recognized in Rwanda. In discussions with the local population with regard to the shape of the new land registration system, women and men indicated that land records should state who are the owners but should also include those people who have "interests" in the land (such as children of owners and common-law wives or husbands). In this way, people felt that women—particularly those who are not legally married—would be able to protect their rights to land (Daley et al. 2010). A recent evaluation of this relatively low-cost land tenure regularization scheme in Rwanda found that it improved land access for legally married women (76 % of married couples) and prompted recording of inheritance rights without gender bias (Ali et al. 2011). The analysis also found a large impact on the adoption and maintenance of soil conservation measures, particularly for female heads of households, suggesting that the program reduced their land tenure insecurity. Finally, it appears that the program did not cause distress sales of land or widespread land loss by vulnerable people.

In India, Landesa has been working with the Department of Revenue and Disaster Management in Odisha State to allocate state land to rural families, in most cases also regularizing the land that families are already occupying (Middey and Fletschner 2010). The land regularization program promotes including women's names on the *pattas* (land documents).

6.4.1.3 Strengthening Women's Rights in Community and Common Property

In addition to efforts to strengthen women's rights over private property, initiatives to strengthen women's rights to community property and the commons also play an important role in providing secure resource tenure for women. Community surveying, demarcation and collective titling, and land use planning exercises to secure community ownership over their land should ensure women's full participation. However, this is not a straightforward matter, and there may be tensions between efforts to secure women's land rights and involvement of customary authorities, who are often partriarchal. Lavigne Delville (2010) reports that the Rural Land Maps (*plans fonciers ruraux*) (PFRs) in West Africa have not recorded the secondary land rights of women or youth.

Efforts to secure collective rights over common property through comanagement have also often overlooked gender issues, but there are efforts to ensure women's voices in joint forest management in India and Nepal (Agarwal 2010). Colfer (2005) reports that adaptive collaborative management using participatory action research in 11 countries in Africa, Asia, and Latin America improved the ability of women to participate in decisionmaking and negotiation within the community and with policymakers. Other approaches, such as leasing land collectively for women's groups, are another avenue to strengthen women's collective land rights (Agarwal 2003). Programs such as leasing degraded forests to women's groups in Nepal or flooded lands to women for collective fishponds in Bangladesh have provided land for landless women, but require explicit attention to including women and to overcoming collective action problems to be effective (IFAD 2003; Hallman et al. 2007).

6.4.2 Interventions That Transfer Land to Women

6.4.2.1 Redistributive Land Reform

Since the 1980s, women's movements and organizations have demanded that land programs explicitly recognize and protect women's equal land rights. In several of the recent land reform programs, such as the Nicaraguan one, the land reform law had explicit language including women as equal beneficiaries; nevertheless, mostly male heads of household received land.[8]

In Brazil, the state agency distributing land reform parcels (INCRA) disregarded the legal option of joint ownership for husbands and wives mandated in the Federal Constitution.[9] The language of the land reform legislation and the procedures followed by the land reform agency (INCRA) was to adjudicate land to the household head.[10] As a result, as of 1996, only 12.6 % of land reform beneficiaries were women (Guivant 2003). In addition, there was little institutional, political, or grassroots pressure for INCRA to recognize women as legitimate land reform beneficiaries or their rights to joint ownership. According to Deere (2003), a decade after legal recognition of joint ownership, the national leadership of the national agricultural workers' union, CONTAG, finally took up the recommendations of its women's commission in 1998 and approved the specific demand that the names of both spouses be included in the land registry. Over the next few years, CONTAG participated in a number of national women's events, such as International Women's Day in 2000 and the Marcha das Margaridas in August of that same year. As a result, INCRA finally acknowledged the legal norm of joint property established in the 1988 constitution by announcing that it would change its titling procedures in order to include the names of both spouses on property documentation (Deere 2003).

The majority of redistributive land reforms in the last few decades, nevertheless, have continued to allocate land to household heads, the great majority of them men, ignoring national legislation and/or constitutional law that explicitly recognize women's equal rights to land (see Deere and León (2001) for data on Latin America). Recent land-titling programs have attempted to rectify this injustice by mandating that land received through the land reform (and other state allocations) be jointly titled to husbands and wives, or by recognizing all women's land rights, irrespective of their civil status. However, joint titling has not necessarily resulted in improvements

[8] Deere and León (2001, 96) state that by 1989, eight years after the start of the land reform, only 9.6 percent of the beneficiaries were women.

[9] According to Article 189 of the Federal Constitution (1988), "The beneficiaries of distribution of rural land through agrarian reform shall receive title deeds or concessions of use. These shall be granted to the man or woman, or to both, irrespective of their marital status, according to the terms and conditions set forth by law."

[10] For example, until 2000, the agrarian reform registration form (*cadastro*) did not have space for writing in the names of two persons. INCRA also insisted on naming men as beneficiaries—if a woman applied, INCRA officials would ask the whereabouts of the spouse, and in the absence of a spouse, they would try to grant beneficiary status to a son.

in women's land rights. In a recent study of land reform in Nicaragua, where joint titling was mandated, men had land titled in their and their brother's or son's names, rather than include their wife's name on the title (Agurto and Guido 2002).

6.4.2.2 Microplots

Traditional redistributive land reform projects are predicated on the assumption that sufficiently large parcels of land (1 ha or more) should be allocated to the rural poor. The quantities of land under question are part of the reason these programs are difficult to operate and/or run the risk of generating controversy. Indeed, there are very few governments that are currently engaged in large-scale land redistribution—whether gender-equitable or not. A promising alternative to this traditional model is the development of microplot allocation programs (Mitchell et al. 2010). In such programs, the state allocates small plots of land (1 acre or less) that include or are near to the homestead to encourage the rural poor to engage in vegetable gardening, small livestock cultivation, tree planting, construction of secure shelter and engage in other small-scale income-generation activities.

Mitchell et al. (2010) assert that the development of microplots is particularly beneficial for women for a number of reasons. First of all, women who are typically tied to the homestead in their role as caregivers are not required to leave the homestead to care for the produce or livestock on the microplots. As such, they may be able to assert more control over how produce from the microplot is used and may gain control over new assets or new sources of income. Furthermore, microplots allow women to increase dietary diversity of their household by growing a variety of fruits and vegetables and tending to small livestock while fulfilling their domestic and childcare responsibilities. This is exemplified in the homestead food production model of the NGO Helen Keller International (Ianotti et al. 2009).

Other initiatives that involve direct allocation or distribution, coupled with rules that allow both sons and daughters to inherit, are ongoing. For example, Landesa has been working with the Department of Land and Land Reforms in Coochbehar (formerly Bihar, in India) to prioritize the allocation of homesteads to girl-only households, and to include a clause in the *pattas* (legal document of landownership rights) that includes all daughters and sons as co-inheritors (Middey and Fletschner 2010). Landesa has been working with local NGOs to hold "community conversations" among families, to explore the negative implications of providing dowries and of early marriage for girls. The intervention also includes the formation of "Boys and Girls Groups" in these communities to provide information, education, and social support to develop strategies to deal with issues surrounding inheritance. The curriculum for these groups includes sessions on land rights. An important aspect of this approach is that it goes beyond legislation or assignment of rights, recognizing the need for accompanying programs to address local norms in order to ensure that women's land rights are secured.

6.4.2.3 Collective Plots

Another alternative to traditional land distribution is for women's groups to acquire land collectively to cultivate. In this alternative, the state or communities can allocate land to women's groups, or women can pool funds to buy land, particularly if women acquiring land individually through purchase is deemed too controversial by their families. Group-based land acquisition may also provide women with seed funds with which to purchase land on their own. For example, an NGO in Bangladesh that targeted poor women for the adoption of polyculture fishpond technology made arrangements for women's groups to lease land, to excavate the fishpond through a food-for-work program, and then to cultivate the fishponds jointly. Years later, some of the more successful members had begun to purchase their own land for fishponds, which was then titled in their own names (see Hallman et al. 2007; Quisumbing and Kumar 2011). In an HKI homestead vegetable production program in Burkina Faso, concerns that husbands would take over wives' gardens once these would become profitable were addressed by first holding community meetings that explained the rationale for providing program support through women. This paved the way for community agreements that ceded land for the village model farms to women's groups for the duration of the project, and potentially beyond (Nielsen 2010).

6.4.3 Interventions That Improve the Implementation of Reforms

6.4.3.1 Increasing Gender Inclusiveness of Land Administration Bodies

Land administration bodies carry out and enforce the land rights (and rules and regulations) mandated by legislation that protect women's land rights. These include the land registry, cadastral office, titling agency, and land magistrates and courts that are often staffed by male technical, legal, or administrative personnel who are unlikely to be knowledgeable of women's land rights or sensitive to the constraints women face in claiming their rights.

The need for gender training of project or program personnel has been evident in numerous titling programs in Latin America and Asia. As the Nicaragua land titling project revealed, it is not sufficient for national program officials to receive gender training—titling personnel on the ground who are actually collecting property owner information and making land measurements also need to be aware of the program's gender objectives and the laws behind them (Agurto and Guido 2002). The titling project in Laos greatly improved the level of women's participation when the Lao Women's Union (LWU) became an active and integrated member of the adjudication teams (GRID 2000). The LWU has integrated gender into the education, training, and information dissemination activities at the village level; it produces and distributes information, education, and communication materials such as calendars, posters, TV and radio spots, and distributes song tapes in villages in their efforts to educate titling beneficiaries with regard to the risks and benefits of land titles. The Global Land Tools Network has developed training materials for land professionals to integrate gender perspectives in the design and implementation of land tools and policies (UN-HABITAT 2011).

In addition to gender training of program personnel, increasing the number of women personnel employed in these land institutions is an important priority. Government sends an important message by hiring as many competent and qualified women staff as possible for the different components and activities. Women paralegals, surveyors, adjudicators, technicians, lawyers, registrars, judges, and public relations specialists will also encourage women to approach program staff to make their property claims or request assistance. Where there is strict separation of women and men, it will be necessary to have female staff at all levels in order to provide services for women; separate entrances and rooms for women clients may also be necessary.

While it is ideal to strive for gender parity in program personnel, the reality may be that there are fewer women trained in the legal and technical professions. In order to achieve a significant and representative number of women staff members, it may be necessary to include a disproportionate number of women in training programs. In Mozambique, a gender program was initiated in the Technical-Professional School of Geodesy and Cartography in 1996 and by 1998, 40 % of students were women. By 2000, 32 % of all staff members were women, although this dropped to 13 % at professional and management levels and 15 % at provincial and district levels (Norfolk and Tanner 2007). The shortage of trained women may require training community members to assist in the fieldwork, for example, as paralegals to assist adjudicators and technicians to work with surveyors in the titling teams. In Rwanda and Tanzania, for example, legislation mandates that local land committees throughout the country and local government management committees be composed of at least 30 % women, in an effort to increase the voices and visibility of rural women (Daley et al. 2010; Walker 2002). Special efforts and procedural accommodations may be necessary to encourage community women to participate in these activities and for them to feel comfortable during training and in the carrying out of their project responsibilities.

The experience with Ethiopia's land registration program suggests that having women in local land administration committees encourages participation of female-headed households in the land registration process and does not decrease male-headed households' participation. Having female members in the land administration committees also increases attendance at meetings relating to the land registration (Kumar and Quisumbing 2012). Gender-equitable land registration processes also reinforce awareness of reforms in family law. Kumar and Quisumbing (2012) find that awareness about the land registration process is positively correlated with the shift in perceptions toward equal division of land and livestock upon divorce. This is especially true for male-headed households, indicating that interventions to improve gender equality can reinforce each other. Moreover, the presence of female members in the land administration committees has a positive effect on the changes in perceptions toward a more equal distribution of assets upon divorce.

This effect persists even after controlling for local norms regarding the distribution of assets upon divorce. This suggests that the presence of females in an important village-level committee may provide support to women and also may be a source of information regarding new family codes that strengthen women's property rights.

6.4.3.2 Legal Literacy Programs

In many countries, a major impediment to women's participation in land programs is widespread ignorance among women and, to a lesser extent, men of their rights to land due to illiteracy and lack of dissemination of information and legislation at the local level. This lack of access to information is often compounded by restricted access to formal legal professionals and general inability of poor rural populations to pay for such services.

One promising initiative that aims to combat precisely these problems is grassroots community-based paralegal programs focused on land and housing issues. While paralegal programs in general have been used for decades, the specific focus on gender and land rights issues is relatively recent. Although each program varies in strategy and objectives, a number of different programs have been implemented in the last 5 years, including those in India (Landesa), Lesotho (Federation of Women Lawyers), Tanzania (EnviroCare), and Rwanda (Women's Land Link Africa). The scale and structure of programs vary from large umbrella organizations with formalized training protocol and payment structures, to networks of community volunteers who undergo rapid training to be information sources in their communities. The Uganda Land Alliance (ULA) recently launched a pilot project to build the capacity of 20 grassroots paralegals to serve as sources of legal information on topics including women's and children's rights, landlord-tenant relations, marriage rights, and will writing in the Luwero district. Half of the paralegals were women, and all paralegals undertook a basic and intensive gender training course.[11] This model, as well as others implemented by grassroots organizations, aims to improve women's knowledge of their legal rights and enhance their capacity to act when their rights are challenged.

Although paralegal programs present a promising approach to enhancing women's land rights, there is thus far little evidence on how effective these programs are at improving knowledge, attitudes, and practices on gender and land issues. Part of the problem is the lack of a monitoring and evaluation culture within implementing organizations. Without systematic monitoring and evaluation, programs are modified in an ad hoc manner. Without evaluation, it is difficult to recommend what programs can be scaled up. Likewise, it is difficult to know what design features can be modified for local conditions without adversely affecting the overall outcome of the intervention (Quisumbing and Pandolfelli 2010).

More recent initiatives have taken steps to resolve this issue. A monitoring and evaluation component was included in the ULA pilot project that conducted gender

[11] For more information, see http://www.icrw.org/where-we-work/training-grassroots-paralegals-help-women-exercise-their-property-rights

sensitizations in 64 villages, reaching over 2,500 men and nearly 3,000 women in a 9-month period in 2009/10 (Jacobs et al. 2011). Monitoring data show that land disputes over property boundaries were the most common type of conflict encountered by paralegals, and that although attitudes of both men and women are generally positive toward women's land rights, there are large gaps in knowledge surrounding legal rights. In addition, qualitative interviews point to success stories, both in client conflict resolution, and in empowerment of the paralegals themselves. Challenges and recommendations included the need for improved coordination between land rights organizations and institutions as well as the need to address paralegal workload burdens. The documentation by ICRW is extremely informative, as few other organizations have documented in a systematic way the successes and challenges of their programs beyond publication of toolkits and training manuals (Barnes et al. 2009; FIDA-LESOTHO 2006).

6.5 Conclusions

Whether women's land rights are justified as contributing to greater productivity, poverty reduction, or human rights, empowering women through land rights is a key mechanism to achieve gender equity and to increase household welfare. But strengthening women's land rights is not a simple matter of reforming land laws. Property rights are influenced by a host of other factors, including inheritance patterns, marital regimes, customary and religious law and practices, as well as the practical implementation of legal reforms, and achieving gender equity in land rights requires addressing all of these factors. The state and civil society are both needed to specifically address and implement both statutory reforms and administrative structures as well as reduce constraints such as patriarchal norms that prevent the realization of these rights. Whether the state is redistributing land, resettling rural households on improved agricultural land, or titling the land of rural households, we have seen that civil society and the international community need to demand state action and resources and to collaborate with the state in implementing gender-equal programs to improve women's legal and *de facto* land rights.

The support provided by donor and development agencies that cooperate with and provide funding for state land programs and civil society initiatives has often gone beyond the productivity impacts of financial investments to catalyze broader societal change in support of gender equity. State land programs and civil society initiatives themselves reflect both internal and external pressure to pay attention to gender equity in their land programs. But progressive legal or policy reforms and programs to strengthen land rights or to transfer land to women are not enough: to change rights on the ground requires working with the implementing agencies to ensure that they understand and implement the reforms in a gender-equitable manner. It may also involve a change in deeply rooted community norms that restrict women's property rights. Achieving gender equity in land rights requires the participation of women (and men) at all levels: national and local government, local civil society, women's organizations, and local women's groups.

One difficult challenge is protecting women's rights to land in indigenous or communal lands. Many states recognize these territories as belonging to the community or indigenous groups and often title the community as the owner. Legal language usually stipulates that internal land rights and land use practices are regulated by local customary tenure rules. The community authorities who make land management and allocation decisions are often men and usually only male household heads participate in community meetings. Women have access to household land through their father, brother, or husband, and to communal land as members of the household. Increasing women's participation in community meetings and decisionmaking processes and having them acknowledged as full members of the community are first steps in recognizing that they are full members of the community and have the same rights to land as men. This recognition will protect their rights to land as land use patterns change and the customary tenure rights and rules also change.

A recent phenomenon in land tenure is the selling and leasing of large tracts of agricultural land, particularly in Sub-Saharan Africa and even in Asia. These land deals are usually portrayed either as large land grabs facilitated by corrupt state officials or as unprecedented opportunities to increase agricultural productivity and offer employment for rural populations. Both pictures have some truth to them, of course. Studies that support either argument have thus far not considered the gendered effect of these agri-business enterprises on women's rights to land or their livelihood strategies. One recent paper that cites two case studies suggests that the design and implementation of these transactions repeat the old, tired strategies of previous land tenure programs: it is men who are consulted and sought out and men who are assumed to be the landholders (Behrman et al. 2011). Neglecting the issue of women's rights to land and to livelihood as governments enter into these land deals will only result in further erosion of their rights (and possibly those of local populations) and their ability to support their households.

In addition to the different land rights, therefore, tenure security—not simply legal landownership—is essential in rural contexts, and for rural women it is often tenure security that is most lacking. State agencies and communities need to recognize and enforce land rights for women. It may fall to communities to enforce women's land rights because state agencies often have a weak presence in rural areas. Customary institutions and norms are all the more important in such cases, highlighting the need for more than legislative changes in order to strengthen women's property rights. However, when land becomes more valuable and outsiders (and insiders) try to stake claims, then state-recognized rights become especially important for tenure security and ensuring gender equity in documentation of land rights is crucial.

Acknowledgments This chapter draws heavily from a working paper by Susana Lastarria-Cornhiel, Ambreena Manji, and Zoraida Garcia commissioned by the Gender, Equity, and Rural Employment Division of the Food and Agriculture Organization of the United Nations (FAO). We acknowledge insightful comments from Eve Crowley, AnaPaola De La OCampo, Clara Park, and Martha Osorio.

References

Agarwal B (1994) A field of one's own: gender and land rights in South Asia. Cambridge University Press, Cambridge

Agarwal B (2003) Gender and land rights revisited: exploring new prospects via the state, family, and market. J Agrar Change 3(1–2):184–224

Agarwal B (2010) Gender and green governance: the political economy of women's presence within and beyond community forestry. Oxford University Press, Oxford

Agurto S, Guido A (2002) Nicaragua: Diagnostico rapido sobre el impacto de la titulacion mancomunada. Fundación Internacional para el Desafío Económico Global, Managua

Aldashev G, Chaara I, Platteau J-P, Wahhaj Z (2009) Using the law to change the custom. University of Namur, Namur

Alden Wily L (2011) The tragedy of public lands: the fate of the commons under global commercial pressure. International Land Coalition, Rome

Ali DA, Deininger K, Goldstein MP (2011) Environmental and gender impacts of land tenure regularization in Africa: pilot evidence from Rwanda. World Bank policy research working paper series. World Bank, Washington, DC. http://ssrn.com/abstract=1912357

Asian Development Bank (2002) Sociolegal status of women in Indonesia, Malaysia, Philippines, and Thailand. Manila, the Philippines. Available at http://www.adb.org/Documents/Studies/Sociolegal_Status_Women/default.asp

Barnes S, Hersh M, Stricker M, Takigawa J (2009) Measuring the impact of a paralegal program on women's empowerment in rural Uganda. Uganda Community-Based Association for Child Welfare (UCOBAC), Kampala, and the School of International and Public Affairs (SIPA), Columbia University, New York. http://new.sipa.columbia.edu/sites/default/files/UCOBAC_FINAL_REPORT.pdf

Behrman J, Meinzen-Dick R, Quisumbing A (2011) The gender implications of large-scale land deals. IFPRI discussion paper 01056. International Food Policy Research Institute, Washington, DC

Besley T (1995) Property rights and investment incentives. J Pol Econ 103(5):913–937

Bomuhangi A, Doss C, Meinzen-Dick R (2011) Who owns the land? Perspectives from rural Ugandans and implications for land acquisitions. IFPRI discussion paper 01136. International Food Policy Research Institute, Washington, DC

Bruce J (2006) Land law reform: achieving development policy objectives. Law, justice, and development series. World Bank, Washington, DC

Brulé R (2011) When legal reform matters: findings from India on women's benefits from property rights reform. Presented at the annual South Asia conference, University of Wisconsin-Madison, 21–23 October

Byres TJ (2004) Introduction: contextualizing and interrogating the GKI case for redistributive reform. J Agrar Change 4(1 & 2):1–16

Colfer CJP (2005) The complex forest: communities, uncertainty and adaptive collaborative management. Resources for the Future, Washington, DC

Daley E, Dore-Weeks R, Umuhoza C (2010) Ahead of the game: land tenure reform in Rwanda and the process of securing women's land rights. J East Afr Stud 4(1):148–169

Deere CD (2003) Women's land rights and rural social movements in the Brazilian Agrarian Reform. J Agrar Change 3(1–2):257–288

Deere CD, Contreras Diaz J (2011) Acumulación de activos: Una apuesta por la equidad. Facultad Latinoamericana de Ciencias Sociales, Quito

Deere CD, Doss C (2006) The gender asset gap: what do we know and why does it matter? Fem Econ 12(1–2):1–50

Deere CD, León M (2001) Empowering women: land and property rights in Latin America. University of Pittsburgh Press, Pittsburgh

Deere CD, Alvarado GE, Twyman J (2011) Gender inequality in asset ownership in Latin America: female owners vs. household heads. Unpublished manuscript, University of Florida, Gainsville, FL

Deininger K, Ayalew Ali D, Holden S, Zevenbergen J (2008) Rural land certification in Ethiopia: process, initial impact, and implications for other African countries. World Dev 36(10):1786–1812

Deininger K, Goyal A, Nagarajan HK (2010) Inheritance law reform and women's access to capital: evidence from India's Hindu Succession Act (June 1, 2010). World Bank policy research working paper series. World Bank, Washington, DC

Dercon S, Krishnan P (2000) In sickness and in health: risk sharing within households in rural Ethiopia. J Pol Econ 108(4):688–727

Di Gregorio M, Hagedorn K, Kirk M, Korf B, McCarthy N, Meinzen-Dick R, Swallow B (2008) Property rights, collective action, and poverty: the role of institutions for poverty reduction. CAPRi working paper 81. International Food Policy Research Institute, Washington, DC

Fafchamps M, Quisumbing AR (2002) Control and ownership of assets within rural Ethiopian households. J Dev Stud 38(6):47–82

Fafchamps M, Kebede B, Quisumbing AR (2009) Intrahousehold welfare in rural Ethiopia. Oxf Bull Econ Stat 71(4):567–599

FAO (Food and Agriculture Organization of the United Nations) (2002) Gender and law—women's rights in agriculture, Chapter II (FAO Legislative Study). Rome

FIDA (International Federation of Women Lawyers)–LESOTHO (2006) Paralegal training manual. In: Collaboration with Community Legal Resource Centre (CLRAC). Supported by USAID Women's Legal Rights Initiative. Maseru, Lesotho

Fortmann L, Antinori C, Nabane N (1997) Fruits of their labors: gender, property rights, and tree planting in two Zimbabwe villages. Rural Sociol 62(3):295–314

Giovarelli R (2006) Overcoming gender biases in established and transitional property rights systems. In: Bruce J, Giovarelli R, Rolfes L Jr, Bledsoe D, Mitchell R (eds) Land law reform: achieving development policy objectives. Law, justice, and development series. World Bank, Washington, DC, pp 67–106

Giovarelli R (2009) Gender and land tenure reform. In: Mitchell R, Hansted T (eds) One billion rising: land law and the alleviation of global poverty. Leiden University Press, Leiden, pp 195–233

Gopal G (2001) Law and legal reform. In: Quisumbing AR, Meinzen-Dick R (eds) Empowering women to achieve food security, 2020 focus 6 brief 12. International Food Policy Research Institute, Washington, DC

GRID (Gender Resource Information & Development Center) (2000) Gender and land documents: how do society's perceptions of gender affect women? Gender Resource Information & Development Center (GRID), Vientiane

Guivant JS (2003) Agrarian change, gender and land rights: a Brazilian case study. Social policy and development programme paper no. 14. United Nations Research Institute for Social Development, Geneva

Haddad L, Hoddinott J, Alderman H (eds) (1997) Intrahousehold resource allocation in developing countries: methods, models, and policy. Johns Hopkins University Press for the International Food Policy Research Institute, Baltimore

Hallman K, Lewis D, Begum S (2007) Assessing the impact of vegetable and fishpond technologies on poverty in rural Bangladesh. In: Adato M, Meinzen-Dick R (eds) Agricultural research, livelihoods, and poverty: studies of economic and social impacts in six countries. Johns Hopkins University Press for the International Food Policy Research Institute, Baltimore, pp 103–148

Hatcher J, Meggiolaro L, Santonico Ferrer C-I (2005) Cultivating women's rights for access to land. Action Aid International, Rome

Ianotti L, Ruel M, Cunningham K (2009) Improving diet quality and micronutrient nutrition: homestead food production in Bangladesh. IFPRI discussion paper 928. International Food Policy Research Institute, Washington, DC

IFAD (International Fund for Agricultural Development) (2003) Kingdom of Nepal Hills Leasehold Forestry and Forage Development Project interim evaluation. Rome

Ikdahl I, Hellum A, Kaarhus R, Benjaminsen TA, Kameri-Mbote P (2005) Human rights, formalization, and women's land rights in southern and eastern Africa. Studies in women's law no. 57. Institute of Women's Law, University of Oslo, Oslo, Norway

Jacobs K, Saggers M, Namy S (2011) How do community-based legal programs work? Understanding the process and benefits of a pilot program to advance women's property rights in Uganda. International Center for Research on Women, Washington, DC

Kevane M (2004) Women and development in Africa: how gender works. Lynne Rienner, Boulder/London

Kumar N, Quisumbing AR (2012) Policy reform toward gender equality in Ethiopia: little by little the egg begins to walk. IFPRI discussion paper 1226. International Food Policy Research Institute, Washington, DC

Lastarria-Cornhiel S (1997) Impact of privatization on gender and property rights in Africa. World Dev 25(8):1317–1333

Lastarria-Cornhiel S (2010) Land tenure, titling, and gender in Bolivia. Saint Louis Univ Public Law Rev 29(1):193–242

Lavigne Delville P (2010) Registering and administering customary land rights: can we deal with complexity? In: Deininger K, Augustinus C, Enemark S, Munro-Faure P (eds) Innovations in land rights, recognition, administration, and governance. Joint organizational discussion paper—issue 2. World Bank, Global Land Tool Network, International Federation of Surveyors, and Food and Agriculture Organization of the United Nations, Washington, DC, pp 28–42

Mbilinyi MJ (1972) The 'new woman' and traditional norms in Tanzania. J Mod Afr Stud 10(1):57–72

Meinzen-Dick R, Mwangi E (2008) Cutting the web of interests: pitfalls of formalizing property rights. Land Use Policy 26(1):36–43

Meinzen-Dick R, Brown LR, Sims Feldstein H, Quisumbing AR (1997) Gender, property rights, and natural resources. World Dev 25(8):1303–1316

Middey G, Fletschner D (2010) Security for girls through land. Presentation at the inception workshop on evaluating the impacts of agricultural development programming on gender inequalities, asset disparities, and rural livelihoods. International Food Policy Research Institute and International Livestock Research Institute, Nairobi, Kenya, 5–7 Nov

Mitchell R, Hanstad T, Nielsen R (2010) Microplots for the world poor. In: Prosterman RL, Mitchell R, Hanstad T (eds) One billion rising: land law and the alleviation of global poverty. Leiden University Press, Leiden, pp 153–194

Nielsen J (2010) Enhanced homestead food production for improved food security and nutrition in Burkina Faso. Presentation at the inception workshop on evaluating the impacts of agricultural development programming on gender inequalities, asset disparities, and rural livelihoods. International Food Policy Research Institute and International Livestock Research Institute, Nairobi, Kenya, 5–7 Nov

Norfolk S, Tanner C (2007) Improving tenure security for the rural poor: Mozambique—country case study. Food and Agriculture Organization of the United Nations, Maputo

Peterman A (2011a) Women's property rights and gendered policies: implications for women's long-term welfare in rural Tanzania. Special section on gender and assets. J Dev Stud 47(1):1–30

Peterman A (2011b) Widowhood and asset inheritance in Sub-Saharan Africa: empirical evidence from 15 countries. CPRC working paper 183. Chronic Poverty Research Centre, Manchester, UK

Platteau J-P (2008) The evolutionary theory of land rights as applied to Sub-Saharan Africa: a critical assessment. Dev Change 27(1):29–86

Quisumbing AR, Hallman K (2005) Marriage in transition: evidence on age, education, and assets from six developing countries. In: Lloyd CB, Behrman JR, Stromquist NP, Cohen B (eds) The changing transitions to adulthood in developing countries: selected studies. National Academies Press, Washington, DC, pp 200–269

Quisumbing AR, Kumar N (2011) Does social capital build women's assets? The long-term impacts of group-based and individual dissemination of agricultural technology in Bangladesh. J Dev Eff 3(2):220–242

Quisumbing AR, Maluccio J (2003) Resources at marriage and intrahousehold allocation: evidence from Bangladesh, Ethiopia, Indonesia, and South Africa. Oxf Bull Econ Stat 65(3):283–328

Quisumbing AR, Otsuka K (2001) Land inheritance and schooling in matrilineal societies: evidence from Sumatra. World Dev 29(12):2093–2110

Quisumbing AR, Pandolfelli L (2010) Promising approaches to address the needs of poor female farmers: resources, constraints, and interventions. World Dev 38(4):581–592

Rajagopal D (2007) Rethinking current strategies for biofuel production in India. Paper presented at the international conference on linkages between energy and water management for agriculture in developing countries, Hyderabad, India, 29–30 January

Razavi S (ed) (2003) Agrarian change, gender, and land rights. Blackwell Publishing, Oxford

Rossi A, Lambrou Y (2008) Gender and equity issues in liquid biofuels production: minimizing the risks to maximize the opportunities. Food and Agriculture Organization of the United Nations, Rome

Saito K, Mekonnen H, Spurling D (1994) Raising the productivity of women farmers in Sub-Saharan Africa. Africa Technical Department series no. 230. World Bank, Washington, DC

Swaminathan H, Suchitra JY, Lahoti R (2011) KHAS: measuring the gender asset gap. Indian Institute of Management Bangalore, Bangalore

Thomas D, Contreras D, Frankenberg E (2002) Distribution of power within the household and child health. Mimeo/RAND, Santa Monica

Toulmin C (2008) Securing land and property rights in Sub-Saharan Africa: the role of local institutions. Land Use Policy 26(1):10–19

Udry C (1996) Gender, agricultural production, and the theory of the household. J Pol Econ 104(5):1010–1046

UN-HABITAT (United Nations Human Settlements Programme) (2005) Shared tenure options for women: a global overview. UN-HABITAT (United Nations Human Settlements Programme), Nairobi

UN-HABITAT (United Nations Human Settlements Programme) (2011) Designing and evaluating land tools with a gender perspective: a training package for land professionals. UN-HABITAT (United Nations Human Settlements Programme), Nairobi

UN-OCHA (United Nations Office for the Coordination of Humanitarian Affairs) (2002) Zimbabwe: focus on women's lack of access of land. Integrated Regional Information Network (IRIN), December 4. Retrieved at http://www.irinnews.org/report.aspx?reportid=40021

Walker C (2002) Land reform in Southern and Eastern Africa: key issues for strengthening women's access to rights in land. Unpublished manuscript prepared for the Food and Agriculture Organization of the United Nations, Rome

World Bank (2003) Land policies for growth and poverty reduction. World Bank Policy Research Report. World Bank, Washington, DC

World Bank (2008) World development report 2008: agriculture for development. World Bank, Washington, DC

World Bank/FAO/IFAD (Food and Agriculture Organization of the United Nations/International Fund for Agricultural Development) (2009) Gender in agriculture sourcebook. World Bank, Washington, DC

Zwarteveen M, Meinzen-Dick R (2001) Gender and property rights in the commons: examples of water rights in South Asia. Agric Hum Values 18(1):11–25

Chapter 7
A Review of Empirical Evidence on Gender Differences in Nonland Agricultural Inputs, Technology, and Services in Developing Countries

Amber Peterman, Julia A. Behrman, and Agnes R. Quisumbing

Abstract Empirical research on gender dimensions in agricultural inputs has focused on land. This chapter reviews existing microeconomic empirical literature from the past 10 years on gender differences in use, access, and adoption of nonland agricultural inputs in developing countries. The review focuses on three key areas: (1) technological resources, (2) natural resources, and (3) human resources. In general, there has been more empirical research on inorganic fertilizer, seed varieties, and extension services than on tools and mechanization and life-cycle effects, and most of the studies are from Sub-Saharan Africa. A consistent finding is that, across different types of inputs, men generally have higher input measures than women, and that this input gap is responsible for observed productivity differences between men and women; however, this finding is often sensitive to the use of models that control for other background factors, as well as the type of

A. Peterman (✉)
Department of Public Policy, University of North Carolina, Chapel Hill, Chapel Hill, NC, USA
e-mail: amberpeterman@gmail.com

J.A. Behrman
Department of Sociology, New York University, 295 Lafayette Street, 4th floor, New York, NY 10012, USA
e-mail: jab965@nyu.edu

A.R. Quisumbing
Poverty, Health, and Nutrition Division, International Food Policy Research Institute (IFPRI), 2033 K Street, N.W., Washington, DC 20006, USA
e-mail: a.quisumbing@cgiar.org

A.R. Quisumbing et al. (eds.), *Gender in Agriculture: Closing the Knowledge Gap*, DOI 10.1007/978-94-017-8616-4_7,

gender indicator implemented in the analysis. The final section presents future directions, opportunities, and recommendations for microeconomic gender analysis of nonland agricultural inputs.

Keywords Gender • Agriculture • Access to farm inputs • Assets • Women

7.1 Introduction

Since the 1990s, policymakers and development practitioners have highlighted the critical importance of gender in the implementation, evaluation, and effectiveness of programs across a range of social and economic sectors.[1] *Gender in Agriculture,* a recent sourcebook produced by the World Bank, the Food and Agriculture Organization of the United Nations (FAO), and the International Fund for Agricultural Development (IFAD) (2009, 2), warns that the "failure to recognize the roles, differences and inequities [between men and women] poses a serious threat to the effectiveness of the agricultural development agenda." Similarly, IFAD states that although female farmers are primary contributors to the world's food production and security, they are "frequently underestimated and overlooked in development strategies" (UN News Center 2010). In short, there is agreement that gender inequalities and lack of attention to gender in agricultural development contribute to lower productivity, lost income, and higher levels of poverty as well as undernutrition. This recent and renewed interest in gender and agriculture has produced several new initiatives, calls for action, and commitments from the international development community since 2005 (see, for example, IFAD 2003; IFPRI 2007; World Bank 2007). In addition, guides, tool kits, and other resources on theory and practice of gender integration and promising programmatic approaches have been developed to streamline gender-specific agricultural development initiatives (Doss 1999; Mehra and Rojas 2009; Quisumbing and Pandolfelli 2010; UN-HABITAT 2006; World Bank et al. 2009). Despite these advancements, there is a lack of consensus on the actual magnitude and effects of gender differences in access to agricultural inputs. Where information is available, it is generally focused on access to land or based on dated and region-specific research. Given the importance of producing evidence-based policies, this chapter proposes to update the current knowledge on household-level microeconomic effects and levels of gender differences in access to nonland agricultural inputs through review of published and unpublished literature between 1999 and 2009.

This chapter contributes to the literature in several ways. First, we focus *strictly* on empirical household or plot-level data from program evaluations and agricultural and socioeconomic research in order to summarize and bound parameters for estimates in a reasonable range. We include only articles that are based on quantitative indicators, reasonable measurement of outcomes, and attention to econometric

[1] Here, *gender* represents a social construction of what it means to be of the male or female sex, including cultural, ethnic, economic, religious, and ideological influences.

evaluation techniques.[2] We therefore do not review studies based on aggregate cross-country data or cluster means generated from census data, because such data do not adequately capture the intercluster variation and heterogeneity of the agricultural sector. We review studies that focus explicitly on gender as well as those that include gender as an explanatory indicator in evaluations of other outcomes. This assessment will be conducted with the knowledge that percentages and effect sizes are not strictly comparable because of the diverse technological products, crop varieties, program designs, and empirical techniques from which results are derived.[3] Therefore, although we discuss and include outcome measures in the review, the common theme across all studies included is the provision of sex-disaggregated input data. Second, as previously mentioned, we focus on papers published between 1999 and 2009 to update the literature, given the rapidly evolving environmental, technological, and demographic trends in that period. A body of rigorous and significant literature from the 1980s and 1990s has provided empirical evidence on gender differences in access to inputs. However, this literature has been reviewed sufficiently in past studies (Quisumbing 1994, 1996; Schultz 2001; Kevane 2004), and there is little value in continuing to revisit this material. Finally, although we attempt to make regional comparisons to help identify how women farmers face similar or diverging constraints according to their geographic region of origin (Asia, Sub-Saharan Africa, the Middle East, Eastern Europe, or South/Latin America), our ability to do so is limited by data availability, since most studies on gender differences in access to inputs (with the exception of labor) come from Sub-Saharan Africa.[4]

The review is focused on access to agricultural inputs in three main areas: (1) technological resources (including inorganic fertilizer, insecticide, improved seed varieties, and equipment), (2) natural resources (including water and soil fertility), and (3) human resources (including labor and extension services). The review is compiled by online searches of published material as well as inclusion of working papers and forthcoming evaluations from researchers working in gender and agriculture.[5] Each section is summarized in a table with key components and effect sizes

[2] We do not use a strict sample size cut-off per se but include only studies that generate descriptive statistics across gender-disaggregated subgroups.

[3] Although we attempt to compare and contrast findings, please note that we do not conduct a meta-analysis. Meta-analysis would necessitate a substantial number of studies examining the same types of inputs (and associated outputs), which is not an appropriate analysis because of diversity of inputs.

[4] As noted, the regions we compare include Asia, Sub-Saharan Africa, the Middle East (including North Africa), Eastern Europe, and South/Latin America. When we refer to "region-specific" trends, we lump areas of the world into these five regional categories. Throughout the chapter we sometimes refer to subregions within these five categories (for example, South Asia) or within specific countries (for example, the southern region of Zimbabwe); these instances will be specifically noted.

[5] We started by reviewing original research on gender inequalities in agriculture, followed by papers that cite these studies. We then conducted online searches using keywords for various inputs in each category (Google Scholar, peer-reviewed journals, and websites of the Consultative Group on International Agricultural Research) and publication searches of websites of agricultural research organizations. We also conducted "snowball" citation techniques and sent emails to researchers in the field working on gender and agriculture within various institutions.

as a method of organizing and comparing inputs and outcomes. As we mentioned earlier, we do not explicitly include access to land because it has traditionally been the focus of other reviews, and is also covered in Lastarria-Cornhiel et al. (Chap. 6), although we will inevitably touch on linkages between land access and access to other inputs. In addition, although we acknowledge the importance of bargaining power, women's status, cultural and religious beliefs surrounding agriculture, and community norms, we do not explicitly include how these are determined, but rather focus on how these factors affect the distribution of inputs between men and women. We conclude by making recommendations to address the research gaps in measuring gender differences in nonland agricultural inputs, to highlight the policy implications of the reviewed empirical work, and suggest directions for future research.

Before we present our review of the three focus areas, it is useful to more clearly articulate the range of definitions implicitly or explicitly assigned to the term *women's use* of various resources. When generalizing about gender differences for a given input (such as fertilizer or seed varieties), we often use the terms *use*, *access*, or *adoption* interchangeably; however, within a specific study or framework, these terms may connote entirely different. These distinctions are critically important, as differences across studies may in fact be the result of variations in definitions of terms rather than magnitude of gender differences. The literature on property rights and collective action defines bundles of rights, which refer to gradients of control over a given resource, usually applied to land or other natural resources. For example, bundles of rights for land can be divided into the right to use the asset (including the right to access, the right to extract resources), the right to appropriate the return from the asset (including earnings and income), the right to change its form, substance, and location (including decisionmaking rights such as management, and the exclusion of other users), and alienation (including transfer of rights to others) (Di Gregorio et al. 2008).[6] These bundles of rights are applied at different levels (individuals, families, groups, the state), and actors often overlap in their levels of rights. Although this framework is a useful starting point for thinking about women's control of agricultural inputs, we limit our review to production, and thus concepts of transfer or exclusion will not typically apply. Therefore, we define *use* of an input as the actual application of that resource in productivity-producing outputs, specifically, at the individual or household level, whether the input was obtained through extraction, purchase, or barter. The use of inputs is generally straightforward and can be operationalized for both technological inputs, such as fertilizer or seed varieties, and natural and human resources. We define *access* to an input as the availability or potential for use at the individual, household, or community level. Access implies the right or ability to use a resource or input, but is not an actual use measurement. We define *adoption* as the initial use of an input or method by an individual, household, or community that often, but not always, occurs in the context of an established program or scheme. Finally, we make use of the additional term

[6] For a detailed presentation of property rights and collective action framework, including measurements, institutional actors, and linkages to poverty reduction, see Di Gregorio et al. (2008); for a review of implementation of this framework in evaluation work, see Mwangi and Markelova (2008).

participation, which we define as the ability to freely and fully partake in and engage with a social or political group or network. Although not all the inputs and studies we review relate directly to these definitions, they will serve as a general guide throughout this chapter. Where deviations from these terms are necessary, they will be noted in the text.

In part, the levels and appropriateness of use, access, and adoption of inputs are determined by the setting, farming systems, and context of the study in question. A number of rudimentary generalizations can be made about the differences in farming systems across regions. In Asia, where monogamous extended or nuclear families dominate, and where families jointly farm agricultural land, men serve as the primary agricultural decisionmakers and laborers.[7] In many African societies, where polygamous families are common, access to resources and decisionmaking is divided between household members (Dey 1985). While African women play a large role in agricultural production, there is often a gendered division of labor that links women to the production of food crops and men to cash crops (Boserup 1970). In Latin and South America, where the monogamous family structure is dominant, there is a gender division of labor in both industrialized crop production and peasant farming (Ashby 1985). In general, women's agricultural participation in family farming systems is much more important in the Andean countries and Central America than in the southern region of South America (Deere and Leon 1987). These regional differences will be further explored in the discussion section.

7.2 Evidence of Gender Inequalities

7.2.1 Technological Resources: Inorganic Fertilizer, Insecticide, Improved Seed Varieties, and Mechanical Power

Advancements in technological resources have positively had an impact on farmers in developing countries by providing a means to improve soil fertility and increase land productivity and overall crop yields. Female farmers, who are more likely to be asset poor and subsistence oriented than their wealthier male counterparts, stand to benefit significantly from such technology (World Bank et al. 2009). In this chapter, we examine four main categories of inputs of particular importance to small-scale female farmers: (1) inorganic fertilizer, (2) insecticides, (3) improved seed varieties, and (4) mechanical power. Inorganic fertilizer refers to a nitrogen-based chemical mixture used to improve soil fertility. Inorganic fertilizer is differentiated from organic fertilizer (such as animal manure, compost, or other living mulch) by its manufacture, chemical modification, and external purchase. Insecticides and pesticides (also called farm chemicals, agrochemicals) are primarily synthetic

[7] Polygamy exists in Asia, but not to the same extent as in Sub-Saharan Africa.

spray-applied agents used to expand agriculturally productive land and increase crop yields through pest, bacteria, and weed destruction or control.[8] Improved and genetically modified seed varieties are artificially produced by cross-pollination to increase yield, uniformity, and resistance to disease. By *mechanization* we mean the introduction of mechanized farming tools or other equipment (tractors, plows, seeders, and weeders) into the farming practice. For the purposes of gender analysis, technology inputs are unique in that they typically (but not always) imply a monetary purchase as a prerequisite to use, in contrast to other categories, which may require time or natural resource endowment.

Table 7.1 summarizes the 24 studies reviewed that contain statistics on gender differences in access to technological resources. Articles are listed in alphabetical order of the first author's surname (column 1) and therefore do not represent importance or significance of studies. Column 2 lists the country or countries or region of the indexed study and the crop, if applicable. Column 3 reports the sample size and unit of analysis in the study. Columns 4 through 7 indicate differential access or mean values of a specified input type (column 4, for example, shows fertilizer or seed varieties) reported for women (column 5) and for men (column 6) in a specified unit of disaggregation (column 7). Where additional analysis was conducted, columns 8, 9, and 10 list stratifying variables, outcome variables, and effect sizes (coefficients and standard errors in parenthesis) for each study. Comments on relevant findings, including methods or caveats, interpretations of use operationalized by the study, and indicator of peer-reviewed publication status follow in the remaining columns.

Much of the research on gender differences in access to technological inputs focuses on inorganic fertilizer, which perhaps reflects the important role fertilizer continues to play within debates about agricultural productivity and poverty reduction. In the literature on inorganic fertilizer, an important theme is that, given equal access to fertilizer (controlling for other inputs and background factors), female farmers adopt fertilizer at the same rates as male farmers. Such findings suggest that accessibility of inputs, not propensity to use inputs, is a key issue for many female farmers. A highly cited paper is Doss and Morris's (2001) study of 420 maize farmers in Ghana, which found that once researchers controlled for access to complementary inputs (land, education, labor), they found no significant difference in rates of adoption between male and female farmers. Similarly, Thapa (2009) found little evidence for gender differences in value of farm output in 2,360 Nepalese households after controlling for access to inorganic fertilizer and other key inputs. Gilbert et al. (2002) analyzed a cropping system trial survey in Malawi and found a significant gender difference in fertilizer use among the 1,385 farmers selected to participate in the trial. Following a treatment period in which all participants were supplied with inorganic fertilizer inputs, the authors found no significant gender difference in maize yield. Jagger and Pender (2006) examined the effects of the presence of local organizations that promote improved technology use in rural Uganda and found female heads of household are significantly more likely to adopt inorganic fertilizer than their male counterparts.

[8] Pesticides may also be organic or organic compounds synthesized in a laboratory.

Table 7.1 Gender differences in access to technological inputs: fertilizer, insecticide, seed varieties, and other technological inputs

(1)	(2)	(3)	(4)	(5)	(6)	(7)	(8)	(9)	(10)	(11)	(12)
			Use of access to input								
Authors (year)	Country (crop)	Sample size	Input type	Women	Men	Unit	Gender indicator	Outcome measure	Effect size	Comments	Published
Atreya (2007)	Nepal	434 hhlds	Pesticides	–	–	–	Male respon-dents	Decisions of pesticide use in a household	0.425*** (nr)	Partial correlations among individual characteristics and pesticide use-knowledge, attitudes and practices (but not actual use) find almost all respondents of both genders were aware of negative impacts of pesticide use on human health and environment; however, females were at higher risk due to lower level of awareness of safe pesticide use practices	x
				–	–	–		Proper practice: wind direction	1.03** (nr)		
				–	–	–		Knowledge of safety precautions	0.117*** (nr)		
Babatunde et al. (2008)	Nigeria	60 hhlds	Farm tools	5,946	75,706	Value in naira	Female heads	–	–	Bivariate analysis finds male-headed households have significantly higher valued farm tools (access) as compared to female-headed households; however, this study found no differences in farm output by gender	x
Bourdillon et al. (2002)	Zimbabwe (maize)	136–200 hhlds[a]	Maize hybrid	–	–	–	Not specified[b]	Adoption	NS	Probit analysis finds that gender of household head has no significant impact on adoption rates; however, statistics are not reported (sensitivity analysis discussed in text only)	

(continued)

Table 7.1 (continued)

(1)	(2)	(3)	(4)	(5)	(6)	(7)	(8)	(9)	(10)	(11)	(12)
			Use of access to input								
Authors (year)	Country (crop)	Sample size	Input type	Women	Men	Unit	Gender indicator	Outcome measure	Effect size	Comments	Published
Chirwa (2005)	Malawi	156 hhlds	Fertilizer	–	–	–	Female plot owners	Adoption	−0.146 (−0.58)	Probit analysis finds that gender of plot owner farmer has no significant association with adoption rates	x
			Maize hybrid	–	–	–			0.096 (0.37)		
			Fertilizer	–	–	–	Female heads	Adoption	−7.3*** (−2.57)	Probit analysis finds female-headed households associated with lower adoption rates	
			Maize hybrid	–	–	–			−0.23 (−0.85)	Probit analysis finds gender of household head insignificant in predicting adoption rates	
Doss and Morris (2001)	Ghana (maize)	420 farmers	Modern seed varieties[c]	39	59	%	Female farmers	Adoption	−0.085 (0.200)	Two-stage probit models, find no significant difference in adoption rates between male and female farmers once access to complementary inputs (land, education, labor) are controlled for	x
			Inorganic fertilizer	16.2	22.5	%			0.093 (0.225)		
Enete et al. (2001)	Nigeria (cassava)	62 hhlds	Inorganic fertilizer	19	14	%	Female plot owners	Cassava yields	–	Female-owned plots have significantly higher mean cassava yields; however, no multivariate analysis presented to attribute to inputs	x
			Improved seed varieties	5	0	%			–		
Fisher et al. (2000)	Senegal	60 hhlds	Stabling technique[d]	–	–	–	Age of wives		−1.77** (0.70)	Logistic regression models factors related to the bargaining power of wives (proxied by age, number of wives and number of children of first wife) in household decision to adopt stabling (which is an intensive labor technique)	x
				–	–	–	Number of wives	Adoption	3.24* (1.68)		
				–	–	–	Number of children of 1st wife		−0.67* (0.38)		
Freeman and Omiti (2003)	Kenya	399 hhlds	Fertilizer	–	–	%	Male heads	Adoption and intensity of use	2.48 (10.212)	Tobit regression model finds no significant differences in adoption and intensity of use after controlling for other inputs	x

Gilbert et al. (2002)	Malawi (maize)	1,385 farmers	Inorganic fertilizer	36.0	66.2	kg	Gender stratified: female farmers	Use (pretreatment, high altitude zone)	−3.68*** (nr)	Descriptive statistics disaggregated by gender and agroecological zone show that before treatment there were significant differences in fertilizer use based on the gender of farmer	x
				20.4	28.9	kg		Use (pretreatment, low-med altitude zone)	−2.03* (nr)		
				2,460	2,470	kg ha^{-1}		Yield (posttreatment, high altitude zone)	NS (nr)	Descriptive statistics disaggregated by gender and agroecological zone show that when female farmers were provided with seed and fertilizer inputs (access) for the trial, their farm management efforts (use) were equally as productive as the male in farmers'	
				2,540	2,560	kg ha^{-1}		Yield (posttreatment, low-med altitude zone)	NS (nr)		
Horrell and Krishnan (2007)	Zimbabwe (primarily maize)	300 hhlds	Inorganic fertilizer	55 (de jure); 67 (de facto)	63	%	Female heads (distinction between de jure and de facto)	Usage	−0.03 (0.03) de jure; −1.90 (1.30) de facto	Sample of 300 households distinguishes between de jure female-headed (widow headed, n=52) and de facto headed (n=17). Tobit regression analysis finds no significant difference in maize yields achieved or fertilizer usage by female-headed households. However, further analysis finds de facto female heads of household receive low prices for their output and lack access to selling consortiums; thus disadvantages persist	x
				137 (de jure); 175 (de facto)	156	kg/ha (among users)		Inputs/ha	NS (nr)		
			Seeds	13.4 (de jure); 17.7 (de facto)	14	kg/ha (among users)			NS (various)	Tobit regression models find no significant differences in use among de jure or de facto female-headed households and male-headed households for both maize and all crop samples	
			Machinery	4.2 (de jure); 3.6 (de facto)	5.2	Ownership index		–	S** (nr)	Bivariate t-tests show significant differences in use between male-headed and de facto female-headed households only	

(continued)

Table 7.1 (continued)

(1)	(2)	(3)	(4)	(5)	(6)	(7)	(8)	(9)	(10)	(11)	(12)
Authors (year)	Country (crop)	Sample size	Use of access to input: Input type	Women	Men	Unit	Gender indicator	Outcome measure	Effect size	Comments	Published
Jagger and Pender (2006)	Uganda	451 hhlds	Inorganic fertilizer	–	–	–	Female heads	Adoption	0.136*** (nr)	Two-stage probit models used to look at impacts of programs and organizations on technology adoption. Control factors also include the number of males and females in the household. Number of males is weakly associated (10 % level) and number of females is significantly (5 % level) associated with fertilizer use, and both are insignificant in predicting pesticide use	
			Pesticides	–	–	–			–0.087 (nr)		
Kinking-ninhoun-Mêdagbé et al. (2008)	Benin (rice)	45 farmers	Inorganic fertilizer	406	361	kg/ha	Female farmers	Rice yield	0.062 (0.105)	Production function estimates indicate gender and quantity seeds insignificant but quantity fertilizer (access, use) significant in predicting yields after controlling for other inputs	x
			Insecticide	0.84	0.95	l/ha					
			Seeds	134	215	kg/ha					
Nkedi-Kizza et al. (2002)	Uganda	18 farmers; 90 soil samples	Fertilizer	–	–	–	Female plot owners	Soil fertility indicators	NS (nr)	Chemical analysis of soil fertility indicators across male and female plots suggests that females are not allocated plots of inferior quality. Therefore, lower yields in female-owned plots are likely due to other socioeconomic factors (lack of access to fertilizer, extension, etc.). Actual differences in fertilizer use were not directly explored	x

Oladele and Monkhei (2008)	Botswana	See note "e" below	Cattle (draft)	8.402	24.796	Number of holdings	Gender stratified: male farmers	Ownership	2.88 S (0.05)	Bivariate t-tests show that across Botswana, males own (access, use) significantly more draft animals than females do	x
			Donkey (draft)	6.809	18.405				2.40 S (0.05)		
			Goat (nondraft)	11.177	23.514				2.16 S (0.05)		
Ouma et al. (2002)	Kenya (maize)	127 farmers	Improved seed varieties	–	–	–	Male farmers	Adoption	2.21* (1.06)	Logit analysis finds gender has a significant association with adoption of improved seed, while OLS regression finds gender is insignificantly associated with fertilizer use	
			Fertilizer					Use (kg/ha)	10.5 (39.9)		
Pender and Gebremedhin (2006)	Ethiopia	500 hhlds	Draft animal power (oxen)	–	–	–	Female heads	Input use	−0.207*** (nr)	In a study to inform sustainable land management practices, OLS regression finds female-headed households use significantly less draft animal power	x
			Improved seed varieties	–	–	–			−0.002 (nr)		
			Fertilizer	–	–	–			−0.050 (nr)		
Sanginga et al. (2007)	Nigeria (soybeans)	203 hhlds	Improved seed varieties	–	–	–	Female farmers	Usage	−25.122* (nr)	Tobit model finds gender of farmer negatively associated with usage of improved seed. Qualitative evidence is also presented by gender on welfare and distributional effects	
Shankar and Thirtle (2005)	South Africa (cotton)	91 smallholders	Improved seed varieties[f]	–	–	–	–	–	−0.19 (0.30)	Probit model of Bt. adoption finds no significant differences by gender	x
SOAS et al. (2008)	Malawi	2,491 hhlds	Fertilizer subsidy coupon	–	–	–	–	Distribution of coupon by household head gender	–	Study of coupon distribution finds female-headed households less likely to receive (access) fertilizer coupons of all types, and to receive fewer per household	

(continued)

Table 7.1 (continued)

(1)	(2)	(3)	(4)	(5)	(6)	(7)	(8)	(9)	(10)	(11)	(12)
			Use of access to input								
Authors (year)	Country (crop)	Sample size	Input type	Women	Men	Unit	Gender indicator	Outcome measure	Effect size	Comments	Published
Thapa (2009)	Nepal	2,360 hhlds	Inorganic fertilizer	1,428	2,119	Value in rupee	Female heads	Value of farm output	–0.018 (0.056)	Production function estimates give weak/little evidence for gender differences after controlling for other inputs (access)	
Tiruneh et al. (2001)	Ethiopia	180 hhlds	Inorganic fertilizer	4,776	6,456	Birr per ha	Gender stratified: female heads	Gross output value	–	Production function analysis suggests gender differences in output partly related to lower quantities of inputs used by females (fertilizer, extension, and others)	
			Improved wheat technology	30	14	%	Not specified	Adoption	5.7 S (0.05)	Logit analysis shows a significantly higher proportion of male households than female households grew improved wheat varieties	
Uttaro (2002)	Malawi	60 farmers	Inorganic fertilizer	62 spouses; 45 heads	67	%	Female farmers	Adoption	–	Study finds as a group, married women are more likely to have access to some fertilizer than are female-headed households	x
			Maize hybrid	39 spouses; 43 heads	69	%		Use of hybrid maize	–	Decision tree modeling finds women (both household heads and spouses) more constrained in access to fertilizer and seeds (partly due to price) than men. All three groups show little difference in preferences and beliefs regarding input use	

Van de Fliert et al. (2001)	Indonesia (sweet potato)	123 farmers	Fertilizer	–	–	–	Female partici-pants	Profitability	NS (nr)	Study compares knowledge, skills, practices, input and output usage, and profitability of participants to nonparticipants of integrated crop management Farmer Field Schools. Estimation of sweet potato profit function finds female indicator is not significant (and therefore is excluded from table results)
World Bank and Government of Malawi (2007)	Malawi	11,280 hhlds[g]	Fertilizer	–	–	–	Female farmers	Decisions about input use and planting	–	Poverty vulnerability analysis finds women, on average, make half of the decisions on crops not requiring fertilizer, while only 10 % of the time with crops requiring fertilizer

Source: Authors' compilation

Notes: Articles listed in alphabetical order of first author's last name; Effect size refers to gender indicator coefficients with standard errors in parentheses unless otherwise noted. *S* significant, *NS* not significant, *nr* not reported, *hhlds* households, *ha* hectare(s). * Significant at 10 %; ** significant at 5 %; *** significant at 1 %. Outcome measures of adoption refer to rate of adoption of corresponding input type; published indicates the study is published in a peer reviewed journal

[a]Analysis is stratified by village (Mupfurudzi village or other) and year (1994–1995, 1995–1996, and 1996–1997), and sample sizes range from 136 to 200

[b]Report discusses whether sex of household head affects adoption but does not report corresponding variable in statistical tables or explanations; it was not possible to ascertain whether gender indicated male or female

[c]Modern varieties are improved open-pollinating varieties and hybrids developed by a formal breeding program

[d]Stabling is a technological package consisting of a stable, a food supplement, an animal health-care program, and an improved method of producing manure. A major benefit of stabling is increased milk production

[e]Data used in this study come from the 2007 Agricultural Census; livestock ownership across six regions was compiled. Although exact sample size was not provided, it is assumed to be large, covering the entire rural agricultural population of the country

[f]The *Bacillus thuringiensis* (Bt) gene in Bt varieties of cotton produces a natural insecticide

[g]Sample sizes are not reported in World Bank and Government of Malawi (2007); however, the referenced IHS2 survey Extract of Findings provides this information

Findings from several additional studies contradict initial expectations that female household heads are disadvantaged in their fertilizer usage and adoption rates. Freeman and Omiti (2003) and Bourdillon et al. (2002) found that the sex of household head has no significant effect on adoption and intensity of use of inorganic fertilizer in 399 households in Kenya and among stratified samples of 136–200 households in Zimbabwe. In a sample of 156 households in Malawi, Chirwa (2005) found men and women plot owners do not differ significantly with respect to fertilizer adoption. However, in a parallel analysis using the same sample but using headship as an indicator of gender, he found that female-headed households are less likely to adopt fertilizer (note, however, the sample size is only 156 households). Horrell and Krishnan (2007) found no significant difference in maize yields achieved or fertilizer usage by female household heads in Zimbabwe. However, further analysis found de facto female heads of household do receive lower prices for their output and lack access to selling consortiums; thus, disadvantages persist.

Many of the same studies that examine fertilizer use also analyze gender differences in seed varieties. The Doss and Morris (2001) study in Ghana found that once they controlled for access to complementary inputs (land, education, labor), there were no significant difference in rates of modern seed variety adoption between male and female farmers. Similarly, Horrell and Krishnan (2007) found no significant difference in maize yields achieved or seed usage per acre by female heads of household. Tiruneh et al. (2001) in their study of households in Ethiopia found that a significantly higher proportion of male than female heads of household use improved wheat.[9] Logit analysis stratified by sex shows that in male-headed households, farm size and extension service contact significantly and positively affected adoption, whereas farm size and asset ownership are associated with adoption in female-headed households. Sanginga and colleagues (2007) found female farmers less likely to use improved soybean seeds in Nigeria, at least in part, because male farmers continue to have more money to spend on hiring extra labor and have better market access opportunities. However, Sanginga and colleagues also found that more and more women are growing soybeans, a traditionally male crop, thus blurring presumed cropping norms. The studies by both Chirwa (2005) and Bourdillon et al. (2002) found that the sex of household head has no significant effect on adoption of improved seed in Malawi and Zimbabwe, respectively, although the authors of neither study provides an explanation for why this might be the case, and, as previously mentioned, sample sizes are relatively small (N = 156–200).

We found fewer relevant studies that examined inequities in pesticide use by gender. Jagger and Pender (2006) used a two-stage model to examine program effects on pesticide adoption among 451 Ugandan households and found female headship is insignificant in predicting adoption. Atreya's (2007) exploration of pesticide knowledge, attitudes, and practices (but not actual use) among 434 households in Nepal found that almost all respondents were aware of negative impacts of pesticide use on human health and environment; however, females were at higher risk of

[9] Improved wheat seed is artificially produced by cross-pollination to improve yield, uniformity, and resistance to disease.

incorrect usage because they had less knowledge of how to use pesticide safely. Kinkingninhoun-Mêdagbé and colleagues' (2008) study of 45 rice farmers in Benin found significant gender differences in farmers' use of pesticide, which they largely attribute to gender-based discrimination. This lack of information may be indicative of the relatively low importance placed on pesticide use by agriculture-based research and programs.

Only two qualifying empirical studies were reviewed that found gender differences in use of production tools and equipment; we return to the paucity of research in this area in the discussion section. In the Zimbabwe study of agricultural differences in productivity, Horrell and Krishnan (2007) included an index of farm machinery as a control indicator and found significant bivariate differences between male and de facto female heads of household but not between male and de jure female heads of household. Babatunde and colleagues (2008) also found significant bivariate differences between male and female heads of household in value of farm tools owned in a sample of 60 Nigerian households. However, several related studies looked at gender-based differences in access to/ownership of draft animals. Draft animals are essential for the operation of manual plows and are an important source of manure; some studies cite ownership of draft animals as a key factor in increasing agricultural productivity among the rural poor (Smith 2008). Oladele and Monkhei (2008) found significant differences in the populations of animals owned by men and women in Botswana; men are significantly more likely to own cattle, donkeys, and horses, whereas women are significantly more likely to own goats. Pender and Gebremedhin (2006) found that female heads of households are negatively associated with the use of draft animals (oxen) in Ethiopia. This study also found that female heads of household achieve 42 % lower crop yields than male heads of household with similar use of labor, ox power, and other inputs, thereby indicating a further gender-based disadvantage in productive use of inputs. Fisher et al. (2000) examined the role of women's bargaining power among Senegalese cattle owners in the decision to adopt a bundle of "stabling technology" and found that the more bargaining power a wife has, the more likely the household is to reject adoption of this labor-intensive technique.[10] This may be because stabling leads to an increase in labor for women and a concurrent loss in income (when milk becomes more lucrative, men take on the traditional women's role of selling milk). Further analysis reveals that adoption of the practice does lead to a loss of income for women but an overall improvement in household welfare that may benefit women in the long run.

In summary, we reviewed 24 studies of technological input use, access, and adoption that fit our criteria. The majority examine more than one technological input, including 18 measures of fertilizer, 13 measures of seed varieties, 7 measures of tools, and 3 measures of pesticide use, access, and adoption. Sixteen of 24 studies were published in peer-reviewed journals. Overall, where descriptive statistics for inputs were provided (for 24 input indicators), 19 (79 %) found men have higher

[10] Stabling is a technological package consisting of a stable, a food supplement, an animal healthcare program, and an improved method of producing manure. A major benefit of stabling is increased milk production.

mean access and 5 (21 %) found women have higher mean access to the given resource. Where further bivariate or multivariate analysis was conducted (for 39 input indicators), 23 (59 %) found gender indicators are not significant with respect to outcome measures when other factors are controlled for, while 15 (38 %) found differences persist and men have higher outcome measures; one study (3 %) found that women have higher outcome measures. The lack of significant differences is driven by the studies on inorganic fertilizer, where key background factors accounting for differences are education, wealth/asset stores, and land indicators. Many of these studies, however, identify alternative channels, through which gender disparities persist, such as receiving lower prices for yields or through poor access to markets. However, since these channels are outside the main focus of these studies, they are only described and not analyzed in great detail.

7.2.2 Natural Resources: Water and Soil Fertility

The importance of natural resources is a growing concern in agricultural production as population pressures expand and stress the finite provision of environmental resources. Water is a supremely valuable resource not only for agriculture but also for domestic and household work, small business, commercial use, and general health and hygiene. It is therefore not surprising that there are social constructs concerning decisions about policy, access and allocation, and pricing of water and that gender has been high on the policymaking and programmatic water agenda (Singh et al. 2006; UNDP 2006; von Koppen 2002; World Bank et al. 2009, Module 6). Because access to water can refer to a wide range of provision types, not all of which are appropriate for our review, we only include studies that specifically analyzed water for agricultural or mixed garden and household use.[11] We therefore included studies on soil fertility that use sex-disaggregated data on any natural soil improvement technique, including, but not limited to, use of manure and compost, application of fallow periods, or other intercropping techniques, such as hedgerow or alley farming, that have the ability to improve soil fertility.

Table 7.2 summarizes the 13 studies that examine gender differences in access to natural resources and follows the format described for Table 7.1 on technological resources. Despite the importance of irrigation and access to water for agricultural outcomes, comparatively few empirical micro-level studies examine gender differences.[12] Using a sample of 1,131 households from the 2000 China National Rural Survey, de Brauw et al. (2008) found no difference in the percentage of irrigated

[11] Studies that examine drinking water or domestic use only are therefore not included. However, women might use drinking water for kitchen gardens or small plots for home consumption, even if it is not noted or analyzed in the study. Because the literature on any type of water use is so large, we decided to exclude these studies.

[12] Since irrigation often relies on water schemes or centralized infrastructure, there have been more case studies and other institutional analyses since 2000.

Table 7.2 Gender differences in access to natural resources: water and soil fertility

(1)	(2)	(3)	(4)	(5)	(6)	(7)	(8)	(9)	(10)	(11)	(12)
			Use of access to input								
Authors (year)	Country (crop)	Sample size	Input type	Women	Men	Unit	Gender indicator	Outcome measure	Effect size	Comments	Published
Adesina et al. (2000)	Cameroon	255 farmers	Alley farming[a]	–	–	–	Male plot owners	Adoption	1.08** (0.61)	Using logit regression and controlling for other inputs. men are more likely to adopt alley farming. perhaps due to lack of women's land and tree rights	x
de Brauw et al. (2008)	China	1.131 hhlds	Irrigated land	66.4	65.2	%	Female managers	Plot revenue	0.0019 (0.041)	OLS regression controlling for village-level fixed effects. authors find no evidence of female differences in productivity (results unchanged with use of female heads or female share of hours worked)	x
Enete et al. (2001)	Nigeria (cassava)	62 hhlds	Manure	19	71	%	Female owners	Cassava yields	–	Female-owned plots have significantly higher mean cassava yields: however. no multivariate analysis presented to attribute to inputs	x
Goldman and Heldenbrand (2001)	Uganda	80 plots	Fallow period	15[b]	37.5	%	Female plot owners	Change in per capita output	–	Comparison of mean differences in production indicates women (especially single women) are disadvantaged and have lower outputs as compared to men	x
			Manure	70[b]	62.5	%			–		

(continued)

Table 7.2 (continued)

(1)	(2)	(3)	(4)	(5)	(6)	(7)	(8)	(9)	(10)	(11)	(12)
			Use of access to input								
Authors (year)	Country (crop)	Sample size	Input type	Women	Men	Unit	Gender indicator	Outcome measure	Effect size	Comments	Published
Hope et al. (2003)	South Africa	539 hhlds	Private piped water	21.8[b]	31.8	%	Female heads	Number of garden crops grown	S (nr)	Using bivariate analysis, the relationship between access to water and number of garden crops is significant at the 1 % level	
Horrell and Krishnan (2007)	Zimbabwe (primarily maize)	300 hhlds	Manure	1,014 (de jure) 1,094 (de facto)	1,380	kg/ha	Female heads	–	NS (various)	Tobit models predicting logged values of kg/ha of manure inputs among maize and all crops show that headship variables are insignificant	x
				42 (de jure) 67 (de facto)	57	%		–			
Jagger and Pender (2006)	Uganda	451 hhlds	Animal manure	–	–	–	Female heads	Adoption of land management practice	0.106 (nr)	Study was conducted to evaluate the impacts of programs and organizations on technology adoption. In a two-stage probit analysis, although headship indicator is insignificant (as well as control variable of number female household members), control of number of males is associated (1 % level) with adoption of crop residues and manure	
			Crop residues	–	–	–			–0.024 (nr)		
			Mulching	–	–	–			–0.073 (nr)		

Kazianga and Masters (2002)	Burkina Faso	258 farmers	Field bunds[c]	–	–		Female farmers	Adoption and intensity	−0.093** (0.054)	Approximately 16 % of cropland is farmed by women. Analysis uses multivariate tobit regression; however, results are sensitive to choice of model	x
			Microcatchments	–	–				−0.256*** (0.061)		
Kinkingninhoun-Mêdagbé et al. (2008)	Benin (rice)	45 farmers	Distance to main irrigation channel	2.55	2.7	Meters	Female farmers	Rice yield	0.062 (0.105)	Production function estimates indicate gender insignificant, but irrigation level is significant in predicting yields after controlling for other inputs	x
Pender and Gebre-medhin (2006)	Ethiopia	500 hhlds	Manure and composting	–	–	–	Female heads	Input use	−0.087***(nr)	Study conducted to inform sustainable land management practices and uses probit regression. Female-headed households make up 21.8 % of the sample and average use of manure/composting is 22.8 % and of burning is 11.0 % in total	x
			Burning to prepare fields	–	–	–			0.025 (nr)		
Phiri et al. (2004)	Zambia	218 hhlds	Improved tree fallows[d]	36	23	%	Female heads	Adoption	NS (nr)	Using log-linear models controlling for wealth, no significant differences in mean rates of adoption by gender were found	x

(continued)

Table 7.2 (continued)

(1)	(2)	(3)	(4)	(5)	(6)	(7)	(8)	(9)	(10)	(11)	(12)
			Use of access to input								
Authors (year)	Country (crop)	Sample size	Input type	Women	Men	Unit	Gender indicator	Outcome measure	Effect size	Comments	Published
Somda et al. (2002)	Burkina Faso	116 hhlds	Composting	40[b]	65	%	Female farmers	Adoption	–0.504 (0.83)	Logit models show women farmers equally likely to adopt composting when controlling for other inputs. However, when stratifying by region (two regions), gender is significant in both, one positive and one negative, suggesting regional cultural or crop differences	x
Swinkels et al. (2002)	Kenya	45 hhlds	Alley farming[e]	28	72	%	Female heads	–	–	All households participated in trial; low mean acceptance rates among women are attributed in part to reluctance to trim hedges due to physical strength and the view of hedges as men's crops	x

Source: Authors' compilation

Notes: Articles listed in alphabetical order of first author's last name; *Effect size* refers to gender indicator coefficients with standard errors in parentheses unless otherwise noted. *S* significant, *NS* not significant, *nr* not reported, *hhlds* households. * Significant at 10 %; ** significant at 5 %; *** significant at 1 %. Outcome measures of *adoption* refer to rate of adoption of corresponding input type; *published* indicates the study is published in a peer reviewed journal

[a]Alley farming (or hedgerow cropping) involves planting of food crops between hedgerows of nitrogen-fixing leguminous hedgerow species, which have deep roots for nutrient capture and recycling. The technique requires the occasional trimming of hedgerows for application as mulch

[b]Mean differences are not presented in the paper but were calculated by authors from disaggregated statistics

[c]Field bunds (barriers to soil and water runoff) and microcatchments (small holes in which seeds and fertilizers are placed) are conservation techniques

[d]Two-year tree fallows, mainly *Sesbana sesban* (requiring nursery) and *Tephrosia vogelii* (directly seeded)

[e]Types included *L. leucocephala, Leucaena diversifolia, Calliandra calothyrus*, or *Gliricidia sepium* planted from inoculated seedlings

land under female management (66.4 %) and under male management (65.2 %). The absence of differences in water use is consistent with a study of 45 rice growers in Benin that found average distance of female farmers to the main irrigation channel is slightly greater than that of male farmers (2.7 m versus 2.55 m); however, the sample size is very small (Kinkingninhoun-Mêdagbé et al. 2008). Findings from a Limpopo household survey (N=552) in South Africa linking poverty and water supply found female-headed households are significantly disadvantaged in their access to piped water (22 %; 32 % of male-headed households have such access), and bivariate methods show that access is significantly associated with an increase in kitchen garden crops (Hope et al. 2003). The mixed findings for gender differences in water use and access may be in part obscured by the fact that women are often responsible for fetching water for household domestic use, which may also be used for small-scale farming for household consumption.

A range of improved soil fertility methods has been the focus of many interventions, partially because of the gain in productivity realized without the provision of infrastructure or costly technology. Although the sample sizes of the studies included are relatively small, results generally indicate men are more likely to have access to or implement soil fertility improvement techniques than women. For example, in Cameroon, male plot owners are significantly more likely to adopt alley farming techniques, controlling for other inputs, which the authors attribute to potential disincentives to invest because of lack of land and tree property rights for women (Adesina et al. 2000). Low acceptance rates also were found among Kenyan women heads of household for alley farming, which the authors speculate may owe to the view of hedges as men's crops and women's reluctance to trim hedges, a task that involves heavy physical labor (Swinkels et al. 2002). Although mean differences indicate female heads of household are actually more likely to adopt tree fallows in Zambia in a sample of 218 households, the difference is insignificant, once other factors are controlled for (Phiri et al. 2004). While one may suspect women would have comparatively more access to natural products, like manure and compost, than to purchase fertilizer products, the few studies we reviewed give mixed results. Horrell and Krishnan (2007) found no significant differences in use of manure between female and male heads of household in Zimbabwe. In Uganda, a study of 80 plots found female owners report higher use of manure in comparison to male owners (70 % versus 62.5 %); however, in Nigeria, among 62 cassava-producing households, female farmers applied manure on 19 % of plots, whereas manure was applied to 71 % of male-owned plots (Goldman and Heldenbrand 2001; Enete et al. 2001). Jagger and Pender (2006) evaluated the effect of a program for natural resource management on 451 households in Uganda and found no differences between male- and female-headed households in their adoption of animal manure, mulching, and crop residue. Using probit regression, Pender and Gebremedhin (2006) found that female heads of households in Ethiopia are no different than their male counterparts in burning to prepare fields; however, women are less likely to use manure and composting to increase productivity. Finally, in a sample of 116 households in Burkina Faso, gender analysis of composting techniques found mixed results by region, suggesting that cultural or cropping differences may effect adoption (Somda et al. 2002).

In summary, we reviewed 13 studies of natural resource input use, access, and adoption that fit our criteria. The majority of studies examine measures of soil fertility (14 measures), while the minority examine water measures (three measures). Eleven of 13 studies were published in peer-reviewed journals. Overall, where descriptive statistics for inputs were provided (for 11 input indicators), eight (72 %) found men have higher mean values, and three (27 %) found women have higher mean values for the given resource. Where further bivariate or multivariate analysis was conducted (for 14 input indicators), nine (64 %) found gender indicators are not significant with respect to outcome measures when other factors are controlled for, while five (36 %) found differences persist and men have higher outcome measures. None of the reviewed studies found women have higher outcome measures in further bivariate or multivariate analysis. The factors accounting for the differences in significance vary, ranging from regional and market variations to quality and quantity of land. We hypothesize that this is, in part, the result of the diverse nature of inputs (ranging from soil improvement techniques to formal irrigation schemes) and because sample sizes in this section are relatively smaller than in other sections.

7.2.3 Human Resources: Agricultural Labor and Extension Services

The effect of human resources on agriculture is a broad and extensive topic, ranging from health and nutrition to education and labor contributions (see Chaps. 11 by Harris and 14 Termine and Dey de Pryck and Termine). The process through which intrahousehold allocations of human resources are determined may, in fact, reflect the distribution of agriculture-specific inputs. However, because other studies have reviewed many of the relationships with these broader categories of human resources, we chose to limit our examination of human resources to two main proximate and definitive inputs: (1) agricultural labor and (2) extension or agricultural knowledge services. Agricultural labor refers not only to women's own ability to produce outputs (own labor) but also to the quantity and quality of supplemental labor they are able to access (hired or outside labor), which is often nonpaid labor allocated within the household. Note that this evidence is strictly differentiated from macro-estimates of women's contribution to the total agricultural workforce or the percentage of output produced by women farmers. Extension services (also known as agricultural advisory services) refer to the range of information, training, and agriculture-related knowledge provided by government, nongovernmental organizations (NGOs), and other sources that increase farmers' ability to improve productivity. Extension services are delivered on the ground by extension agents or livestock officers who are charged with information dissemination. Extension services may take the form of individual field visits, technical advice at organized meetings, visits to demonstration plots and model farms, or Farmer Field Schools (FFS) (for reviews of gender and agricultural extension frameworks, systems, policy, and programs, see Ragasa, Chap. 17, and Davis et al. 2007; World Bank/IFPRI 2010; World Bank et al. 2009, Module 7).

Table 7.3 summarizes the 17 studies that examine gender differences in access to human resources, following the format described for Table 7.1 on technological resources. By far the most research has been conducted on various forms of extension services. A comprehensive and extensive review of primary survey data in Ghana, Ethiopia, and India, completed by a "gender and governance" team of more than 16 researchers for the World Bank and IFPRI (2010), found large gender inequalities in access to extension services. Although the type of extension varies by county, mean differences are especially prominent in Ghana, where, on average, less than 2 % of female heads of households and female spouses in male-headed households have contact with extension agents, whereas nearly 12 % of men do. In Karnataka, India, 20 % of female household heads, but 27 % of male household heads, report extension service visits at home or on the farm in the past year. The authors not only included measures of access by sex but also analyzed measures of farmer satisfaction with services, gendered aspects of service provision, and institutional frameworks by country and validated by using qualitative research. Interestingly, in conducting multivariate analysis to explain contact with agents, gender variables become insignificant across countries; this is true in India and Ghana, due to inclusion of asset/wealth variables, and in Ethiopia, due to local fixed effects. This dynamic perhaps speaks to the tendency of female heads of household to be asset poor and/or to variation in the supply-side characteristics/policies of extension services, which may be more women friendly by region within Ethiopia.

It is also possible that results reflect the diminished power of the female headship variable to produce statistically significant results because of low percentages of women reporting contact with extension services. It is of note that in the World Bank/IFPRI findings (2010), women's access to livestock-related extension services are slightly better than for agricultural extension. For example, in Ghana, 0–24 % of female heads of household and 0–15 % of female spouses have access to livestock-related extension services compared with 5–34 % of male household heads who have such access.[13] Interestingly, evidence from Ghana, Ethiopia, and India indicates that the public sector provides the majority of extension services. The World Bank/IFPRI (2010) study found that NGOs, private-sector enterprises, and community-based organizations (CBOs) all play a relatively limited role in delivery of extension services.

Another recent comparative study by Davis and colleagues (2010) examined FFS in Kenya, Tanzania, and Uganda, using a longitudinal quasi-experimental impact evaluation design. Findings suggest that female community members in Kenya and Tanzania have equal access to services, while women in Uganda are less likely to participate in FFS. A promising finding of the Davis et al. (2010) study is that women who participate in FFS are more likely to adopt nearly all other major technologies, including improved seed varieties, soil fertility management, and pest control techniques. All other reviewed studies on extension services report mean values of access that are lower for women than men: 19 % versus 81 % in Malawi (Gilbert et al. 2002), 1.13 contacts versus 2.03 contacts in Uganda (Katungi et al. 2008), 7 % versus 13 % in Malawi (World Bank/Malawi 2007). The only study with

[13] The varieties in percentages refer to the differences in percentages between different zones surveyed.

Table 7.3 Gender differences in access to human resources: labor, extension services, and life cycle

(1)	(2)	(3)	(4)	(5)	(6)	(7)	(8)	(9)	(10)	(11)	(12)
			Use of access to input								
Authors (year)	Country (crop)	Sample size	Input type	Women	Men	Unit	Gender indicator	Outcome measure	Effect size	Comments	Published
Babatunde et al. (2008)	Nigeria	60 hhlds	Labor	2,077	3,060	Hours	Female heads	–	–	Using bivariate analysis, male-headed households have significantly more hours of labor inputs than female-headed households; however, there were no mean differences in farm output by gender	x
Davis et al. (2010)	Kenya	300 farmers	Farmer Field Schools (FFS, individual member-ship)	66.3	33.7	%	Female heads	Participation in FFS	–0.143 (nr)	Participation in FFS is equally available to female community members in Kenya and Tanzania. In Uganda, female-headed households are less likely to participate. Main reasons given for nonparticipation were lack of time, information, and distance. In addition, results suggest that FFS have a higher impact in terms of productivity, crop, and livestock income for female-headed than for male-headed households	
	Tanzania	284 farmers		31.3	68.7	%			0.25 (nr)		
	Uganda	267 farmers		50.2	49.8	%			–3.470*** (nr)		

de Brauw et al. (2008)	China	1,131 hhlds	Labor	1,081	942	Annual hours worked	Female household members	–	–	Using the last-round (2000) statistics, women are shown to work, on average, more hours than men; however, this ratio or their role in management is not changing significantly over time. In addition, there are no productivity differences found between female-headed or -managed farms and those run by males	x
Doss and Morris (2001)	Ghana (maize)	420 farmers	Extension services	50.43[a]	43.5	% with ≥ one contact	Female farmers	–	–	Authors use two-stage probit models to predict technology use, and use number of extension services as a control variable, which is a consistent positive predictor of use (note the different construction of extension as compared to mean statistics calculated here)	x
Enete et al. (2001)	Nigeria (cassava)	62 hhlds	Hired labor	76	57	%	Female owners	Cassava yields	–	Female-owned plots have significantly higher mean cassava yields; however, no multivariate analysis is presented to attribute to labor inputs	x

(continued)

Table 7.3 (continued)

(1)	(2)	(3)	(4)	(5)	(6)	(7)	(8)	(9)	(10)	(11)	(12)
			Use of access to input								
Authors (year)	Country (crop)	Sample size	Input type	Women	Men	Unit	Gender indicator	Outcome measure	Effect size	Comments	Published
Fletschner (2008)	Paraguay	210 hhlds[b]	Labor	–	–	–	Additional male adults	Technical efficiency[c]	0.068*** (nr)	Individual-level OLS (among spouses) finds households with more male labor exhibit higher technical efficiency, whereas additional female labor is not associated with increased technical efficiency	x
				–	–	–	Additional female adults		0.01 (nr)		
Gilbert et al. (2002)	Malawi (maize)	1,385 farms	Extension services	19	81	%	Female farmers	–	–	Mean values show that female farmers are a disproportionately low percentage of those contacted by extension agents to conduct intercropping trial	x
Horrell and Krishnan (2007)	Zimbabwe (primarily maize)	300 hhlds	Labor (economically active members of household)	3.97 (de jure); 3.12 (de facto)	4.14	People	Female heads	–	S** (nr)	Bivariate t-tests show significant differences between male- and de facto female-headed households only. These results are consistent with multivariate OLS results predicting log of household labor availability both for maize and all other crops	x

Katungi et al. (2008)	Uganda (banana)	352 hhlds	Extension services	4.11	7.78	Percent with contact	Male heads	Information exchange	0.079** (0.029)	Multinomial logit model suggests that female-headed households are disadvantaged in formal/informal information exchange, and extension services contribute to informal exchange, controlling for other characteristics, and this effect is larger for women than for men	x
Kinkingninhoun-Mêdagbé et al. (2008)	Benin (rice)	45 farmers	Labor	118	95	Man days/ha	Female farmers	Rice yield	0.062 (0.105)	Production function estimates indicate gender and labor inputs are insignificant in predicting yields after controlling for other inputs	x
Moore et al. (2001)	Senegal	694 husbands and wives	Knowledge nursery techniques	1.15	1.12	1–3 knowledge scale	Wives	–	–	Mean values show that knowledge surrounding natural resource management is influenced by access to extension services in different ways for husbands and wives. Women are most responsive to female extension services, and men have more access to informal networks for information sharing	x
			Knowledge composting	0.25	0.71						
			Knowledge forestry practices	2.28	2.87						

(continued)

Table 7.3 (continued)

(1)	(2)	(3)	(4)	(5)	(6)	(7)	(8)	(9)	(10)	(11)	(12)
			Use of access to input								
Authors (year)	Country (crop)	Sample size	Input type	Women	Men	Unit	Gender indicator	Outcome measure	Effect size	Comments	Published
Oladeebo and Fajuyigbe (2007)	Nigeria (rice)	100 farmers	Extension services	5.18	6.28	Number contacts	Female farmers	Stratified by gender: Productivity	0.176 (11.5)	Regression coefficients are reported from female farmer regressions. Productivity analysis finds that female farmers have higher technical efficiency than male farmers	x
			Family labor	6.25	9.01	Man days			0.006** (1.810)		
			Hired labor	21.33	13.30	Man days			0.0003* (3.384)		
Paolisso et al. (2002)	Nepal	264 hhlds	Time fruit and vegetables	20.93	43.35	Min/12-h day	Male farmers	Cultivation and care activities	Various	Evaluates effects of a vegetable and fruit cash crop program by gender and finds that households with a preschooler allocate more time to productive activities but decrease childcare, while households with more than one child do not face this trade-off	x
			Time cereal and livestock	228.13	244.2						
			Time under age 5 childcare	33.24	10.76						
Pender and Gebremedhin (2006)	Ethiopia	500 hhlds	Labor	–	–	Person days/ha	Female heads	Input use	−0.415*** (nr)	Study conducted to inform sustainable land management practices and uses OLS regression. Female-headed households made up 21.8 % of the sample, and average person-days per hectare was 86.4 in total	x

Rola et al. (2002)	Philippines	68 farmers	Farmer Field Schools (FFS)	57.35	42.65	%	Female farmers	FFS graduate	S** (nr)	Females are a higher proportion of FFS graduates using bivariate z-tests. Qualitative components of the study indicate that women are more likely to attend FFS because they have more free time and more patience to sit through classes, not necessarily because they make farming decisions	x
Thapa (2009)	Nepal	2,360 hhlds	Own male labor (> 16 years)	1,450	3,923	Hours	Female heads	Value of farm output	−0.018 (0.056)	Production function estimates give weak/ little evidence for gender differences after controlling for other inputs	
			Own female labor (>16 years)	6,858	5,105						
			Hired labor	227	226						
World Bank (2007)	Malawi	11,280 hhlds	Extension services	7	13	%	Female heads	–	–	Summary of key gender differences show that females are disadvantaged in access to extension, possibly due to smaller average farm size	

(continued)

Table 7.3 (continued)

(1)	(2)	(3)	(4)	(5)	(6)	(7)	(8)	(9)	(10)	(11)	(12)
			Use of access to input								
Authors (year)	Country (crop)	Sample size	Input type	Women	Men	Unit	Gender indicator	Outcome measure	Effect size	Comments	Published
World Bank and IFPRI (2010)	Ethiopia	1,753 hhlds	Extension services	20	27	%	Male farmers	Contact with agent in last year	0.158 (0.121)	Probit analysis suggests that female-headed households are not disadvantaged in access to services when controlling for other factors, specifically regional variation	x
	Ghana	861 hhlds	Extension services	0–2.1 (by zone)[d]	10.9–12.3 (by zone)	% with ≥ one contact	Male heads	Contact with agent	0.044 (0.030)	Probit analysis suggests that female households are not disadvantaged in access to services while controlling for other factors; however, coefficient on male head is larger than other control factors (none of which is significant with the exception of asset indexes in variations of main model)	

India	676 hhlds	Extension services	20	27	Percent with ≥ one contact	Female heads	Contact with agent in last year	1.099 (0.716)	Probit analysis suggests that female-headed households are not disadvantaged in access to services when controlling for other factors, specifically, assets

Source: Authors' compilation

Notes: Articles listed in alphabetical order of first author's last name; *Effect size* refers to gender indicator coefficients with standard errors in parentheses unless otherwise noted. *S* significant, *NS* not significant, *nr* not reported, *hhlds* households, *ha* hectares. * Significant at 10 %; ** significant at 5 %; *** significant at 1 %. Outcome measures of *adoption* refer to rate of adoption of corresponding input type; *published* indicates the study is published in a peer reviewed journal

[a]Mean differences are not presented in paper but were calculated by authors from disaggregated statistics

[b]Within each household, both husbands and wives were interviewed

[c]By definition, a household is considered technically efficient if no other household (or combination of households) produces more output with a similar level of inputs (Paris 1991)

[d]Percentages are by zone: male-headed households (11.7 in forest, 12.3 in transition, and 10.9 in savannah); female-headed households (0 in forest, 2.1 in transition, and 0 in savannah)

somewhat mixed results is from Senegal and looks at husband-wife pairs. It found that women's knowledge of various agricultural techniques is less than men's, with the exception of nursery techniques, in which they are approximately equal (Moore et al. 2001). In general, sample sizes in the extension literature are much larger (for example, 1,385 farms in the Gilbert et al. [2002] study, 11,280 in the World Bank study using the Malawian Integrated Household Survey) as compared to sample sizes in studies examining other inputs, perhaps reflecting use of household and other survey data not collected specifically for the evaluation of agricultural extension or other agricultural interventions.

One interesting, yet relatively unexplored, avenue of research is whether there are gender-based differences in the quality of information received by men and women. One factor that may influence quality and quantity of information is the gender of the extension agents or livestock officers. The World Bank/IFPRI (2010) study found extension agents and livestock officers in Ghana, Ethiopia, and India to be predominantly male; in Ghana, only 10 of 70 extension agents interviewed were female; in Ethiopia, agents were almost exclusively male; in Karnataka, India, none of the 41 agricultural extension workers was female, 1 of 41 junior engineers was female, and 4 of 40 veterinary assistants were female. Gender imbalances may cause problems in disseminating information. For example, in Ethiopia, researchers note that male extension agents are prevented from interacting with female farmers by strict cultural taboos. Another issue noted is that male extension officers may be more likely to subscribe to the common misconception that women are not farmers and overlook women in the household when delivering information. On the other hand, researchers in Senegal found that female extension agents can have a positive impact on dissemination of knowledge among both men and women (Moore et al. 2001). Another factor that may influence both quality and quantity of information available to women is access to information and communication technologies (ICTs—telecommunications, computer, and Internet use). While ICTs are increasingly becoming important tools in information dissemination, women often have limited access to ICTs. For example, a recent study found women in Africa, the Middle East, and South Asia to be, respectively, 23, 24, and 37 % less likely than their male counterparts to own a mobile phone, a key communication technology (GSMA Development Fund 2010).

The disparities in male and female access to extension services, noted throughout the literature, are particularly troubling, given that evidence from the World Bank/IFPRI (2010) study indicates that access to extension services is a key determinant of adoption of new information and use of new technologies and farming practices. For example, in the case of Ghana, multivariate analyses indicate that an extension agent visit was the only variable positively and significantly associated with adoption of new agricultural technology. Even if men and women are given equal access to extension services and information of equal quality, gender-based differences in use or adoption of new practices often persist because of lack of complementary knowledge or necessary inputs. A key example of this comes from the Doss and Morris (2001) study, which found gender-based differences in adoption of modern maize varieties and chemical fertilizer result from gender-based differences in access to necessary complementary inputs.

Comparatively fewer studies discuss gendered labor differentials. The most rigorous examples come from de Brauw et al. (2008), who compare large-scale panel surveys from China, and from a working paper that uses the Nepal Living Standards Survey of a population-level sample of 2,360 households (Thapa 2009). In the China study, de Brauw and colleagues specifically examined the "feminization" of Chinese agriculture by measuring labor allocation decisions within the household on labor use, welfare, and productivity over time, and found little evidence to support the hypothesis for the feminization of agriculture. Findings are robust to use of alternative survey data and construction of gender indicators at different units of analysis. Results from Nepal show that female heads of household report higher commitments of female labor (6,857 h) than male labor (1,450 h), whereas male household heads also report more female labor, although they claim a more equitable ratio (5,105 h of female labor to 3,922 h of male labor). Interestingly, female-headed households report slightly more hours of hired labor, although these values are relatively low and quite similar (227 h for female and 217 h for male heads of household). Although, in subsequent production function estimates, being in a female-headed household does not seem to matter for productivity, all labor indicators are highly significant (at the 1 % level) and contribute positively to the value of farm output, indicating that differences in productivity are explained by differences in access to inputs (including labor, land, and technology). Also in Nepal, Paolisso and colleagues (2002) evaluated the effect of the Vegetable and Fruit Cash Crop Program (VCP) in a sample of 264 households, stratifying results by gender of respondent. Findings indicate that men and women spend roughly the same average time in cereal and livestock production (228 and 244 min per 12-h day for men and women, respectively); however, women spend more time caring for children under 5, while men spend more time in fruit and vegetable production (women spend 33 min on childcare, whereas men spend 11 min; women spend 21 min and men 43 min on fruit and vegetable production). Interestingly, Paolisso et al. (2002) find differential program impacts both by gender and by family type. The VCP had a greater impact on shifting men's time use to vegetable and fruit production; however, men, and especially women in households with one preschooler, reduced the time they spent caring for the child (this result was not found for households with more than one preschooler). The trade-offs between productive and reproductive labor is clearly important in determining household labor allocation. For example, Quisumbing and Yohannes (2004) found nearly 27 % of women cite childcare as a reason for not applying to public works (typically food-for-work) programs in rural Ethiopia in contrast to approximately 3 % of men.

Additional studies generally show that females have larger labor constraints. In regression analysis, Pender and Gebremedhin (2006) found that female heads of household are significantly associated with lower labor participation, as measured by person days per acre, using a 500-household sample from Ethiopia. Again, the Horrell and Krishnan (2007) study included the number of working-age adults in the household as an indicator of labor availability and, by using bivariate methods, found that differences exist between male household heads and de facto female household heads—male-headed households are larger, on average, by one person (4.14 versus 3.12 people). Fletschner's (2008) study of 210 households in Paraguay found that households with more male labor exhibit higher technical efficiency, whereas additional female labor has no impact on technical efficiency.[14] The two remaining studies on labor inputs in Nigeria (Enete et al. 2001) and Benin (Kinkingninhoun-Mêdagbé et al. 2008) both report higher labor inputs for female-owned plots and female farmers than men. However, because studies are limited in their sample sizes and crop diversity, results should be regarded with caution (62 cassava-farming households in Nigeria and 45 rice-farming households in Benin).

In summary, we reviewed 18 studies of human resource input use, access, and adoption that fit our criteria. These include 15 measures of extension services and other educational services and 16 measures of labor inputs. Fifteen of 18 studies were published in peer-reviewed journals. Overall, where descriptive statistics for inputs were provided (for 28 input indicators), 15 (53 %) found men have higher mean access and 13 (46 %) found that women have higher mean access to the given resource. Where further bivariate or multivariate analysis was conducted (for 17 input indicators), nine (53 %) found that gender indicators are not significant with respect to outcome measures when other factors are controlled for, while five (29 %) found that differences persist and men have higher outcome measures, with three (18 %) finding that women have higher outcome measures. Assets and geographical variations seem to be key factors in accounting for differences across studies where gender differences were found previously. In comparison with labor inputs, analysis of extension services is especially well developed and increasingly has considered alternative gender dimensions, including gender of extension agents, quality of information, and time constraints of participation in trainings.

7.3 Discussion and Policy Implications

What value does this review add to the overall knowledge of gendered access to nonland farm inputs? We focus the discussion on three key aspects of the review and finish with a summary, suggestions for future research, and policy implications of our findings. First, we offer some conjectures and speculations as to why we find (and do not find) differences in women's access between and across studies. Second, we try to note some general regional similarities and differences across research on gender and nonland inputs throughout Asia, Latin/South America, Sub-Saharan Africa, and the Middle East. Third, we discuss briefly some issues and promising work in mechanization in which we find few studies fitting our inclusion criteria. Fourth, we discuss the new challenges and opportunities in high value, organic, and fair-trade agriculture for female farmers and how this may have repercussions for and interact with women's access to inputs in the developing world.

[14] By definition, a household is considered technically efficient if no other household (or combination of households) produces more output with a similar level of inputs (Paris 1991).

It is hard to generalize why gender differences are or are not found across inputs, study designs, and regions. However, a common theme throughout the literature reviewed is that crop choices and division of labor differ by gender within disparate regional and cultural contexts. For example, throughout Sub-Saharan Africa, lucrative cash crops are often perceived to be "male crops," and crops for home consumption are perceived to be "female crops" (Kasante et al. 2001; World Bank/Malawi 2007). Related to this issue, Doss (1999) notes that there may be differences in choices of inputs by gender, based on whether the crop is produced for home or for the market. For example, yield may be the most important consideration in market-targeted crops, while other factors, such as taste, storability, and ease of processing (such as drying, fermenting, pounding), may be important in determining crops for home consumption. However, Doss's (2002) study of nationally representative household survey data from Ghana found few crops can be defined as men's crops, and none is obviously a women's crop. Therefore, this and other evidence suggests that, in some settings, boundaries between male and female crops may be less rigid than they initially appear (Quisumbing et al. 2001).

Concerning division of labor, within Sub-Saharan Africa, males are often responsible for the physically intensive task of clearing the land, and women are responsible for weeding and postharvest processing (Guyer 1991; Kasante et al. 2001). In Asian systems, men typically provide the labor in land preparation, and women provide labor in planting, cultivation, and crop care such as weeding (Quisumbing and McClafferty 2006). In future research it is worth further exploring the impact of technology adoption on the traditional gendered division of labor. For example, Fisher et al. (2000) find that the adoption of the stabling technique in rural Senegal makes milk more profitable by improving production; as a result, the marketing of milk shifts from the female to the male domain. In reality, studies that examine one input in isolation capture only a partial picture of realities in which synergies exist between farm inputs and relative outputs. Therefore, it would be expected that as inequalities in access to technology and services are reduced, the potential for increased productivity and output will increase across sectors.

On a methodological note, throughout the reviewed studies, authors make use of (mainly) two very different units of analysis when assessing inequalities in use, adoption rates, or outputs. For example, in examining fertilizer and seed varieties within the technological section, Enete et al. (2001), Freeman and Omiti (2003), the World Bank/IFPRI (2010) study on Ethiopia, and Gilbert et al. (2002), studied the sex of the individual farmer/plot owner; whereas Bourdillon et al. (2002), Jagger and Pender (2006), Tiruneh et al. (2001), the World Bank/IFPRI (2010) studies on Ghana and India, and Thapa (2009), examined the sex of the household head. Only Chirwa (2005) and Doss and Morris (2001) examined both. In their sensitivity analysis, Doss and Morris (2001) point out that using the sex of the farmer allows for examination of female farmers in both male- and female-headed households. This is significant because, as Bourdillon and colleagues (2002) point out, even in female-headed households of rural Zimbabwe, men (such as adult sons) are expected to make agricultural decisions. Because sex of household head is not always a perfect indicator of female access or decisionmaking, there is a need for more studies that conduct sensitivity analysis between measures of female management and female headship. Horrell and Krishnan (2007) make a further distinction between female de jure and female de facto households and find differences persist mainly among de facto households. However, it should be noted that, because the full sample size is 300 households, this stratification results in small sample sizes, especially among the de facto female-headed households (N = 17). The heterogeneity of women or men within these categorizations is important, as they may differ significantly with respect to background characteristics, as shown by different technology adoption rates when interactions between headship and literacy are included to predict adoption rates (World Bank/IFPRI 2010). Uttaro (2002) makes another pertinent differentiation among men, married women, and single women when looking at differential access to inorganic fertilizer in Malawi, finding that married women access inorganic fertilizer at a higher rate than single household heads. In short, the specific gender indicator used seems to matter, and further research is needed to conduct these types of sensitivity analyses (Deere et al. 2010).

The overwhelming bulk of evidence we reviewed is from studies in Sub-Saharan Africa (more than 75 %, depending on inclusion of cross-country studies). In the Latin and South American, Eastern European, and Middle Eastern regions, we found few qualifying studies. This may be a reflection of regional or cultural differences in households and farming practices that, in turn, influence research questions and methods. For example, outside Sub-Saharan Africa, where there are clearly demarcated men's and women's plots, it is harder to measure differences in men's and women's nonland inputs, perhaps with the exception of labor inputs. In addition, this may be driven by regional differences in research funding streams, policy interest, and donor programmatic focus. While there has been a larger body of research with a regional focus on Asia in the past few decades, these studies typically use a different kind of gender disaggregation. For example, in general, labor and other inputs are disaggregated by sex (male-hired labor, female-hired labor, male family labor, female family labor), but outputs are not. This is likely the result of the joint nature of Asian family farming and the relatively low incidence of female headship. Ultimately, the percentage of female-headed households in most studies has been so small that it does not necessarily warrant separate estimation by sex of household heads. Some recent exceptions to this trend are the study by de Brauw et al. (2008), which found little support for the hypothesis of Chinese feminization of agriculture, and the studies by Thapa (2009) and Paolisso and colleagues (2002) in Nepal.

The regional disparities in evidence may also be a function of the percentage of women engaged in agriculture in the Sub-Saharan region; however, we should not assume that this is a driving force. For example, according to International Labor Organization (2009) estimates, agriculture accounted for 65.1 % of the sectoral share of employment for women in Sub-Saharan Africa in 2007; however, this percentage is identical to that of South Asia, followed by Southeast Asia and the Pacific (43.9 %), East Asia (41.2 %), North Africa (38.9 %), and the Middle East (32.0 %). In fact, in comparison with men, women in the Middle East have the higher regional proportion of agricultural workers (agriculture accounts for only 13 % of the sectoral share of employment for men). *Women in Agriculture in the*

Middle East reviews published and unpublished work and compared the state of women working in agriculture in Palestine, Israel, Egypt, and Jordan, including the gender effects of the resettlement process (Motzafi-Haller 2005). The compilation of research emphasizes the importance of discriminating and oppressive political factors, especially in the context of civil conflict, that determine women's ability to obtain and successfully use agricultural inputs. Given the importance of context and cultural influences on the underlying ability of women to secure and use inputs, there is a great need for regionally diversified microempirical work on women and agriculture.

We found few studies that focus on or include mechanization, tools, and other farming equipment disaggregated by gender.[15] This may be, in part, because modern farming equipment, such as tractors and tillers, are not commonly available to either gender or used in rural agricultural work, especially in Sub-Saharan Africa. Several studies from the late 1980s and early 1990s point to gender differences in tool ownership and access. In a Gambian irrigated rice scheme, less than 1 % of women owned a weeder, seeder, or multipurpose cultivation implement, while 12 % of men owned a weeder, 27 % of men owned a seeder, and 18 % of men owned a multipurpose cultivation implement (von Braun et al. 1989). Further, only men (8 %) owned any type of plow. In a household survey, the value of farm tools and equipment owned by Kenyan women across three districts was 18 % of the value of the same implements owned by male farmers (Saito et al. 1994). In a more recent study of productivity differences by gender in a rice irrigation scheme in central Benin, researchers did not explicitly control for access to tools; however, Kinkingninhoun-Mêdagbé and colleagues (2008) note that equipment such as motor-cultivators used for plowing and transport is managed by groups. Since women's groups were not provided with operators, they could not start plowing until the drivers for men's groups completed work on the men's fields, thus delaying women's plowing and subsequent planting (Kinkingninhoun-Mêdagbé et al. 2008). In addition, in a review of gender and agriculture inputs and productivity, Quisumbing (1994) concluded that farmers who use tools and other equipment may be more likely to adopt other technologies, which speaks directly to the interactive and synergetic aspects of agricultural inputs.

Looking forward, several key issues are ripe for research, program implementation, and policy. First, we reiterate the need to collect and analyze sex-disaggregated data in agricultural research. If possible, data disaggregation at the plot level is preferred to disaggregation at the household or farm level, which may obscure intrahousehold dynamics. We also recommend the collection of several indicators of gender to provide more robust results (for example, female heads of household, female-owned plots, female-owned assets, female-managed plots, and so on). While the attention to sex-specific data is improving, some recent publications still do not disaggregate, analyze, or even control for gender indicators in their analyses. Providing descriptive statistics or controlling for gender often involves fairly simple calculations and

[15] However, there is more research on mechanization and technology applied to postharvest labor. See, for example, Mulokozi et al. (2000), Paris et al. (2001), and Singh et al. (1999).

has the potential to build a more robust body of work identifying gender differences in access to agricultural inputs. Second, while a fair amount of attention has been paid to differential access to inputs in some areas (for example, seed varieties, inorganic fertilizer, fallow techniques, extension services), comparatively little evidence exists about several other inputs (such as mechanization). Third, there is a lack of evidence of gender differences in input use from Middle Eastern, Latin/South American, and Eastern European regions, perhaps because of underlying assumptions regarding farm and family organization, such as the assumption that all farm output is pooled. Even in Asia, where there is a wealth of sex-disaggregated data on labor inputs, there is relatively little evidence from outputs on male- and female-managed plots because of the assumption that farming is conducted jointly and output is shared. But even in Asia, there may be homestead plots or livestock that are women's exclusive responsibility.

As the success and sustainability of many interventions reflect, gaining access to productive resources is not just a legal, political, or economic issue; it is a matter of changing gender relations, views, and social institutions in many settings. Having adequate information to inform policy decisions across a variety of settings is crucial. In fact, without attention to the larger scope of gender relations, interventions to provide equal access to inputs and resources have in certain cases led to increased conflict (see, for example, Lastarria-Cornhiel 1997; Tripp 2004; Whitehead and Tsikata 2003). It is our hope that attention to gender in agricultural research, program implementation, and policy will gain increased attention and become standard practice in the coming decade.

Acknowledgments The authors are grateful to Jere Behrman, Kristin Davis, an anonymous reviewer, and participants of *The State of Food and Agriculture* contributors' workshop for helpful comments on an earlier draft and to all authors who provided drafts of forthcoming work. The results presented do not reflect the institutional views of the FAO or IFPRI; all errors are our own.

References

Adesina AA, Mbila D, Nkamleu GB, Endamana D (2000) Econometric analysis of the determinants of adoption of alley farming by farmers in the forest zone of southwest Cameroon. Agric Ecosyst Environ 80(3):255–265

Ashby J (1985) Women and agricultural technology in Latin America and the Caribbean. In: Ashby J, Gomez S (eds) Women, agriculture, and rural development in Latin America. International Fertilizer Development Center/Centro Internacional de Agricultura Tropical, Cali, pp 1–38

Atreya K (2007) Pesticide use, knowledge, and practices: a gender differences in Nepal. Environ Res 104(2):305–311

Babatunde RO, Omotesho OA, Olorunsanya EO, Owotoki GM (2008) Gender differences in resources allocation among rural households in Nigeria: implications for food security and living standard. Eur J Soc Sci 5(4):160–172

Boserup E (1970) Woman's role in economic development. St. Martin's Press, New York

Bourdillon M, Hebinck P, Hoddinott J, Kinsey B, Marondo J, Mudege N, Owens T (2002) Assessing the impact of HYV maize in resettlement areas of Zimbabwe. Summary report. International Food Policy Research Institute, Washington, DC

Chirwa EW (2005) Adoption of fertilizer and hybrid seeds by smallholder maize farmers in southern Malawi. Dev S Afr 22(1):1–12

Davis K, Ekboir J, Mekasha W, Ochieng CMO, Spielman DJ, Zerfu E (2007) Strengthening agricultural education and training in Sub-Saharan Africa from an innovation systems perspective: case studies of Ethiopia and Mozambique. IFPRI discussion paper 00736. International Food Policy Research Institute, Washington, DC

Davis K, Nkonya E, Kato E, Ayalew Mekonnen D, Odendo M, Miiro R, Nkuba J (2010) Impact of farmer field schools on agricultural productivity and poverty in East Africa. IFPRI discussion paper 00992. International Food Policy Research Institute, Washington, DC

de Brauw A, Li Q, Liu C, Rozelle S, Zhang L (2008) Feminization of agriculture in China? Myths surrounding women's participation in farming. China Q 194(June):327–348

Deere CD, León M (1987) Introduction. In: Deere CD, León M (eds) Rural women and state policy: feminist perspectives on Latin American agricultural development. Westview, Boulder, pp 1–20

Deere CD, Alvarado GE, Twyman J (2010) Poverty, headship, and gender inequality in asset ownership in Latin America. Working paper 296. Center for Gender in Global Context, Michigan State University, East Lansing

Dey J (1985) Women in African rice farming systems. In: International Rice Research Institute (ed) Women in rice farming: proceedings of a conference on women in rice farming systems. Gower Publishers, Brookfield, pp 419–444

Di Gregorio M, Hagedorn K, Kirk M, Korf B, McCarthy N, Meinzen-Dick R, Swallow B (2008) Property rights, collective action, and poverty: the role of institutions for poverty reduction. CAPRi working paper 81. International Food Policy Research Institute, Washington, DC

Doss CR (1999) Twenty-five years of research on women farmers in Africa: lessons and implications for agricultural research institutions; with an annotated bibliography. Economics program paper 99-02. Centro Internacional de Mejoramiento de Maiz y Trigo, Mexico City

Doss C (2002) Men's crops? Women's crops? The gender patterns of cropping in Ghana. World Dev 30(11):1987–2000

Doss CR, Morris ML (2001) How does gender affect the adoption of agricultural innovations? The case of improved maize technology in Ghana. Agric Econ 25(1):27–39

Enete AA, Nweke FI, Achike AI, Tollens E (2001) Differentiated gender ownership of cassava fields and implications for root yield variations in smallholder agriculture of South Nigeria. Tropicultura 19(3):105–109

Fisher MG, Warner RL, Masters WA (2000) Gender and agricultural change: crop-livestock integration in Senegal. Soc Nat Resour 13(3):203–222

Fletschner D (2008) Women's access to credit: does it matter for household efficiency? Am J Agric Econ 90(3):669–683

Freeman AH, Omiti JM (2003) Fertilizer use in semi-arid areas of Kenya: analysis of smallholder farmers' adoption behavior under liberalized markets. Nutr Cycl Agroecosyst 66(1):23–31

Gilbert RA, Sakala WD, Benson TD (2002) Gender analysis of a nationwide cropping system trial survey in Malawi. Afr Stud Q 6(1 & 2):223–243. http://web.africa.ufl.edu/asq/v6/v6i1a9.htm. Accessed 7 Sept 2009

Goldman A, Heldenbrand, K (2001) Gender and soil fertility management in Mbale District, Southeastern Uganda. Afr Stud Q 6(1 & 2):45–76. http://www.africa.ufl.edu/asq/v6/v6i1a3.htm. Accessed 21 Apr 2010

GSMA Development Fund (2010) Women and mobile: a global opportunity: A study on the mobile phone gender gap in low and middle income countries. Global System Mobile Association. http://www.gsma.com/mobilefordevelopment/wp-content/uploads/2013/01/GSMA_Women_and_Mobile-A_Global_Opportunity.pdf. Accessed 18 Mar 2010

Guyer JI (1991) Female farming in anthropology and African history. In: di Leonardo M (ed) Gender at the crossroads of knowledge: feminist anthropology in the postmodern era. University of California Press, Berkeley, pp 257–277

Hope RA, Dixon P-J, von Maltitz G (2003) The role of improved domestic water supply in livelihoods and poverty reduction in Limpopo Province, South Africa. In: International symposium on water, poverty and productive uses of water at the household level, Muldersdrift, South Africa, 21–23 January. www.irc.nl/page/8060. Accessed 1 Aug 2009

Horrell S, Krishnan P (2007) Poverty and productivity in female-headed households in Zimbabwe. J Dev Stud 43(8):1351–1380

IFAD (International Fund for Agricultural Development) (2003) Mainstreaming a gender perspective in IFAD's operations: plan of action 2003–2006. IFAD (International Fund for Agricultural Development), Rome

IFPRI (International Food Policy Research Institute) (2007) Proceedings of the consultation on strengthening women's control of assets for better development outcomes. IFPRI (International Food Policy Research Institute), Washington, DC

ILO (International Labor Organization) (2009) Global employment trends for women. Geneva. http://www.ilo.org/wcmsp5/groups/public/@dgreports/@dcomm/documents/publication/wcms_103456.pdf. Accessed 5 Aug 2009

Jagger P, Pender J (2006) Influences of programs and organizations on the adoption of sustainable land management technologies in Uganda. In: Pender J, Place F, Ehui S (eds) Strategies for sustainable land management in the East African Highlands. International Food Policy Research Institute, Washington, DC, pp 277–308

Kasante D, Lockwood M, Vivian J, Whitehead A (2001) Gender and the expansion of nontraditional agricultural exports in Uganda. In: Razavi S (ed) Shifting burdens: gender and agrarian change under neo-liberalism. Kumarian, Bloomfield, pp 35–66

Katungi E, Edmeades S, Smale M (2008) Gender, social capital, and information exchange in rural Uganda. J Int Dev 20(1):35–52

Kazianga H, Masters WA (2002) Investing in soils: field bunds and microcatchments in Burkina Faso. Environ Dev Econ 7(3):571–591

Kevane M (2004) Women and development in Africa: how gender works. Lynne Rienner, Boulder/London

Kinkingninhoun-Mêdagbé FM, Diagne A, Simtowe F, Agboh-Noameshie AR, Adegbola PY (2008) Gender discrimination and its impact on income, productivity, and technical efficiency: evidence from Benin. Agric Hum Values 27(1):57–69

Lastarria-Cornhiel S (1997) Impact of privatization on gender and property rights in Africa. World Dev 25(8):1317–1333

Mehra R, Hill Rojas M (2009) Food security and agriculture in a global marketplace: a significant shift. International Center for Research on Women, Washington, DC. www.icrw.org/docs/2008/a-significant-shift-women-food%20security-and-agriculture%20FINAL.pdf. Accessed 2 Aug 2009

Moore KM, Hamilton S, Sarr P, Thiongane S (2001) Access to technical information and gendered NRM practices: men and women in rural Senegal. Agric Hum Values 18(1):95–105

Motzafi-Haller P (ed) (2005) Women in agriculture in the Middle East. Ashgate Publishing, Burlington

Mulokozi G, Mselle L, Mgoba C, Mugyabuso JKL, Ndossi GD (2000) Improved solar drying of vitamin A-rich foods by women's groups in the Singida District of Tanzania. Research report series 5. International Center for Research on Women, Washington, DC. http://www.icrw.org/files/publications/Improved-Solar-Drying-of-Vitamin-A-rich-Foods-by-Womens-Groups-in-the-Singida-District-of-Tanzania.pdf. Accessed 6 Aug 2009

Mwangi E, Markelova H (2008) Collective action and property rights for poverty reduction: a review of methods and approaches. CAPRi working paper 82. International Food Policy Research Institute, Washington, DC

Nkedi-Kizza P, Aniku J, Awuma K, Gladwin C (2002) Gender and soil fertility in Uganda: a comparison of soil fertility indicators for women and men's agricultural plots. Afr Stud Q 6(1):27–43. http://web.africa.ufl.edu/asq/v6/v6i1a2.htm. Accessed 6 Aug 2009

Oladeebo JO, Fajuyigbe AA (2007) Technical efficiency of men and women upland rice farmers in Osun State, Nigeria. J Hum Ecol 22(2):93–100

Oladele O, Monkhei M (2008) Gender ownership patterns of livestock in Botswana. Livest Res Rural Dev 20(10). www.lrrd.org/lrrd20/10/olad20156.htm. Accessed 11 Aug 2009

Ouma JO, Murithi FM, Mwangi W, Verkuijl H, Gethi M, de Groote H (2002) Adoption of maize seed and fertilizer technologies in Embu District, Kenya. Centro Internacional de Mejoramiento de Maiz y Trigo, Mexico. http://webcache.googleusercontent.com/search?q=cache:7dQsu-

A6z2AJ:www.researchgate.net/publication/237629169_Adoption_of_Maize_Seed_and_Fertilizer_Technologies_in/file/60b7d524bc93f2b408.pdf+&cd=2&hl=en&ct=clnk&gl=us. Accessed 27 Sept 2009

Paolisso M, Hallman K, Haddad L, Regmi S (2002) Does cash crop adoption detract from childcare provision? Evidence from rural Nepal. Econ Dev Cult Change 50(2):313–337

Paris Q (1991) An economic interpretation of linear programming. Iowa State University Press, Ames

Paris T, Feldstein HS, Duron G (2001) Technology. In: Quisumbing AR, Meinzen-Dick RS (eds) Empowering women to achieve food security. 2020 Focus 6. Policy brief No. 5. International Food Policy Research Institute, Washington, DC

Pender J, Gebremedhin B (2006) Land management, crop production and household income in the Highlands of Tigray, Northern Ethiopia: an econometric analysis. In: Pender J, Place F, Ehui S (eds) Strategies for sustainable land management in the East African Highlands. International Food Policy Research Institute, Washington, DC, pp 107–140

Phiri D, Franzel S, Mafongoya P, Jere I, Katanga R, Phiri S (2004) Who is using the new technology? the association of wealth status and gender with the planting of improved tree fallows in Eastern Province, Zambia. Agr Syst 79(2):131–144

Quisumbing AR (1994) Improving women's agricultural productivity as farmers and workers. Education and Social Policy Department discussion paper series 37. World Bank, Washington, DC. www-wds.worldbank.org/external/default/WDSContentServer/WDSP/IB/2001/03/20/000094946_0103030540312/Rendered/PDF/multi_page.pdf. Accessed 28 Aug 2009

Quisumbing A (1996) Male-female differences in agricultural productivity: methodological issues and empirical evidence. World Dev 24(10):1579–1595

Quisumbing AR, McClafferty B (2006) Using gender research in development. Food security in practice technical guide series 2. International Food Policy Research Institute, Washington, DC

Quisumbing AR, Pandolfelli L (2010) Promising approaches to address the needs of poor female farmers: resources, constraints, and interventions. World Dev 38(4):581–592

Quisumbing AR, Yohannes Y (2004) How fair is workfare? Gender, public works, and employment in rural Ethiopia. Policy research working paper 0-3039. World Bank, Washington, DC

Quisumbing AR, Payongayong E, Aidoo JB, Otsuka K (2001) Women's land rights in the transition to individualized ownership: implications for the management of tree resources in Western Ghana. Econ Dev Cult Change 50(1):157–182

Rola AC, Jamias SB, Quizon JB (2002) Do farmer field school graduates retain and share what they learn? An investigation in Iloilo, Philippines. J Int Agric Ext Educ 9(1):65–76

Saito K, Mekonnen H, Spurling D (1994) Raising the productivity of women farmers in Sub-Saharan Africa. Africa technical department series No. 230. World Bank, Washington, DC

Sanginga PC, Adesina AA, Manyong VM, Otite O, Dashiell KE (2007) Social impact of soybean in Nigeria's southern Guinea savanna. International Institute of Tropical Agriculture, Ibadan, Nigeria. http://www.hubrural.org/IMG/pdf/iita_soybean.pdf. Accessed 27 Sept 2009

Schultz TP (2001) Women's role in the agricultural household: bargaining and human capital investments. In: Gardner B, Rausser G (eds) Agricultural and resource economics handbook, vol 1. Elsevier, Amsterdam, pp 383–456

Shankar B, Thirtle C (2005) Pesticide productivity and transgenic cotton technology: the South African smallholder case. J Agric Econ 56(1):97–116

Singh G, Singh G, Kotwaliwale N (1999) A report on agricultural production and processing technologies for women in India. Gender Technol Dev 3(2):259–278

Singh N, Gunnar J, Bhattacharya P, Gustafsson J-E (2006) Gender and water management: some policy reflections. Water Policy 8(2):183–200

Smith R (2008) Cattle's effect on land and labour productivity: evidence from Zambia. J Int Dev 20(7):905–919

SOAS (School of Oriental and African Studies), Wadonda Consult, Michigan State University, and Overseas Development Institute (2008) Evaluation of the 2006/7 Agricultural Input Subsidy Programme, Malawi. Final report. London

Somda J, Nianogo AJ, Nassa S, Sanou S (2002) Soil fertility management and socioeconomic factors in crop-livestock systems in Burkina Faso: a case study of composting technology. Ecol Econ 43(2):175–183

Swinkels RA, Shepherd KD, Franzel S, Ndufa JK, Ohlsson E, Sjogren H (2002) Assessing the adoption potential of hedgerow intercropping for improving soil fertility, Western Kenya. In: Franzel S, Scherr SJ (eds) Trees on the farm: assessing the adoption potential of agroforestry practices in Africa. CABI Publishing/International Centre for Research in Agroforestry, New York, pp 89–110

Thapa S (2009) Gender differentials in agricultural productivity: evidence from Nepalese household data. University of Trento, Department of Economics, Trento, Italy. Electronic paper

Tiruneh A, Tesfaye T, Mwangi W, Verkuijl H (2001) Gender differentials in agricultural production and decisionmaking among smallholders in Ada, Lume, and Gimbichu Woredas of the Central Highlands of Ethiopia. Centro Internacional de Mejoramiento de Maiz y Trigo, Ethiopian Agricultural Research Organization, and the European Union, Mexico City

Tripp AM (2004) Women's movements, customary law and land rights in Africa: the case of Uganda. Afr Stud Q 7(4):1–19

UNDP (United Nations Development Programme) (2006) Mainstreaming gender in water management: resource guide, version 2.1. Gender and Water Alliance (GWA). New York

UN-HABITAT (United Nations Human Settlements Programme) (2006) Mechanism for gendering land tools: a framework for delivery of women's security of tenure. Strategies and outline adopted at the High Status Round Table on Gendering Land Tools, 21 June, Nairobi, Kenya. www.unhabitat.org/downloads/docs/4223_75221_gltn.doc. Accessed 3 Aug 2009

UN News Centre (2010) Rural women key to economic growth, says UN official. http://www.un.org/apps/news/story.asp?NewsID=34165&Cr=agriculture&Cr1=. Accessed 21 Apr 2010

Uttaro P (2002) Diminishing choices: gender, small bags of fertilizer, and household food security decisions in Malawi. Afr Stud Q 6(1):77–110. www.africa.ufl.edu/asq/v6/v6i1a4.htm. Accessed 3 Aug 2009

Van de Fliert E, Johnson N, Asmunati R, Wiyanto X (2001) Beyond higher yields: the impact of sweetpotato integrated crop management and farmer field schools in Indonesia. In: Scientist and farmer: partners in research for the 21st Century. Program report 1999–2000. International Potato Center, Lima, pp 331–342

von Braun J, Hotchkiss D, Immink M (1989) Nontraditional export crops in Guatemala: effects on production, income and nutrition. Research report 73. International Food Policy Research Institute, Washington, DC. www.ifpri.org/pubs/abstract/73/rr73.pdf. Accessed 3 Aug 2009

von Koppen B (2002) A gender performance indicator for irrigation: concepts, tools, and applications. Research report 59. International Water Management Institute, Colombo

Whitehead A, Tsikata D (2003) Policy discourses on women's land rights in Sub-Saharan Africa: the implications of the return to the customary. J Agrar Change 3(1&2):67–112

World Bank (2007) Global monitoring report 2007: millennium development goals—confronting the challenges of gender equity and fragile states. World Bank, Washington, DC

World Bank/FAO/IFAD (Food and Agriculture Organization of the United Nations/International Fund for Agricultural Development) (2009) Gender in agriculture sourcebook. World Bank, Washington, DC

World Bank/IFPRI (International Food Policy Research Institute) (2010) Gender and governance in rural services: insights from India, Ghana, and Ethiopia. Gender and Governance Author Team. International Bank for Reconstruction and Development/World Bank, Washington, DC

World Bank/Malawi (2007) Malawi poverty and vulnerability assessment (PVA): investing in our future. Synthesis report: main findings and recommendations. Poverty reduction and economic management 1, report 36546-MW. World Bank, Washington, DC

Chapter 8
Rural Women's Access to Financial Services: Credit, Savings, and Insurance

Diana Fletschner and Lisa Kenney

Abstract This chapter reviews rural women's access to financial services, a key factor underlying many successful rural development strategies. Designing appropriate financial products for women to be able to save, borrow, and insure is essential to strengthen women's role as producers and widen the economic opportunities available to them. Context-specific legal rights, social norms, family responsibilities, and women's access to and control over other resources play an important role in shaping rural women's needs for capital and their ability to obtain it. The chapter argues that interventions that improve rural women's *direct* access to financial services—not mediated through their husbands—can be beneficial on two fronts. First, by addressing the constraints women face, these interventions enhance women's productive capacity. Second, by improving women's relative power in their households, these interventions can lead to both a more efficient allocation of resources and to improved health, nutrition, and education in their families, all of which are expected to improve long-term production capabilities. The products and service delivery models introduced to address some of the constraints faced by women include technical innovations that improve access to existing financial services, changes in product design to better tailor products to women's preferences and constraints, and the development of new products such as micro-insurance.

Keywords Women • Savings • Credit • Insurance • Financial capital

---

D. Fletschner (✉)
Director of Research, Monitoring and Evaluation, Landesa,
1424 Fourth Ave., Suite 300, Seattle, WA 98101, USA
e-mail: DianaF@landesa.org

L. Kenney
PhD Student, Department of Geography, University of Calgary,
2500 University Dr. NW, Calgary, Alberta T2N 1N4, Canada
e-mail: lmkenney@ucalgary.ca

A.R. Quisumbing et al. (eds.), *Gender in Agriculture: Closing the Knowledge Gap*, DOI 10.1007/978-94-017-8616-4_8,

8.1 Introduction

Ensuring that farmers have adequate access to financial resources is a key tenet of successful rural development strategies. Policymakers have long understood that rural producers who cannot meet their needs for capital must settle for suboptimal production strategies. When producers are unable to make the necessary upfront investments or cannot bear additional risk, they have to forgo opportunities to boost their productivity, enhance their income, and improve their well-being (Besley 1995; Boucher et al. 2008; World Bank 2008). Furthermore, without adequate access to loans or insurance, producers who face negative shocks, such as droughts, illness, or a significant drop in the prices they receive, can lose some of the few assets they do have (Diagne and Zeller 2001). Conversely, producers who have access to well-designed credit, savings, and insurance services can avail themselves of capital to finance the inputs, labor, and equipment they need to generate income; can afford to invest in riskier but more profitable enterprises and asset portfolios; can reach markets more effectively; and can adopt more efficient strategies to stabilize their food consumption (Zeller et al. 1997). In the aggregate, broader access to financial services provides opportunities for improving the agricultural output, food security, and economic vitality of entire communities and nations.

Despite this widely accepted notion, rural financial programs have been largely designed, crafted, and implemented with the male head of household as the intended client and fail to recognize that women are active, productive, and engaged economic agents with their own financial needs and constraints. Women constitute approximately half of the rural labor force and, while not always counted, they are economically active in each subsector of the rural economy. Even though millions of women throughout the world contribute to national agricultural output and family food security, detailed studies from Latin America, South Asia, and Sub-Saharan Africa consistently indicate that rural women are more likely to be credit constrained than men of equivalent socioeconomic conditions (Fletschner 2009; Diagne et al. 2000).

Well-designed products that enable women to adequately save, borrow, and insure against unexpected shocks are therefore essential in any efforts to strengthen women's role as producers and expand the set of economic activities they can undertake, the scale at which they can operate and their ability to benefit from economic opportunities. Yet, with the notable exception of a number of prominent microfinance programs, the vast majority of rural credit, savings, and insurance programs do not take into account that women's legal, social, and economic position in their communities differ from men's.

8.2 Rural Financial Markets Are Not Gender Neutral

To understand how commercial and state-owned development banks, cooperatives, traders, and processors can improve their outreach to women, it is fundamental to identify how context-specific legal rights, social norms, family responsibilities, and women's access to and control over other resources shape their need for capital and their ability to obtain it.

8.2.1 Property Rights and Control over Assets

Legal regulations and customary rules often restrict women's access to and control over assets that can be accepted as collateral such as land or livestock. Women are much less likely to have land titled under their name, even when their families own land, and are less likely than men to have control over land, even when they do formally own it (also see Lastarria-Cornhiel et al., Chap. 6). Biased inheritance rights often bestow land to male relatives, leaving both widows and daughters at a disadvantage (Agarwal 2003). Neither the state mandated agrarian reforms of past decades that granted much of the land to "household heads," who were typically men, nor the more recent market-assisted land reforms have led to significant improvements in women's access to and control over land (Deere and León 1997; Bezner Kerr 2008). Even in countries where laws do protect women's land rights, these laws tend to be loosely regulated and implemented (Parada 2008; Morrow Richardson 2004; USAID 2003).

Women's control over their families' livestock varies by culture (Tipilda and Kristjanson 2008, also see Kristjanson et al., Chap. 9). Yet, typically, men are responsible for the purchase, sale, or pawning of large animals, such as cows, horses, and oxen, while women tend to claim control over small animals such as goats, sheep, poultry, and pigs (World Bank et al. 2009; IFAD 2004; Miller 2001).

Finally, in settings where men are portrayed and perceived as the main breadwinner, women's ability to offer family assets as collateral and their incentives to invest in productive activities are influenced by family dynamics that are likely to prioritize men's investments (Ospina 1998).

8.2.2 Cultural Norms and Family Responsibilities

Socially accepted norms of behavior and the roles women play in their families can have profound effects on the type of economic activities in which women can engage, the technologies available to them, the people and agencies with whom they can interact, the places they can visit, the time they have available, and the control they can exert over their own capital.

In settings where sociocultural norms restrict women's mobility, their interactions with members of the opposite sex or their ability to attend trainings or receive formal education, women's access to information, institutions, and markets is compromised. This is the case when women are not allowed to use public transportation, when they cannot afford to pay for it, or when they cannot get away from their household responsibilities (Primo 2003). It is also the case when women are prevented from interacting directly with men other than close relatives, or when they feel awkward doing so, limiting their participation in agricultural or financial training and their ability to benefit from working with extension agents and veterinarians, most of whom are male and primarily address other men (Aina 2006; Esenu et al. 2005). As a result of these constraints, rural women tend to get their information from informal networks of women,[1] reinforcing the gender gap in access to information. The gap can be substantial: recent work to quantify it using data from Paraguay compares husbands' and wives' knowledge of financial markets and finds that rural women are 15–21 % less likely than men to have basic information about the financial institutions in their communities (Fletschner and Mesbah 2011).

Even when they have access to information on the financial services and market opportunities available to them, women may be less equipped to process the information. Their lower levels of literacy and lack of exposure to other languages, especially relative to male family members, hampers women's ability to benefit directly from information that is provided in writing or in languages other than those they speak at home (UNDP 2007; Ngimwa et al. 1997) and to fully understand the conditions of complex financial products available to them (Brown 2001). This matters as demonstrated by Cole et al. (2009) experimental work in India and Indonesia, which finds that financial literacy is a strong predictor of demand for financial services.

Social norms also define the type of economic activities in which women can engage, the amount of time they can invest in them, and the markets they can access. In most rural communities, activities tend to be sharply segregated by gender (Kevane 2004; Roberts 1998; Schroeder 1996). Women are typically responsible for cooking, childcare, laundry, cleaning, and the collection of water and fuelwood (Fletschner 2008a; Bezner Kerr 2008). While the gendered division of labor within agricultural production varies by location, men are typically in charge of tilling, plowing, fumigating, and selling crops to wholesale traders, and women tend to do most of the animal husbandry and the processing of agricultural or animal products (Fletschner 2008a; World Bank et al. 2009). In aquaculture and fishing, men are the primary fishers, while women mend nets, collect shellfish, smoke and dry fish for sale, and sell at local markets (World Bank et al. 2009).

Women's ability to undertake entrepreneurial activities that depart from well-established social norms is influenced by whether or not a sufficiently large group of women engage in comparable enterprises. As a result, each woman's economic opportunities are shaped not just by their own individual access to financial resources but also by whether those other women are able to obtain the capital they need (Fletschner and Carter 2008).

Cultural norms and family dynamics can also limit women's ability to exercise control over the savings they have or the semi-liquid assets they own. Anderson and Balland (2002) and Gugerty (2007) hypothesize that one of the reasons for the high level of female participation in rotating credit and savings associations (ROSCAs) is that this socially accepted strategy to save allows women to protect their savings

[1] As documented, for example, in Aryeetey's (1995) description of seed technology diffusion in Ghana.

from husbands and other relatives. Alternatively, women may choose individual savings programs that allow them to keep details or even knowledge of these savings to themselves to avoid being subjected to pressure from others.

Finally, social traditions can leave women in a particularly vulnerable position since, in addition to the risks associated with pregnancy and childbearing, women are more likely to experience domestic violence, to experience greater hardships in case of divorce, and to lose their assets when their spouses die (Banthia et al. 2009).

8.2.3 Behavioral Differences

Whether as a result of innate psychological characteristics or of attitudes influenced by social conditions, men and women tend to exhibit systematic differences in their behavior. Of particular importance when assessing the adequacy of financial products available to rural women is how men and women differ in their willingness to take risks. Studies in psychology and economics found that, on average, women tend to be more averse to risk than men and that, other things equal, women are more likely to forego activities that offer higher returns if these opportunities require them to bear too much risk (Fletschner et al. 2010; Croson and Gneezy 2008; Browne 2006). Producers who are more risk averse are less likely to adopt new technologies, to undertake projects that are expected to offer higher profits but expose them to more risk, or to apply for loans that may cause them to lose the collateral they own (Liu 2008; Dercon 2006; Boucher et al. 2008; Fletschner et al. 2009). In other words, compared with men, and without adequate insurance, women are more likely to consider borrowing against collateral as a risky transaction and might be less interested in taking out loans even when credit is available to them. The finding that women are, on average, more risk averse than men suggests that women will have a stronger preference for financial products tailored to help them save in a secure environment, insure against risks (particularly risks that may have a larger impact on women), or borrow without risking losing their assets.

8.2.4 Institutional Discrimination

Finally, rural women's access to financial resources is also limited by biased lending practices that emerge when financial institutions in the area consider them smaller, less experienced, and therefore less attractive clients, or when institutions lack the knowledge to offer products tailored to women's preferences and constraints (Fletschner 2009). The extent to which institutions reach out to women and the conditions under which they do vary noticeably, but women are at a disadvantage when an institution does not fund the type of activities typically run by women, when it does not accept female guarantors, when its requirements are not clear or widely known,[2] or when, as it is typically the case, loans to women are smaller than those granted to men for similar activities (Fletschner 2008a; World Bank et al. 2009; Ospina 1998; Baydas et al. 1994).

Combining these elements and acknowledging that, compared with men, women tend to have more limited control over resources accepted as collateral, less access to information, to be more risk averse and face a different set of activity-regulating social norms and family rights and responsibilities, it is easy to see that the type of financial products they need, the conditions under which they are willing to participate in formal financial markets and their ability to meet their financial needs differs dramatically from those of their male partners. Supporting these arguments, quantitative studies in Paraguay, Malawi, and Bangladesh find that rural women are more likely to be credit constrained than men, and the factors affecting whether or not they are able to meet their needs for capital are different (Fletschner 2009; Diagne et al. 2000). Similarly, Floro (2002) finds that women and men differ in their savings patterns, a difference that she attributes to differences in risk attitudes, options, and constraints.

The examples described above illustrate a number of avenues through which legal regulations, social norms, family responsibilities, and behavioral attributes can hinder rural women's ability to meet their financial needs, but one might be tempted to ignore these constraints and assume that women with male partners, the vast majority of adult women, can rely on their husbands to overcome these obstacles. In fact, this is the logic underlying financial market interventions that focus on poor households by targeting the male household head. To the extent they consider women as producers who can have a demand for capital, these interventions assume, implicitly, that when women lack adequate access to financial resources, they can rely on their husbands to intermediate funds to them. This is a very problematic assumption. While, arguably, some women are indeed able to overcome these obstacles with their husbands' assistance, others are not as fortunate. For those women who cannot count on their husbands as financial intermediaries, having direct access to capital becomes a necessity.

Before discussing the possibility of spousal intermediation of funds in more detail, it is important to recognize that a large number of women are members of households that do not fit the dual-headed model of the family to which we have been referring and on which we will concentrate most of the discussion that follows. In fact, polygamous households are common in a number of African countries as well as within Islamic communities, and female-headed households are increasingly capturing the attention of policymakers who acknowledge their prevalence and understand their economic relevance.

A review of the literature finds that compared to dual-headed households, female-headed households tend to be poorer and more risk averse. In Zimbabwe,

[2] This lack of clarity and transparency enables employees responsible for loan approvals to frame them as special favors that women are often unable to repay. The most common forms of repaying such favors—such as inviting loan officials for a drink or for dinner or the giving of bribes—are not considered acceptable behavior for women (Ospina 1998; Lycette and White 1989).

female-headed households reported 50–75 % lower total household income and up to 60 % lower per capita incomes (Horrell and Krishnan 2007). Similarly, female-headed households in Ghana and Bangladesh were more likely to be asset poor (Quisumbing et al. 2001). And, in Ethiopia, female-headed households were more likely to state fear of risk as the main reason for not taking up formal credit (Vilei and Chisholm 2005).

A joint study by FAO and the United Nations Development Programme (UNDP) (2002) carried out in Vietnam indicates female-headed households borrow less, have less access to formal credit, and pay higher interest on loans than dual-headed households. There are similar findings from Nicaragua (Foltz et al. 2000) and from Nigeria (Lawal and Muyiwa 2009), where female-headed households were found to be almost 200 % more credit constrained than dual-headed households.

One should note, however, that, while widespread, female-headed households are quite heterogeneous as a category and that women heads of households are likely to differ in their financial needs and constraints. On the one hand, one can distinguish between *de jure* female-headed households in which the female head is single or widowed; and *de facto* female-headed households in which the male partner does not permanently reside in the household, and while he can influence larger decisions, by and large he is not involved in day-to-day decisions and activities. The financial needs and constraints of women in *de jure* female-headed households are likely to differ from those in households that are *de facto* headed by women. Moreover, the financial needs and constraints of women in female-headed households are also likely to differ, depending on the reasons why they are or became the household head.

Unfortunately, the economic literature documenting these women's ability to meet their financial needs is sparse; consequently, we are unable to shed much light on how polygamous households allocate their resources, or the specific challenges faced by women in households that are female-headed or polygamous. Understanding that their needs for financial resources are likely to differ, we include references to women in these households and document the few results we have been able to find whenever appropriate.

8.3 Can Rural Women Gain Access to Financial Resources with Their Husbands' Assistance?[3]

When spouses have similar goals and priorities, it seems reasonable to expect that women would have the support of their husbands to gain access to financial resources: spouses that agree with each other are likely to pool their resources. That is, they are likely to combine their assets, their labor, their financial resources, and the information

[3] While the male and female partners in dual-headed rural households need not be married, and they often are not, we refer to them as husbands, wives, and spouses to simplify our discussion.

they have in order to produce an agreed upon basket of goods and services, and then consume or invest the profits they obtain according to their shared priorities.

However, and contrary to what is assumed in most development interventions, spouses can disagree and they often do. Family dynamics can be complex. Spouses can differ in how they want to allocate their resources, what they would like to produce, and how they prefer to spend the income they earn (Haddad et al. 1997). Some individuals may be eager to try out new seeds or new technologies, tempted by potentially high returns. Others may feel reluctant to expose their families to the additional risk associated with these new activities. Some may prefer to invest family resources in new irrigation systems, or to acquire a motorbike, and others will be more inclined to invest those funds in their children's schooling or in improving their houses. Some may want to grow commercial crops, while others want to invest in poultry or swine. Some may prefer to specialize, others may want to diversify their production in order to minimize their exposure to risk.

When spouses disagree with each other, they will each try to steer the resources they control toward the allocations they prefer. They may refuse to share information with their partners; they may limit the labor they contribute to their spouses' activities; they may allocate more of their land, fertilizer, or capital to activities in which they have more individual control; or they may choose not to fully pool the income they earn.

Studies have found that spouses often do not have information about each other's savings, private expenditures, ownership of bank accounts or loans (Fletschner and Mesbah 2011; Boozer and Goldstein 2003; Lacoste 2001; Karanja-Diejomaoh 1978); their decisions to consume or save vary depending on what their spouses know (Ashraf 2009); and they could increase their output by reallocating more of the family inputs, such as labor and manure, to women's plots (Udry 1996).

Recent and innovative quantitative studies based on couples from Latin America, South East Asia, and Sub-Saharan Africa suggest spouses do not always pool their financial resources either. Ashraf (2009) finds that spouses in the Philippines sometimes choose to keep savings in accounts their spouses cannot access. Doss (2001) finds that households in Ghana do not fully pool risk, and Duflo and Udry (2004) report that spouses in Côte d'Ivoire do not fully insure each other against rainfall-induced fluctuations in income. Robinson's (2008) field experiment in Kenya suggests that risks are not shared efficiently within the household. And Fletschner (2009) reports that 15 % of the women she surveyed in Paraguay said they were credit constrained even though their husbands claimed to have adequate access to credit.

In light of these results, the standard assumption that women with male partners can count on their assistance to overcome the gender-specific obstacles they face in financial markets is misguided and not supported by evidence. In other words, targeting poor households is not sufficient to ensure that resource-poor female producers are able to meet their financial needs.[4]

[4]The situation can be exacerbated in polygamous households where, as Oni (1996) points out, senior wives can experience reduced security as their husbands allow their favorite and likely more recent wives more access to resources.

Development strategies intended to boost rural women's productive capacity must therefore enhance women's *direct* access to financial resources. Effectively reaching rural women with financial services entails crafting, funding, and implementing policies and programs that take into account the gender- and context-specific conditions that shape their financial needs and constraints.

And while improving women's access to credit is an effective strategy to boost rural production, ensuring they have direct access to credit—which is not mediated through their husbands—has additional benefits. Improving women's *direct* access to financial resources strengthens women's ability to influence their households' behavior—a result that numerous studies have shown is likely to bring about further social and economic benefits, as we discuss below.

8.4 Enhancing Rural Women's Access to Financial Resources to Strengthen Their Ability to Influence Household Decisions

When spouses have conflicting preferences, what the household borrows, invests, produces, consumes, spends, and saves depends on the power each spouse has to influence these decisions. Spouses' relative bargaining position and, as a result, the extent to which they shape their household's behavior depends on how their individual economic and social fallback positions compare (Lundberg and Pollak 1993). Those with considerably more access and control over resources than their partners will be able to exert greater pressure and tilt their households' decisions toward the economic activities and allocations they prefer. Conversely, spouses who, compared with their partners, have very limited access to and control over resources will have very little influence, if any, on their households' economic behavior.

In other words, decisions such as which crops the family grows, whether or not they try a new seed variety, whether they invest in irrigation, the kind of animals they choose to raise, to whom they sell their products, how they allocate their profits, or whether they send their children to school, are affected not only by the resources the family can access, but also by which spouse has access to and control over these resources. As a result, interventions that enhance women's direct access to and control over financial resources will improve their position vis-à-vis their husbands, strengthening their role as decisionmakers and enhancing their ability to influence how their households allocate resources.

Wide-ranging studies from the past couple of decades have consistently found that improvements in women's direct access to and control over resources are associated with positive economic and social outcomes for women, their families, and their communities. There is by now strong evidence suggesting that interventions enhancing rural women's productive capacity by improving their direct access to resources such as land, water, education, fertilizer, paid work, and technical assistance, lead to a more efficient allocation of family resources, enabling families to produce more with the resources they have. For instance, Udry (1996) demonstrates that households in Burkina Faso can increase their production significantly by simply reallocating their labor and fertilizer from the plots owned by men to those owned by women. Jones (1983) shows that families in Cameroon can increase their overall production if women allocate more of their time to rice cultivation, the income from which typically accrues to men. The study by Saito et al. (1994) argues that if rural women in Kenya had the same access to education and agricultural inputs as men, their yields of maize, beans, and cowpeas would increase dramatically. Zwarteveen's (1997) study analyses the economic importance of securing women's direct access to water and irrigation systems.

The findings associated with spouses' inefficient sharing of financial resources are similar: a study based on information about husbands and wives' individual access to credit in Paraguay finds that, on average, and after accounting for men's access to financial resources, rural families produce 11 % less when women cannot meet their credit needs (Fletschner 2008b).

In addition to affecting what and how much is produced, women's direct access to and control over resources have important consequences for their families' investments in human capital. Researchers working in Sub-Saharan Africa, South Asia, and Latin America have consistently found that families benefit when women are in a stronger position[5] (Quisumbing 2003). When women have more influence in their families' economic decisions, their children are better fed (Smith et al. 2003) and their families allocate more of their income to food (Doss 2006; Thomas 1997; Hoddinott and Haddad 1995), to health (Thomas 1997), to education (Doss 2006; Quisumbing and Maluccio 2000; Thomas 1997), and to children's clothing (Bobonis 2009; Quisumbing and Maluccio 2000).

These studies suggest that interventions that improve women's direct access to financial resources will result in higher investments in human capital and have a stronger impact on their children's health, nutrition, and education. In fact, studies from Bangladesh, India, and Malawi find that women's access to credit improves the nutrition of girls, while men's access to credit does not (Hazarika and Guha-Khasnobis 2008); and when women are the ones taking out loans, their children are better fed (Khandker 1998), more likely to be enrolled in school, and more likely to be literate (Holvoet 2004; Pitt and Khandker 1998). One should interpret these results with caution, however, since recent controlled experiments cast doubt on these findings (see, for instance, Roodman and Morduch 2009).

Interventions that improve rural women's access to financial services can therefore be beneficial on two fronts. First, by addressing the constraints women face, these interventions enhance women's productive capacity. That is, increasing women's access to capital can have an impact on production directly. Second, by improving women's relative power in their households, these interventions can lead to both a more efficient allocation of resources, with families producing more

[5] Where women's relative power in the family is approximated by the assets women brought to the marriage, women's share of family assets or income, women's borrowing, women's access to credit, and women's credit limit, relative to men's.

with the same resources, and to better human capital outcomes, such as improved health, nutrition, and education in their families, all of which is expected to improve long-term production capabilities.

8.5 Recent Innovations That Can Improve Women's Access to Financial Resources

In an effort to address some of these constraints, a number of new products and service delivery models have been introduced. These include technical innovations that improve access to existing financial services, changes in product design to better tailor products to women's preferences and constraints, and the development of new products such as micro-insurance.

Technological innovations, such as prepaid cards to distribute loan payments and mobile phone plans to make loan payments and transfer cash, make it easier for women to gain access to capital by reducing the need for women to travel long distances, allowing them to sidestep social constraints that restrict the areas women can visit or the people with whom they can interact (Duncombe and Boateng 2009). In addition, products like the biometric smart cards allow women to have control over who can access their savings accounts (Quisumbing and Pandofelli 2009).

Acknowledging that clients frequently borrow for income-generating activities but end up channeling some of those funds to build or improve their houses and securing their housing needs, several organizations have started offering "housing loans," a product that typically means low-cost mortgages that rely on key features of the microfinance methodology (CUDS 2000). For examples of three different approaches to microcredit for housing finance, see the products offered by the Center for Agricultural and Rural Development in the Philippines, SEWA Bank in India, or FINCA Africa's partnership with Habitat of Humanity in Uganda, Malawi, and Tanzania (CUDS 2000).

Products designed to strengthen women's position include the Grameen Bank's loans for purchasing land or houses that require they be registered in women's names, and the loans offered by Credit and Savings Household Enterprise in India for parents to buy assets for their girls, enabling their daughters to generate income, delay their marriage and have assets they can contribute to their in-laws when they marry (Mayoux and Hartl 2009). Along similar lines, a host of products have been designed to indirectly benefit other women in the community (Mayoux and Hartl 2009). This is the case, for instance, of loans for businesses that employ women, or for businesses that offer services such as childcare that benefit other women.

Following recent trends in microfinance delivery, Come to Save (CTS) Bangladesh now offers a daily collection of voluntary savings and loan repayments to their urban clientele. While this approach might be too onerous to replicate in rural areas, their flexibility in allowing clients to choose loan sizes and repayment schedules according to their capacity and requirements, rather than having to conform to more rigid schemes, can and should be imitated. See Matin et al. (2002) for descriptions of microfinance delivery systems that replace the ideals of discipline and compulsion with those of frequency, convenience, and opportunity to better reach the poor.

In parallel to these developments, the past few years have seen noticeable progress in efforts to extend insurance products to small producers and to rural areas. Weather, livestock, and crop insurance, for instance, are increasingly being provided as safety nets to farmers. By and large, however, these products are designed without paying attention to gender differences and the degree to which women access them is unclear. A notable exception to this pattern is BASIX, a large microfinance institution (MFI) in India, which provides weather insurance to women's self-help group members in drought-prone areas (World Bank 2001).

A number of MFIs and NGOs are, however, offering health insurance to women. Illness can translate into a major income shock for resource-poor households, and women are particularly vulnerable. On the one hand, women face a greater risk of illness due to limited access to healthcare and poorer nutritional status than men. On the other hand, women are more likely to be assigned the role of carers and as a result illness in the family reduces women's ability to engage in income-generating activities and weakens their ability to influence family decisions. Table 8.1 provides an overview of a few insurance plans that target women as the primary policyholder.

Life events such as birth, death, marriage, and other cultural ceremonies also pose as shocks to rural households. Most microinsurance plans described here cover pregnancy and birth-related expenses. Some offer life and funeral insurance (Sriram 2005; Mgobo 2008), but informal safety nets such as burial societies remain important sources of income smoothing for rural households, especially for women, who face losing all assets upon a husband's death (Dercon et al. 2007; Mapetla et al. 2007).

Finally, self-help groups have proven to be an effective avenue for connecting women with financial institutions. These groups of women operate at the village level and typically require that their members meet regularly. Savings are collected from each member and either deposited in rural banks or loaned to other group members. After a group has demonstrated it had the capacity to collect loans, rural banks typically leverage the group's savings and provide additional capital that group members use mostly for agricultural purposes (World Bank et al. 2009).

8.6 Recommendations

The previous sections argued that women's ability to take full advantage of economic opportunities and to influence their families' resource allocation hinge on women's *direct* ability to meet their financial needs and that, in general, the suite of services women prefer and the constraints they face are likely to differ from those of men from similar economic and cultural strata. A broadly-based rural development strategy must therefore give high priority to enhancing women's access to financial services and recognize the need for a multipronged approach that combines corrective measures to remove obstacles that constrain women's participation

Table 8.1 A sample of health insurance products

Provider and country	Target/beneficiaries	Details
BRAC Bangladesh	Originally provided to Bangladesh Rural Advancement Committee members only; since 2007 opened to all community members (poor rural women are policyholders)	Year started: 2001 Number of members: 10,000 as of 2004 (Matin et al. 2005) Results: 55 % did not renew after first year; poorer households less likely to know about program, and better-off households more likely to enroll; some clients found it difficult to pay annual premium, others who did not use services but enrolled found it to be a "waste" (Matin et al. 2005).
SKS Bangladesh	SKS borrowers, who are primarily women (spouse and up to two children covered)	Year started: 2007, expanded in 2009 to cover spouses (usually husbands) Number of members: 210,000 (as of 2008); required for all new borrowers or renewing borrowers (as of 2007) (Chen et al. 2008) Results: women between ages 16–30 are the heaviest users (Chen et al. 2008).
SEWA India	SEWA members and nonmembers (women are policyholders)	Year Started: 1992 Number of members: 110,000 (as of 2003), two-thirds from rural areas (Ranson et al. 2006) Results: found to reduce clients' vulnerability to shocks overall, but slow processing costly to clients; initially, coverage was mandatory for all borrowers, but once became voluntary, 80 % dropped coverage (McCord 2001).
SPANDANA India	Borrowers (compulsory, as part of loan product) (Sriram 2005; CGAP 2004)	Year started: 2003 (Sriram 2005) Number of members: 84,000, including spouses (as of 2004) (CGAP 2004). In 2007, 96.5 % of total borrowers were women (Mix Market 2010).
PASED, LEAP Sudan	Women NGO members (individual low-cost access to state health insurance) (Mayoux and Hartl 2009)	Year started: 2007 (Mayoux and Hartl 2009) Number of members: unknown
KWFT Kenya	Medium and low-income women, with option to cover family members	Year started: 2008 Number of members: unknown, potentially 100,000 (total KWFT members) (Mgobo 2008)
Zurich Financial Services and WWB Various	Women's World Banking (WWB) affiliates (women member MFIs)	Year started: 2009 Number of members: not yet known, but WWB network has 21 million members (WWB 2010)

with deliberate interventions to ensure that rural policies and programs are planned, designed, implemented, monitored, and evaluated in a gender-sensitive way. Recommendations include:

Local- and national-level interventions to address current discrimination and level the playing field for women

- Gender-sensitize staff of all related institutions. Implement efforts to train, raise the gender awareness and develop the skills needed to integrate gender in policy-making, as well as in the planning and delivery of services. For these efforts to be effective, they should reach legislators, technical and financial experts, heads of agencies and their entire staff, especially extension agents and service providers in all sectors. This will help ensure that rural financial programs are designed, budgeted, and implemented, taking into consideration women's needs and what it would take to reach them. It should also lead to the implementation of complementary efforts to enhance women's entrepreneurship and their creditworthiness, such as legal aid programs that strengthen women's land rights, technology adoption programs that are suited to match women's needs and constraints, and the promotion of gender-sensitive value chains.
- Assess and modify laws as needed to ensure women equal protection under the law and their ability to enforce these rights. This is particularly relevant when it comes to rights that affect women's access, use, control, and ownership of natural and physical assets. It is important that women are recognized as landowners, that their names are included in legal land documents, that they have equal rights to offer property as collateral, and that they are able to have bank accounts on their names.
- Improve rural women's access to financial information by developing content that is culturally appropriate and relevant to the needs of women, and by training and mentoring female extension, agricultural, veterinary, and financial agents who can help women identify opportunities and develop financial plans. Ensure that these efforts take into account women's restricted mobility and their time constraints by, for instance, locating sources of information in the communities and scheduling informational meetings or radio programs at times that do not conflict with women's domestic responsibilities.
- Introduce efforts to develop rural women's basic literacy skills, and beyond that, their basic information skills in accessing, processing, and using the information available to them (Fletschner and Mesbah 2011).
- Promote and assist the creation of groups in which women can find a safe space to experiment with new projects, technologies, or economic activities and the support of which women can leverage in their interactions with their families and their communities, as well as with financial institutions, input providers, and potential buyers (United Nations 2009).
- Enhance women's ability to access input and output markets. This can be accomplished by investing in infrastructure and transportation services that enable women to travel safely, in a reasonably priced and culturally appropriate way; by carrying out promotional campaigns aimed at reducing the impact of cultural

norms that discourage women from engaging in entrepreneurial activities; by offering basic trainings designed to boost women's self-esteem and assertiveness, as well as to sharpen their negotiating skills; and by facilitating women's participation in associations or cooperatives.

Interventions in the financial sector that can be classified in three categories: changes to the sector's rules and regulations, reforms at the organization level, and innovations in the design and delivery of financial services

Changes to financial-sector rules and regulations

- Review and revise regulations in the financial sector to support an organization's ability to reach rural women with products that are easy to understand and financially safe; acknowledging, for instance, that restrictions to mobilize savings and accept deposits limit an organization's outreach and that interest rate ceilings often lead to the creation of new and less clear fees.

Reforms at the organization level

- Develop materials and training that effectively communicate to everybody in an organization—from the directors to the credit officers—that female clients are just as creditworthy as male clients and should be treated accordingly. This can typically be done with an in-house analysis of the organization's portfolio: What percentage of the organization's loans is to women? Are the characteristics of those loans different? What are their repayment rates? Some organizations may have computerized information systems that allow them to carry out such analyses; others may have to rely on written or verbal reports from their credit officers. When this is not feasible because, for example, the organization simply has not had enough female clients to make such comparisons, credit officers from other organizations who regard women as good clients may be willing to share their views and their experience. Local information can be supplemented with publicly available data on the financial and social performance of other institutions, such as those reported at www.mixmarket.org.
- Promote a women-friendly and empowering culture throughout an organization that deals with institutional constraints and clearly communicates to female staff and clients that they are welcomed participants (World Bank et al. 2009). Consult women and include them in discussions, decisionmaking, planning, and the provision of services. Ensure that marketing strategies, promotion, and service delivery are gender-sensitive.

Innovations in the design and delivery of financial services

- Engage in market research to understand the financial needs of current and potential female clients as well as the type of products they prefer.
- Conduct a thorough review of the products currently offered and identify features in their design, promotion, or delivery that can affect women's demand for those products or their ability to access them.
- Design products that are flexible and meet women's needs (Matin et al. 2002). Some women may need long-term housing loans; others may be interested in

consumption loans to meet more immediate needs. Some women may be interested in products that require compulsory savings as a strategy that ensures they save; others may be afraid or unable to make that commitment. Some may want their savings to be publicly known to gain status in their families or communities; others may prefer this information to be kept private so that they can retain control over the funds.

- Design products that are convenient and accessible (Matin et al. 2002). Locate services close to where women are and in places they can easily frequent. Disseminate information and promotion materials in places or through channels women can access. Simplify procedures. Make application requirements appropriate for women's literacy and numeracy levels.
- Consider ways in which branchless banking and IT developments can be leveraged to lower transaction costs and to address some of the obstacles women face due to constraints in their mobility or in their social interactions. India, Brazil, Kenya, the Philippines, and South Africa offer examples in which financial institutions have been able to reach rural customers at a lower cost, by relying on post offices, gas stations, stores, and input providers (World Bank 2008). Similarly, Wizzit in South Africa and Globe Telecom and Smart in the Philippines allow their customers to make payments or transfer funds using mobile phones (World Bank 2008).
- Offer financial literature training to ensure women can compare products and make decisions based on a clear understanding of the characteristics and conditions of the products available (Mayoux and Hartl 2009).

(a) Credit

- Review an organization's current requirements to borrow. Consider expanding the menu of acceptable collateral to also include social collateral and the type of physical assets women are more likely to own or control. Offer incremental loans based on individual repayment behavior to reach out to women who might not be able to provide collateral. Ensure women can apply for loans without their husbands' or other male approval.
- Offer a menu of loans that takes into account the diversity of women's needs and constraints: What are they trying to finance? How much will they need? What is a reasonable repayment schedule? Some women will be interested in loan conditions that match seasonal or annual crop cycles; others will demand loans with a long-term horizon so that they can invest in livestock or physical assets; yet others may want short-term funds to purchase business supplies or address temporary consumption needs or health emergencies.
- Design loan packages that support women's engagement in more profitable, but nontraditional, economic activities by bundling credit with additional services, or by helping women connect with agencies or groups where they can obtain that support (Quisumbing and Pandofelli 2009; World Bank et al. 2009).

(b) Savings

- Review an organization's current requirements to save. Consider whether the existing charges and fees combined with any minimum initial deposit requirements are set at a level that poor women can afford.
- Offer a menu of savings programs that take into account the diversity of women's needs and constraints. Some women may prefer a program that makes savings compulsory—where the amounts saved and the timing of deposits are known and agreed—as a commitment device that prevents them from diverting those funds to meet other needs or helps them resist pressure from family and friends. Others probably want more flexible packages that allow them to adjust the timing and amount of their savings to their consumption needs and investment opportunities.
- Similarly, some women may value programs that, by making their participation publicly known, boost their status in their communities. Others may consider it safer or more desirable to keep that information private, retaining control over how much of this information they share with their friends, relatives, and husbands.

(c) Insurance

- Offer a menu of insurance programs that take into account the diversity of women's needs and constraints. Women can differ from men in the economic activities or assets for which they seek insurance: different crops, different livestock investments, different processing enterprises, or different nonfarm activities, for instance. Women can also differ in their demand for health insurance, especially plans that provide reproductive health coverage for them, for flexible life insurance products to cover them were their spouses to die, or to ensure their children benefit in case they pass away (Banthia et al. 2009).
- Simplify insurance contracts and communicate their conditions using language and examples that women can easily understand. Educating prospective clients and reaching them with adequate information is especially important when it comes to insurance products that tend to be new and complex.
- Simplify the process and requirements to make claims, offering women support when assistance is needed. This will ensure that women receive the benefits to which they are entitled and will strengthen other women's confidence in the value of these products.

Government interventions to support a gender-integrated approach

- Support the institutionalization of a gender-sensitive approach to policymaking and program design by
  - Ensuring the collection of sex-disaggregated data and the creation of gender-sensitive indicators that can be easily accessed and used to inform new initiatives as well as to assess, monitor, evaluate, and report progress.
  - Encouraging and supporting the development of networks of practitioners and gender experts that can identify good practices and build capacity (Mayoux and Hartl 2009).

- Promote the creation of new knowledge and the development of adequate technology to better address women's preferences and constraints by supporting research that
  - Identifies women's specific needs for financial services by taking a more in-depth look at the characteristics of the type of investments they want to finance, the type of risks they want to manage or cope with, the product characteristics they prefer, and the specific constraints they face.
  - Helps develop and test technological and institutional innovations that reduce the risk and cost of reaching rural women with the financial services they demand (World Bank 2008).
  - Evaluates the impact of these efforts and identifies successful approaches.
- Sends a clear and consistent message to national and local communities indicating a strong support for efforts that address women's needs and constraints directly, explaining why these efforts are important and backing up these assertions with concrete actions and representative examples. Ensure that women who are leaders in their communities, or experts in gender issues, are involved and participate actively at each stage and at all levels. Such a measure will enhance the credibility of the efforts to reach out to women, will provide them with a more in-depth understanding of the way current constraints operate, and will offer workable insights into how they can be overcome.

References

Agarwal B (2003) Gender and land rights revisited: exploring new prospects via the state, family, and market. J Agrar Change 3(1–2):184–224

Aina L (2006) Information provision to farmers in Africa: the library-extension service linkage. Paper presented at the World Library and Information Congress: 72nd IFLA general conference and council, Seoul, Korea

Anderson S, Balland JM (2002) The economics of Roscas and intrahousehold resource allocation. Q J Econ 117(3):963–995

Aryeetey E (1995) The significance of gender in seed transfer patterns among farmers in South-East Ghana. Paper presented at the OAU/SAFGRAD regional workshop on technology options and transfer systems for foodgrain production in Sub-Saharan Africa, Abidjan, Nigeria

Ashraf N (2009) Spousal control and intrahousehold decision making: an experimental study in the Philippines. Am Econ Rev 9(4):1245–1277

Banthia A, Johnson S, McCord M, Matthews B (2009) Microinsurance that works for women: making gender-sensitive microinsurance programs. Microinsurance paper no. 3, microinsurance innovation facility. International Labour Organization, Geneva

Baydas MM, Meyer RL, Aguilera-Alfred N (1994) Discrimination against women in formal credit markets: reality or rhetoric? World Dev 22(7):1073–1082

Besley T (1995) Savings, credit, and insurance. In: Behrman J, Srinivasan TN (eds) The handbook of development economics, vol 3. Elsevier, Amsterdam, pp 2123–2207

Bezner Kerr R (2008) Gender and agrarian inequality at the local scale. In: Snapp SS, Pound B (eds) Agricultural systems: agroecology and rural innovation for development. Academic, London, pp 281–308

Bobonis G (2009) Is the allocation of resources within the household efficient? New evidence from a randomized experiment. J Pol Econ 117(3):453–503
Boozer M, Goldstein M (2003) Poverty measurement and dynamics. Mimeo/Yale University, New Haven
Boucher S, Carter MR, Guirkinger C (2008) Risk rationing and wealth effects in credit markets: theory and implication for agricultural development. Am J Agric Econ 90(2):409–423
Brown W (2001) Microinsurance—the risks, perils and opportunities. Small Enterp Dev 12(1):11–24
Browne K (2006) Evolved sex differences and occupational segregation. J Organ Behav 27(1):143–162
CGAP (Consultative Group to Assist the Poor) (2004) Improving risk management for the poor. Microinsurance no. 3. CGAP Working Group on Microinsurance, Washington, DC
Chen T, Comfort A, Bau N (2008) Implementing health insurance through microcredit: a case study of SKS Microfinance, India. Center for Microfinance, Chennai
Cole S, Sampson T, Zia B (2009) Financial literacy, financial decisions, and the demand for financial services: evidence from India and Indonesia. Working paper 09–117. Harvard Business School, Cambridge, MA
Croson R, Gneezy U (2008) Gender differences in preferences. J Econ Lit 47(2):448–474
CUDS (Center for Urban Development Studies) (2000) Housing microfinance initiatives. Synthesis and regional summary: Asia, Latin America, and Sub-Saharan Africa with selected case studies. Center for Urban Development Studies, Harvard University, Cambridge, MA
Deere C, León M (1997) Women and land rights in the Latin American Neo-Liberal counter-reforms. Working paper no. 264. Women in International Development, Michigan State University, East Lansing, MI
Dercon S (2006) Risk, growth, and poverty: what do we know? What do we need to know? Queen Elizabeth House working paper series. University of Oxford, Oxford
Dercon S, Hoddinott J, Krishnan P, Wodehanna T (2007) Collective action and vulnerability. Collective Action and Property Rights (CAPRi) working paper no. 83. Consultative Group on International Agricultural Research, Washington, DC
Diagne A, Zeller M (2001) Access to credit and its impact on welfare in Malawi. Research report 116. International Food Policy Research Institute, Washington, DC
Diagne A, Zeller M, Sharma M (2000) Empirical measurements of household's access to credit and credit constraints in developing countries. Food consumption and nutrition division discussion paper 90. International Food Policy Research Institute, Washington, DC
Doss C (2001) Is risk fully pooled within the household? Evidence from Ghana. Econ Dev Cult Change 50(1):101–130
Doss CR (2006) The effects of intrahousehold property ownership on expenditure patterns in Ghana. J Afr Econ 15(1):149–180
Duflo E, Udry C (2004) Intrahousehold resource allocation in Côte d'Ivoire: social norms, separate accounts and consumption choices. Working paper 10498. National Bureau of Economic Research, Inc, Washington, DC
Duncombe R, Boateng R (2009) Mobile phones and financial services in developing countries: a review of concepts, methods, issues, evidence, and future research directions. Third World Q 30(7):1235–1258
Esenu B, Ossiya S, Serunkuma D, Oluka J, Alin D, Owesigire B, Ebiyau G, Olokojo B (2005) Positioning agro-pastoral women in livestock production: the link between socioeconomic factors and improved household food security. Tropicultura 23(1):28–32
FAO/UNDP (Food and Agriculture Organization of the United Nations/United Nations Development Program) (2002) Gender differences in the transitional economy of Vietnam. Food and Agriculture Organization and United Nations Development Programme, HaNoi
Fletschner D (2008a) Rural women's access to capital: intrahousehold bargaining and social effects. VDM Publishing House, Saarbrucken
Fletschner D (2008b) Women's access to credit: does it matter for household efficiency? Am J Agric Econ 90(3):669–683

Fletschner D (2009) Rural women's access to credit: market imperfections and intrahousehold dynamics. World Dev 37(3):618–631
Fletschner D, Carter M (2008) Constructing and reconstructing gender: reference group effects and women's demand for entrepreneurial capital. J Socio-Econ 37(2):672–693
Fletschner D, Mesbah D (2011) Rural women's access to information: do spouses share what they know? World Dev 39(8):1422–1433
Fletschner D, Guirkinger C, Boucher S (2009) Risk, credit constraints and financial efficiency in Peruvian agriculture. J Dev Stud 46(6):981–1002
Fletschner D, Anderson CL, Cullen A (2010) Are women as likely to take risks and compete? Behavioral findings from central Vietnam. J Dev Stud 46(8):1459–1479
Floro M (2002) Mobilizing domestic resources for development: through the lens of poor women. Paper presented at the conference on gender budgets, financial markets, financing for development, Heinrich Böll Foundation, Berlin
Foltz J, Larson B, Lopez R (2000) Land tenure, investment, and agricultural production in Nicaragua. Development discussion paper no. 738. Institute for International Development, Harvard University, Cambridge, MA
Gugerty MK (2007) You can't save alone: commitment and rotating savings and credit associations in Kenya. Econ Dev Cult Change 55(2):251–282
Haddad L, Hoddinott J, Alderman H (1997) Introduction: the scope of intrahousehold resource allocation issues. In: Haddad L, Hoddinott J, Alderman H (eds) Intrahousehold resource allocation in developing countries: models, methods, and policy. Johns Hopkins University Press for the International Food Policy Research Institute, Baltimore, pp 1–16
Hazarika G, Guha-Khasnobis B (2008) Household access to microcredit and children's food security in rural Malawi: a gender perspective. Discussion paper no. 3793. Institute for the Study of Labor, Bonn
Hoddinott J, Haddad L (1995) Does female income share influence household expenditure patterns? Evidence from Côte d'Ivoire. Oxf Bull Econ Stat 57(1):77–96
Holvoet N (2004) Impact of microfinance programs on children's education: do the gender of the borrower and the delivery model matter? J Microfinance 6(2):1–23
Horrell S, Krishnan P (2007) Poverty and productivity in female-headed households in Zimbabwe. J Dev Stud 43(8):1351–1380
IFAD (International Fund for Agricultural Development) (2004) Livestock services and the poor. A global initiative. Collecting, coordinating and sharing experiences. International Fund for Agricultural Development, Rome
Jones C (1983) The mobilization of women's labor for cash crop production: a game theoretic approach. Am J Agric Econ 65(5):1049–1054
Karanja-Diejomaoh W (1978) Disposition of incomes by husbands and wives: an exploratory study of families in Lagos. In: Oppong C, Adaba G, Bekombo-Priso M, Mogey J (eds) Marriage, fertility, and parenthood in West Africa. Changing African family project monograph series no. 4. Australian National University, Canberra
Kevane M (2004) Women and development in Africa: how gender works. Lynne Rienner, Boulder/London
Khandker S (1998) Fighting poverty with microcredit: experience in Bangladesh. Oxford University Press, New York
Lacoste JP (2001) Savings strategies of poor women in Zimbabwe: a socioeconomic perspective. Paper presented at the international conference on livelihood, savings and debt in a changing world. Developing sociological and anthropological perspectives. University of Wageningen, Wageningen, the Netherlands
Lawal J, Muyiwa A (2009) Food security and socioeconomic characteristics of cocoa farming households in Nigeria: support through agricultural biotechnology. In: Proceedings of the 2nd international e-conference on agricultural biosciences, Ibadan, Nigeria
Liu E (2008) Time to change what to sow: risk preferences and technology adoption decisions of cotton farmers in China. Department of Economics, Princeton University, Princeton

Lundberg S, Pollak R (1993) Separate spheres bargaining and the marriage market. J Pol Econ 101(6):988–1010

Lycette M, White K (1989) Improving women's access to credit in Latin America and the Caribbean: policy and project recommendations. In: Berger M, Buvinic M (eds) Women's ventures: assistance to the informal sector in Latin America. Kumarian Press, West Hartford, pp 19–44

Mapetla M, Matobo T, Setoi S (2007) Hunger vulnerability: burial societies as social protection mechanism in Lesotho. Commissioned Study by Regional Hunger Vulnerability Programme of Lesotho

Matin I, Hulme D, Rutherford S (2002) Finance for the poor: from microcredit to microfinancial services. J Int Dev 14(2):273–294

Matin I, Imam N, Ahmed SM (2005) Micro Health Insurance (MHI) pilot of BRAC: a demand side study. BRAC research report. Bangladesh Rural Advancement Committee, Dhaka

Mayoux L, Hartl M (2009) Reaching and empowering women: gender mainstreaming in rural microfinance: guide for practitioners. International Fund for Agricultural Development, Rome

McCord M (2001) Health care microinsurance—case studies from Uganda, Tanzania, India and Cambodia. Small Enterp Dev 12(1):25–38

Mgobo S (2008) Kenya: new scheme for low-cost health insurance launched. http://allafrica.com/stories/200806231599.html. Accessed 27 Jan 2010

Miller BA (2001) Empowering women to achieve food security: rights to livestock. International Food Policy Research Institute, Washington, DC

Mix Market (2010) Profile for Spandana. http://www.mixmarket.org/mfi/spandana/data. Accessed 27 Jan 2010

Morrow Richardson A (2004) Women's inheritance rights in Africa: the need to integrate cultural understanding and legal reform. Human rights brief no. 19. Center for Human Rights and Humanitarian Law, Washington, DC

Ngimwa P, Ocholla DN, Ojiambo JB (1997) Media accessibility and utilization by Kenyan rural women. Int Inf Libr Rev 9(1):45–66

Oni JB (1996) Qualitative exploration of intrahousehold variations in treatment of child illness in polygynous Yoruba families: the use of local expressions. Health Transit Rev 6(1):57–69

Ospina R (1998) Perspectiva de género en la misión rural: Para dotar de poder a las mujeres. Misión Rural Proyecto Proequidad GTZ/CNP, Santa Fé de Bogotá

Parada S (2008) Rural women in Latin America and their access to economic resources. Paper prepared for the 2009 world survey on the role of women in development, Division for the Advancement of Women, Department of Economic and Social Affairs, United Nations, New York

Pitt M, Khandker S (1998) The impact of group-based credit programs on poor households in Bangladesh: does the gender of participants matter? J Pol Econ 106(5):958–997

Primo N (2003) Gender issues in the information society. Paper prepared for the world summit on the information society, UNESCO, Paris, France

Quisumbing AR (ed) (2003) Household decisions, gender, and development: a synthesis of recent research. International Food Policy Research Institute, Washington, DC

Quisumbing A, Maluccio J (2000) Intrahousehold allocation and gender relations: new empirical evidence from four developing countries. International Food Policy Research Institute, Washington, DC

Quisumbing A, Pandofelli L (2009) Promising approaches to address the needs of poor female farmers: resources, constraints, and interventions. IFPRI discussion paper no. 882. International Food Policy Research Institute, Washington, DC

Quisumbing A, Haddad L, Peña C (2001) Are women overrepresented among the poor? An analysis of poverty in ten developing countries. J Dev Econ 66(1):225–269

Ranson MK, Sinha T, Chatterjee M, Acharya A, Bhavsar A, Morris SS, Mills A (2006) Making health insurance work for the poor: learning from SEWA's community-based health insurance scheme. Soc Sci Med 62(3):707–720

Roberts P (1998) Rural women's access to labor in West Africa. In: Stichter S, Parpart J (eds) Patriarchy and class: African women in the home and the workforce. Westview Press, Boulder, pp 97–114

Robinson J (2008) Limited insurance within the household: evidence from a field experiment in Kenya. Working paper. Department of Economics, University of California-Santa Cruz, Santa Cruz
Roodman D, Morduch J (2009) The impact of microcredit on the poor in Bangladesh: revisiting the evidence. Center for global development working paper 174. Center for Global Development, Washington, DC
Saito K, Mekonnen H, Spurling D (1994) Raising the productivity of women farmers in Sub-Saharan Africa. Africa Technical Department series no. 230. World Bank, Washington, DC
Schroeder R (1996) Gone to their second husbands: marital metaphors and conjugal contracts in the Gambia's female garden sector. Can J Afr Stud 30(1):69–87
Smith LC, Ramakrishnan U, Ndiaye A, Haddad L, Martorell R (2003) The importance of women's status for child nutrition in developing countries. Research report 131. International Food Policy Research Institute, Washington, DC
Sriram MS (2005) Expanding financial services access for the poor: the transformation of Spandana. Indian Institute of Management, Vastrapur
Thomas D (1997) Incomes, expenditures, and health outcomes: evidence on intrahousehold resource allocation. In: Haddad L, Hoddinott J, Alderman H (eds) Intrahousehold resource allocation in developing countries: models, methods, and policy. Johns Hopkins University Press for the International Food Policy Research Institute, Baltimore, pp 142–164
Tipilda A, Kristjanson P (2008) Women and livestock development: a review of the literature. ILRI innovation works discussion paper 01–08. International Livestock Research Institute, Nairobi
Udry C (1996) Gender, agricultural production, and the theory of the household. J Pol Econ 104(5):1010–1046
UNDP (United Nations Development Programme) (2007) Human development report 2007. United Nations Development Program, New York
United Nations (2009) Women's control over economic resources and access to financial resources, including microfinance. Paper prepared for the 2009 world survey on the role of women in development. Department of Economic and Social Affairs. Division for the Advancement of Women, New York
USAID (United States Agency for International Development) (2003) Women's property and inheritance rights: improving lives in changing times. Office of Women in Development, Bureau for Global Programs, Washington, DC
Vilei S, Chisholm N (2005) Can credit improve the livelihoods of resource-poor rural households in Ethiopia? Department of Food Business and Development, National University of Ireland, Galway
World Bank (2001) Engendering development through gender equality in rights, resources, and voice. World Bank policy research report. Management 1: report no. 36546-MW. World Bank, Washington, DC
World Bank (2008) World development report 2008. World Bank, Washington, DC
World Bank/FAO/IFAD (Food and Agriculture Organization of the United Nations/International Fund for Agricultural Development) (2009) Gender and agriculture: sourcebook. World Bank/ FAO/IFAD, Washington, DC
WWB (Women's World Banking) (2010) Women's World Banking Fact Sheet. http://www.swwb.org. Accessed 27 Jan 2010
Zeller M, Schrieder G, von Braun J, Heidhues F (1997) Rural finance for food security for the poor. International Food Policy Research Institute, Washington, DC
Zwarteveen M (1997) Water: from basic need to commodity: a discussion on gender and water rights in the context of irrigation. World Dev 25(8):1335–1349

Chapter 9
Livestock and Women's Livelihoods

Patti Kristjanson, Ann Waters-Bayer, Nancy Johnson, Annita Tipilda, Jemimah Njuki, Isabelle Baltenweck, Delia Grace, and Susan MacMillan

Abstract Livestock make substantial contributions to the livelihoods of poor women in Sub-Saharan Africa and South Asia, yet the factors that enhance or constrain livestock-related opportunities for women have received relatively little empirical analysis. This review applies a gender lens to a conceptual framework for understanding the role of livestock in pathways out of poverty, using a livelihoods approach that centralizes the importance of assets, markets, and other institutions. The three hypothesized livestock pathways out of poverty are (1) securing

P. Kristjanson (✉)
CGIAR Program on Climate Change, Agriculture and Food Security (CCAFS),
World Agroforestry Centre, United Nations Avenue, Gigiri, P.O. Box 30677,
Nairobi 00100, Kenya
e-mail: p.kristjanson@cgiar.org

A. Waters-Bayer
ETC Foundation, P.O. Box 64, 3830 AB Leusden, Netherlands
e-mail: ann.waters-bayer@etcnl.nl

N. Johnson
IFPRI, Agriculture for Nutrition and Health (A4NH),
2033 K Street, N.W., Washington, DC 20006, USA
e-mail: n.johnson@cgiar.org

A. Tipilda
Global Forests and Climate Change Programme, Nature Based
Solutions Group, International Union for Conservation of Nature (IUCN),
28 rue Mauverney, CH-1196, Gland, Switzerland
e-mail: Annita.ANNIES@iucn.org

J. Njuki
Senior Program Officer, International Development Research Centre,
P.O. Box 62084 00200, Nairobi, Kenya
e-mail: jemimah.njuki@gmail.com

I. Baltenweck • D. Grace • S. MacMillan
International Livestock Research Institute, P.O. Box 30709, Nairobi 00100, Kenya
e-mail: i.baltenweck@cgiar.org; d.grace@cgiar.org; s.macmillan@cgiar.org

A.R. Quisumbing et al. (eds.), *Gender in Agriculture: Closing the Knowledge Gap*, DOI 10.1007/978-94-017-8616-4_9,

current and future assets, (2) sustaining and improving the productivity of agricultural systems in which livestock are important, and (3) facilitating greater participation of the poor in livestock-related markets. While these three pathways are distinct, with each requiring particular strategies and interventions to be successful, they are not mutually exclusive. The chapter summarizes what is known for each pathway and what these pathways imply for programmatic and policy interventions.

Keywords Gender • Livestock • Livelihoods • Assets • Markets

9.1 Introduction

After several years of relative neglect, the importance of livestock for livelihoods and poverty alleviation is once again being recognized. However, there is also an increasing awareness that certain types of livestock systems are associated with nontrivial consequences such as environmental degradation, greenhouse gas emissions, zoonotic and emerging infectious disease, or food-borne illnesses. There is a need to balance these positive and negative aspects as is made clear by the title of the State of Food and Agriculture 2009 report, "Livestock in the Balance" (FAO 2009). Attention to gender will be central to achieving this balance. Livestock are important in women's livelihoods and asset portfolios. Women do much of the work in livestock systems, whether they own the animals or not; women are differentially exposed to health risks associated with animal production and food processing.

Although two-thirds of the world's 600 million poor livestock keepers are rural women (Thornton et al. 2003), knowledge gaps still exist about rural women's roles in livestock keeping and the opportunities livestock-related interventions could offer them. This is in contrast to considerable research on the roles of women in small-scale crop farming, where the importance of women is widely recognized and lessons are emerging about how best to reach and support women through interventions and policies (e.g., Quisumbing and Pandolfelli 2010; Gladwin et al. 2001). In the 2000s, some researchers have begun to provide evidence of relations between gender and livestock production (e.g., Bravo-Baumann 2000; Deshingkar et al. 2008; Herath 2007; Flintan 2008) but, as this review demonstrates, there remains a dearth of quantitative information on this subject, especially for the mixed crop-livestock systems where most livestock and livestock keepers are found and where the major increases in production will have to occur if the global demand for meat, milk, and other animal products in coming decades is to be met (Herrero et al. 2010). Furthermore, the multiple roles livestock play in livelihoods of the poor make generalizing about women's roles in, and economic contributions to, livestock development problematic, and prioritizing livestock research and interventions for women's development challenging (Niamir-Fuller 1994; LID 2004; Rangnekar 1998; Aklilu et al. 2008; Waters-Bayer and Letty 2010).

9.2 Conceptual Framework

This review applies a gender lens to a conceptual framework for understanding the role of livestock in pathways out of poverty (henceforth "livestock pathways out of poverty") developed by the International Livestock Research Institute (ILRI 2002). This framework takes a "livelihoods approach" that centralizes the importance of assets, markets, and other institutions. The framework has been used to explore different aspects of small-scale livestock production and marketing, such as the impacts of livestock and animal diseases on poverty and poverty dynamics (Kristjanson et al. 2004; Perry and Grace 2009). This is the first time the framework has been used to investigate gender issues.

The three hypothesized livestock pathways out of poverty are (1) securing current and future assets, (2) sustaining and improving the productivity of agricultural systems in which livestock are important, and (3) facilitating greater participation of the poor in livestock-related markets. While these three pathways are distinct, with each requiring particular strategies and interventions to be successful, they are not mutually exclusive. In any particular circumstance, one of these pathways may offer more opportunity than the others for reducing poverty, but livestock keepers, researchers, and developers alike must pay attention to all three pathways if they hope to sustain and optimize development of livestock-based enterprises.

Pathway 1—Helping women secure, build, and safeguard their assets. Recent research on livelihoods and poverty dynamics recognizes the importance of assets to the poor (Carter and Barrett 2006), and in particular, livestock assets to the poor (Kristjanson et al. 2007; Little et al. 2008). While poverty is often measured in terms of income or food security, a household's ability to meet its material needs is determined largely by its assets—the physical, human, social, financial, and natural kinds of capital that determine what livelihood strategies a household can pursue and how well it can cope with risks and shocks (Sparr and Moser 2007; also see Meinzen-Dick et al. Chap. 5). Beyond material wealth, assets provide the basis of agency, or the "power to act, to reproduce, challenge or change the rules that govern the control, use, and transformation of resources" (Sen 1997).

As discussed in other parts of this volume, research on intrahousehold dynamics has shown that interventions that increase women's access to, and control over, assets have been shown to improve household food security and child nutrition and education as well as the well-being of women themselves (Quisumbing 2003; Smith et al. 2003; World Bank 2001). An implication of this research is that development interventions designed to reduce poverty should pay attention to how households accumulate—as well as lose—access to assets. Livestock are an important asset for women because it is often easier for many women in developing countries to acquire livestock assets, whether through inheritance, markets, or collective action processes, than it is for them to purchase land or other physical assets or to control other financial assets (Rubin et al. 2010). However, the relative informality of livestock property rights can be a double-edged sword for women when their ownership of animals is challenged. Interventions that increase women's access and rights to livestock, and then safeguard their stock from theft or untimely death, could help women move along a pathway out of poverty.

Pathway 2—Helping women increase and sustain their livestock productivity. Improvements in the productivity of livestock systems can come in the form of increased outputs of milk, meat, eggs, and surplus animal stock, but such improvements may also take the form of reduced environmental degradation (e.g., less pollution of water sources by livestock excrement) or more efficient use of natural resources (e.g., of water used to grow fodder crops) or lower health risks associated with keeping livestock (e.g., brucellosis). While measuring the productivity of small-scale livestock systems is not straightforward, in part due to the multitude of economic and social roles livestock play in livelihoods, it is generally believed that there is considerable scope for improving the productivity of most small-scale livestock systems in the developing world (FAO 2009; Staal et al. 2009).

The three conventional pillars for improving small-scale livestock productivity lie in improving animal feeds, breeds, and health. Other avenues being explored to improve productivity of livestock systems include improving crop-livestock interactions on mixed smallholder farms, livestock water productivity, carbon sequestration on rangelands, and efficiency of farm animal labor. Despite relatively lower investment by the public sector in livestock research, many technologies exist that appear to be appropriate for smallholder systems, yet adoption rates remain low (FAO 2009). Overcoming problems of appropriateness and access to existing technologies and/or developing new ones could have significant benefits in terms of increased productivity—for sale or for home consumption in the form of nutritious animal-source foods—and in reducing negative environmental and health impacts.

Since both productivity and environmental improvements arise from changes in the way people manage (feed, water, treat, herd, care for) livestock, it is important to understand how these decisions are made, and what factors promote or constrain adoption of new, more efficient technologies and practices. Men and women often manage different types of animals and are responsible for different aspects of animal care. Women and men also typically have different objectives for keeping animals, different authorities and responsibilities regarding animal management, and different abilities to access and use new information and improved technologies. These differences may lead them to have different priorities regarding investments in the adoption of new technologies and practices. To have impact, research and development organizations may need to take these differences into consideration in the types of technologies developed, and the manner in which they are refined, disseminated, and supported.

Pathway 3—Helping women participate in and benefit from livestock markets. The increasing global demand for animal products has been dubbed the "livestock revolution" (Delgado et al. 1999). This demand is expected to provide incentive for adoption of productivity-enhancing technologies and practices for those producers who have access to markets for sales of milk, meat, or eggs. This rising demand could also generate increased employment opportunities along the entire livestock value chain. Because livestock market chains are long and complex, in theory they provide myriad opportunities for the poor to participate in, for example, through the provision of livestock inputs and services or the marketing and processing of livestock products.

Women tend to face more challenges than men in accessing and benefiting from markets, especially more formal markets (see Rubin and Manfre Chap. 12 and Hill and Vigneri Chap. 13). In particular, the indirect consequences for women of "gender-neutral" market development projects need to be carefully examined: where women have insecure rights over livestock or limited control over livestock products and income from their sales, they may have difficulty maintaining control, as livestock become more economically attractive to men.

Poor livestock keepers worldwide face a daily trade-off between selling their (relatively expensive) milk, meat, and eggs to increase their household income and consuming the same (high-quality) foods to increase their household nutrition. Because animal-source foods are so dense in nutrients, including micronutrients that help prevent "hidden hunger," decisions in these matters have potentially large implications for the nutritional as well as economic health of households. Given women's traditional responsibility for household food security, their level of control over decisions about whether to sell or consume the family's animal products, as well as over how to use any income obtained from the sale of animal foods, could greatly determine the nutritional well-being of household members (also see Harris Chap. 11).

9.3 Helping Women Build and Safeguard Their Assets

9.3.1 Women's Ownership of Livestock and the Importance of Livestock Assets to Women

Evidence from many different developing countries and covering many different small-scale livestock and agricultural production systems and livestock species reveals that poor women can and do own livestock. A common perception is that women are more likely to own small stock, such as chickens, sheep, and goats, than larger animals, such as cattle, water buffaloes, and camels. While often the case, studies show that the type of species owned by women varies by region and culture and can be dynamic.

In Asia, for example, analysis of a project involving the Grameen Bank, which provided microcredit loans to women (Todd 1998), showed a clear investment trajectory, with the women given credit investing their new capital in poultry keeping and then moving to goats and eventually to milk cows. In India, Heffernan et al. (2003) found that, despite a common perception that only men own bullocks, they were of particular interest among landless women, who rented them to farmers. In pastoral areas of Ethiopia, a study documented women purchasing bulls (Rubin et al. 2010), while in mixed crop-livestock systems, men and women both own cattle, goats, and sheep, although men own more (Yisehak 2008). In pastoral societies, women frequently own fewer animals than men; however, livestock assets are generally more equitably distributed between men and women than are other assets like land (Flintan 2010). In Uganda, Kenya, and Nigeria, most urban cattle farmers are women. And although women and men in East Africa were found to keep similar numbers of cattle, men in Nigeria own more than ten times as many cattle as women do (Randolph et al. 2007).

In Iraqi Kurdistan, 70 % of both female- and male-headed households own livestock, with female-headed households, on average, owning twice as many animals as male-headed households (Waite 2000). The value of livestock in the female-headed households is also considerably greater than that of livestock in the male-headed households. In this society, where women do not engage in paid labor or other alternative income-earning activities, the care of livestock has traditionally been regarded as a "female activity". In Ethiopia, on the other hand, a study in the Western Shoa region found that women in female-headed households own fewer livestock than men and than women in male-headed households (Torkelsson and Tassew 2008).

Men and women are also likely to differ in the types of breeds they own within a given species, with men more likely to have improved animals than women in dairy areas of Kenya (EADD 2008). While a higher percentage of female-headed households than male-headed households own local cattle, the reverse was observed for (higher-yielding, genetically improved) exotic cattle, with 63 % of male-headed households owning exotic cattle compared to 49 % of female-headed households. These results are consistent with those from Rwanda, where 45 % of male-headed households owned exotic cattle compared to 32 % of female-headed households (EADD 2008). Results from the same study show that in Rwanda and Uganda, female-headed households also owned significantly fewer local cattle (at an average of 4.2 and 5.1 head per household, respectively) than did male-headed households (7.8 and 12.6).

Men and women may also differ in the types of rights they have to livestock. Rights can be divided into user rights, including resource access, rights to withdraw products, rights to exploit commercially, and decisionmaking rights, such as management, exclusion, or alienation (Meinzen-Dick et al. 2004). For example, in many cases, women control cattle milk when it is used for home consumption; however, they cannot sell it and keep the income (Valdivia 2001). Guèye (2000), in a review of backyard poultry in Africa, states that women generally own and care for poultry; however, they can seldom take sole decision over the use of the birds or eggs (consumption, selling, exchange, etc.). McPeak and Doss (2006) found that, among mobile pastoralists in northern Kenya, women had the right to sell milk; however, men were responsible for the overall herd and had the right to decide where the household would camp. If women's marketing objectives conflicted with men's herd management objectives, men used location to limit women's ability to market. In some societies, women may "own" some animals (e.g., having brought them into the family upon marriage or later through inheritance) but have little say about selling or slaughtering them (e.g., among the Massai) (Talle 1988). Yet in other societies, e.g., among the Nandi (Oboler 1996), the women may have a say in sales decisions, even though they do not "own" the animals. Flintan (2008) observed that in some pastoral societies, men cannot sell without approval of women and sometimes also children.

Heffernan et al. (2003) found sharp differences between the sexes in their perceptions of the roles of livestock in Kenya, where women viewed livestock primarily as a means of ensuring food security for the family, while men perceived livestock as a means of making longer-term investments. Rubin et al. (2010) found that livestock are the preferred investment for both men and women in microcredit schemes.

9.3.1.1 How Do Women Acquire Livestock?

Women acquire animals as gifts, they inherit them from family members, they receive them from development projects, and they buy them in markets. The literature indicates that women are more likely than men to acquire livestock through nonmarket rather than market channels; however, this is not always the case. For example, Rao et al. (2002) found in their study in India, that most landless women purchased milking cows out of their own savings coupled with the earnings of their husbands, or depended on moneylenders, or (in the case of Pondicherry) purchased cows through loans taken from the government or private agencies. Only very few (3 out of 57) had obtained the animals as a "family gift" (i.e., a nonmarket channel).

A recent study in Bolivia, India, and Kenya found that, when the data from Kenya were disaggregated by production system and agroecological zones, only a few female-headed but most male-headed pastoral households purchased their animals (Heffernan et al. 2003). Heffernan et al. concluded that women in Kenya appear more able than men to access informal networks to obtain livestock. In India, on the other hand, women had few informal or formal mechanisms for acquiring livestock, whereas men had both.

Zambian women said that they could not buy livestock because income from both livestock and crop agriculture, including their vegetable plots, was controlled by men (Chawatama et al. 2005). This concurs with more widespread evidence of the importance to developing-country women of informal mechanisms for obtaining livestock assets. It also suggests that the reason that these women do not buy more animals in the market is not that they cannot access markets, but rather that they have no cash with which to purchase animals. Removing this constraint, for example through microcredit, can result in more women buying livestock (Todd 1998; Rubin et al. 2010).

Livestock have been freely provided to women by organizations such as Heifer International, FARM-Africa, and Land O'Lakes for many years in many countries in Africa and Asia. In Bangladesh, the Self-Employed Women's Association and Rural Advancement Committees are examples of local initiatives that organize women in the informal economy and facilitate their access to productive resources such as livestock as well as critical services such as health, housing, and childcare. Past livestock development—especially emergency aid initiatives such as restocking—overlooked gendered access issues and, as a result, did not benefit or even had a negative effect on women. Over time, these failures have been documented and lessons identified (Niamir-Fuller 1994; Heffernan et al. 2004), although they are still not always incorporated into programming.

While few rigorous evaluations have been conducted on the impacts of these more recent initiatives, anecdotal evidence of the benefits to women is positive and the projects have done much to bring attention to the value developing-country women place on livestock, and on building assets. For example, in 1998, Heifer International established a Women in Livestock Development (WiLD) initiative to help women use livestock to care for themselves, their families, their environments, and each other. A project is designated "WiLD" if 70 % or more of the participants in the project, including its leaders and decision makers, are women. WiLD projects provide women with cows, goats, water buffaloes, poultry, and other farm animals (Heifer International 2008).

9.3.1.2 Threats to Livestock Assets

Livestock kept by poor people in poor countries face many threats. The animals are typically raised in harsh environments where drought and theft are common, and commercial feed and veterinary services are beyond the means of most people. Women's tolerance for risk may be different from men's, either because they are inherently more risk averse (Rubin et al. 2010) or because they face more difficulties in rebuilding livestock assets that are lost. Preliminary results from an ongoing pilot project on index-based livestock insurance in northern Kenya found that women were more likely than men to purchase insurance for their animals (Mude 2010, personal communication).

Understanding risk preferences and ability to cope with shocks is likely to become more important in the face of increasing climate variability. Turner (1999) found that repeated droughts in Niger strengthened women's control over livestock because they were able to invoke a cultural norm that made men responsible for household food security, with the result that men had to sell their livestock before women did. This led to a change in regional herd composition and an increase in women's relative control. However, another study found that many women in the Sahel felt that they would lose traditional access to resources if competition for rangeland and other livestock resources increased due to increasing climatic vagaries (IFAD 2005).

Another way women lose access to livestock assets is through the dissolution of households, either through divorce or death of a spouse (Mutenje et al. 2008). In such situations, cultural norms often dictate that animals are transferred to other family members (Engh et al. 2000; Goe and Mack 2005; Kanyamurwa and Ampek 2007). While some developing countries have enacted legislation to protect women from loss of property upon the death of their husbands, these laws are rarely implemented, and most widows do not possess the resources to employ legal experts to help them protect their property. While legislation exists to prevent property/asset grabbing in many areas of northern Namibia, for example, it is still common practice for a husband's family to take livestock and other resources from a widow and/or remaining children upon the husband's death (Engh et al. 2000).

Many NGO projects are now addressing this issue. In Zambia, a Heifer International project, through training people, raising awareness, and conducting negotiations at different levels, has enabled women not only to co-own livestock with their husbands, but also to continue their ownership of the animals after their husbands die, animals that otherwise would have been taken away from them by their husbands' relatives. In Thailand, where a high incidence of HIV/AIDS has led to the disintegration of many families, women heads of households are being provided with water buffaloes and training in their management to help the women not only to bring in the rice harvest, but also to generate cash incomes through renting out these valuable animals to other community members (Heifer International 2008).

9.4 Helping Women Increase and Sustain the Productivity of Livestock

Given the general lack of data on productivity of smallholder livestock systems, it is perhaps not surprising that we found few studies comparing productivity of livestock kept by men versus women. A study of an intensive dairy intervention in coastal Kenya found no significant differences in this respect; in fact, female-owned and operated farms performed better than male-owned and operated ones (Mullins et al. 1996). Studies from cropping systems show that, controlling for access to resource such as land and credit, productivity levels are similar between men and women (Alene et al. 2008; Njuki 2001; Smale and Heisey 1994). Perhaps the key issue is not whether the current low levels differ between men and women, but whether the opportunities and constraints to improving productivity differ between men and women, as such improvements are critical to realizing this pathway out of poverty.

The relationship between gender and livestock productivity is not straightforward. Poor men and women keep animals for multiple purposes, both productive (food security, income) and nonproductive (savings, insurance, culture). To the extent that nonproductive reasons predominate and productivity does not correlate highly with asset or cultural value, improving productivity may not be a priority.

Many interventions aimed at intensifying livestock production, such as shifting from grazing to stall-feeding or by keeping potentially higher-yielding but also more demanding breeds, increase the workload of women and girls, because the intensification lies in their traditional tasks (Okali and Sumberg 1985; Mullins et al. 1996; Wangui 2008). To the extent that improvements in productivity require additional labor from women that is not compensated, they may have less incentive to apply the new technology or practice. Helping women contribute to and benefit from improvements in livestock productivity requires careful attention not only to the size but also to the distribution of the costs and benefits associated with improved productivity, especially who benefits from improvements and who makes the investments (financial and time) in generating them.

9.4.1 Role of Women in Livestock Keeping

While there is great variability across systems and socioeconomic contexts, women generally play a major role in managing and caring for animals, even when they are not the owners. Flintan (2008) documents participation of women in every aspect of livestock management in different pastoral systems around the world. In intensive Asian livestock systems, more than three-quarters of livestock-related tasks are the responsibility of women (Niamer-Fuller 1994). Fully 90 % of Nepalese women are engaged in agricultural production (compared to 75 % of men) (Herath 2007).

Indian women play a significant role in livestock-keeping by providing labor; in poorer families, their contribution typically exceeds that of men (George and Nair 1990). In India's tribal, low-rainfall and semi-arid areas, much of the work of managing animals has been transferred to women because the men have left to find jobs elsewhere (a similar phenomena is seen in most of Africa).

In Sub-Saharan Africa, women's roles in crop and livestock production are strongly determined by gender and cultural norms. In Nigeria, Ayoade et al. (2009) report that women feed and manage vulnerable animals (calves, small ruminants, and sick, injured, and pregnant animals), clean barns, milk cows, and make butter and cheese, but are not involved in livestock marketing or managing livestock diseases. These trends are similar to what was found in the Ethiopian highlands, where women clean cowsheds; milk cows; look after calves and sick animals; cut the grass and supervise the feeding and grazing of cows; make dung cakes, butter, and cheese; and sell these products once or twice a week. Men, on the other hand, feed the oxen and take the animals for veterinary treatment when the need arises (Yisehak 2008). Njuki et al. (2004), in a study in central and eastern Kenya, found women were more engaged in feeding of cattle, while men were more involved in watering and disease management. The total time allocation to dairy-related work did not, however, differ significantly between men and women.

9.4.2 Women's Constraints on Technology Adoption

In spite of the central roles they play in small-scale livestock systems, women are severely limited in their ability to make decisions regarding livestock enterprises. In addition, they receive little outside support to help them make better decisions about those enterprises. The agricultural service and input-delivery systems are dominated by men and therefore difficult for women to access (Upadhyay 2005).

Access to land. Although land is not a prerequisite for keeping livestock (if feed can be purchased), grazing lands are key to livestock production in many areas and many traditionally communal grazing areas are being privatized. In agro-pastoral systems in Peru and Bolivia, taking animals to graze is the task mostly of female heads of households. Guillet (1992) documented the benefits of growing alfalfa in fallow fields, a feed resource gaining in importance in the Altiplano region.

The shift to alfalfa has reduced the fallow fields women may use for grazing their sheep or *criollo* cattle, especially those women in poorer households without access to land with appropriate soils for growing alfalfa (Valdivia 2001).

Group ranch, or "block grazing," systems have been tried in various countries, including Kenya, Nigeria, Sierra Leone, Somalia, and Tanzania. In almost all cases, the planners failed to understand the fundamental importance of pastoral reciprocity and alliances in maintaining viable livestock production in ecologically fragile and climatically variable areas. In many regions, this failure led to range wars and a rush for privatization and expropriation of rangelands (Oxby 1987; Mwangi 2005), with particularly negative impacts on women, most of whom were not allowed to join the group ranches, and instead becoming unpaid workers taking care of their husbands' livestock (Talle 1988; Kipuri 1989). With an increasing exodus of men from pastoral to urban areas in search of jobs, the women left behind could not influence decision-making and governance within the group ranches on such important matters as land use and ownership (Mwangi 2005).

Access to extension services, information, and training. A study in the Taurus Mountain villages in Turkey found that most women farmers had little access to information about animal production through public extension services (Budaka et al. 2005). Similar findings have been documented in Cameroon, Ghana, and Madagascar (Salman et al. 1999), in Pakistan (Teufel et al. 1998) and in The Gambia (Jaitner et al. 2001). The reasons given for this lack of access by women to livestock-related extension services included women's long workdays, which precluded them from engaging with, or searching out, extension officers; a neglect of women's needs and circumstances when targeting extension work; and widespread female illiteracy.

Among Maasai pastoralists in southern Kenya, women's access to extension services was restricted by cultural as well as time constraints, with women typically relying on their husbands for information, although delivering extension messages through women's groups was found to be effective (Kimani and Ngethe 2007). Zimbabwean women complained that cattle are generally registered in their husband's names with the Department of Veterinary Services (for the purposes of dipping the animals in acaricides to prevent tick infestation), which serves to exclude the women from livestock initiatives (Chawatama et al. 2005).

Some countries have succeeded in increasing women's access to livestock information and services. For instance, Indian women dairy farmers have been credited with raising the country's milk production levels to among the highest in the world (Herath 2007). Women constitute 93 % of total employment in India's dairy production (World Bank 2001). Starting some two decades ago, India recognized the importance of women in dairying and encouraged their growing participation in the country's large dairy sector. Many dairy cooperative societies were formed across the country, including some specifically for, and run by, women in the states of Andhra Pradesh and Bihar (World Bank 2001).

India has also recruited and trained women extension workers, who are playing crucial roles in disseminating information and technologies. Since the late 1980s, the country's National Dairy Development Board has made women's extension training central to their cooperative development program, which was designed to strengthen the role of women in the control and governance of dairy cooperatives. By1998, 6,000 out of 7,000 dairy cooperative societies in India were women's societies (Patel 1998). Subsequent projects, such as the Women's Dairy Cooperative Leadership Program, have helped Indian women continue to gain more control over the sale of milk and the use of income from it.

Some of the governance-related lessons from India have been applied in other countries such as Tanzania, where the formation of district and regional networks of a self-help initiative proved an ideal platform for linking women dairy producers with the Tanzania Milk Producers' Association and the Tanzania Dairy Board.

In The Gambia, where the proportion of female agricultural extension workers has increased from 5 % in 1989 to more than 60 % today, more attention is being paid to women's livestock information needs and desire for female-led training, especially regarding small ruminant and poultry production (FAO 2003). Similarly, Due et al. (1997) found that in Tanzania, 40 % of women farmers preferred to work with female extension agents and 51 % of the women interviewed mentioned that they wanted to receive information on small ruminant production. Almost all the women (94 %) pointed out that they could attend demonstrations and training courses only if these were conducted in their villages.

Roy and Rangnekar (2007) concluded that participatory and systems approaches applied to development of rural dairy business systems in Andhra Pradesh were particularly useful in understanding the perceptions of women producers, the constraints they faced, and the kinds of training most appropriate for them. An assessment of the impacts of a livestock training course in Kotli, India (Hussain et al. 2004) found that all the women who had received gender-sensitive training thereafter used their new knowledge, particularly regarding vaccination of animals.

Access to animal health services. A promising new trend benefiting women is the linking of public health and veterinary services. While traditionally working independently, the medical and veterinary sectors have recently come together to tackle zoonoses—diseases transmissible between animals and humans, particularly emerging zoonotic diseases such as highly pathogenic avian influenza. In Mongolia, researchers demonstrated that a proposed vaccination effort against brucellosis in livestock was profitable and cost-effective for both livestock and public health sectors (Roth et al. 2003).

Human and livestock health services often fail to serve the poorest livestock keepers, particularly in remote rural settings in Africa and Central Asia, because of financial, logistic, and service-delivery constraints (Heffernan and Misturelli 2008). However, in Chad, between 2000 and 2005, Schelling et al. (2007) demonstrated the feasibility of combining human and animal vaccination programs for nomadic pastoralists and their livestock. By sharing transport and equipment costs, medical doctors and veterinarians reduced their total costs. Joint delivery of human and animal health services is highly valued by hard-to-reach pastoralists. In intervention zones, for the first time, about 10 % of nomadic children were fully immunized annually and more children and women were vaccinated daily in joint human-livestock vaccination rounds than in vaccination campaigns targeting only people.

The literature also points to a need to strengthen institutional links among agricultural research, agricultural extension, and veterinary services. An example of how this can be useful are routine vaccination systems for small ruminants established by agricultural extension services collaborating with veterinary services (Haenlein and Abdellatif 2004; Devendra and Chantalakhana 2002).

Strong producer organizations can also play an important role in efficiently delivering veterinary services to poor livestock-keepers. For example, the Kenya Women's Veterinary Association has partnered with the government to develop the country's semi-arid and arid areas through improvements in livestock-keeping. By building capacity in livestock and disease management skills in local communities, the association has helped improve control of zoonoses and reduce the incidence and costs of tick-borne diseases in cattle and Newcastle disease in poultry. An impact study in Kenya (Kimani and Ngethe 2007) reports that the formation of women's groups has helped improve control of livestock diseases, particularly transboundary diseases.

Several projects in East Africa are experimenting with training villagers, some women, to be animal health workers (also known as paravets, community animal health workers, or community animal first-aid workers) (Allport et al. 2005; Msoffe et al. 2010). An evaluation of the projects attributed their success to the participatory nature of their activities and to their ability to train independent local workers, who were effectively monitored and supported by government services (for medicines, vaccination campaigns, and referrals on serious cases). The evaluation also concluded that women were more heavily involved in the management of ruminants than was previously thought and that, consequently, their participation in the training program should be increased.

In many parts of the world, however, sociocultural barriers continue to hinder women's access to animal health services at the community level. For example, a CARE-led community animal health initiative found that women were generally not allowed to take part in training courses, although the women spent more time than did the men with the animals and were thus in a better position to recognize animal health problems earlier (Rivière-Cinnamond 2005).

Access to credit. Men in developing countries generally have greater and easier access to credit than do women, whose lack of collateral makes them appear not creditworthy (see Fletschner and Kenney, Chap. 8). Women dairy operators in Kenya, for example, typically lack secure titles to property, which prevents them from obtaining credit from formal financial institutions. A survey in Kenya, Rwanda, and Uganda showed that significantly more men than women had applied for loans from financial institutions or local cooperatives (EADD 2008).

In many countries, however, women have developed their own small credit systems. Credit funds and revolving savings women's groups are common throughout Africa. Members of a group each save a certain amount of money monthly, which is then granted in turn to each of the women as a loan, normally at no interest. Most of these loans go toward non-income-earning activities (Place et al. 2004), although some groups allow loans of animals or milk for processing. These systems tend to function best at the village or neighborhood level, where tight social connections ensure that loans are repaid.

Women livestock-keepers have worked together to overcome credit constraints, as in India and Uganda, where they established group bank accounts so the women could access their dairy payments. In another case, a Danish-financed smallholder poultry development project demonstrated the important role that women's groups play in accessing credit in West Africa and Asia. This project took a holistic approach involving capacity building, organization of women into groups, and farmer field schools aimed at giving poor illiterate women farmers and local food vendors the knowledge required to benefit from collective action (Riise et al. 2008).

9.5 Helping Women Participate in and Benefit from Livestock Markets

The actors in livestock value chains include not only livestock producers but also input suppliers, traders, processors, wholesalers, and retailers. Helping women gain access to labor, product, and service markets all along the value chain, and improving their working conditions, can enhance their benefits from participation in livestock markets. While women may play many of these roles along the value chain in many regions, the literature mainly cites their roles as suppliers of livestock products, particularly milk products, and as processors of animal source foods, often street foods.

Women as suppliers of livestock products. Among the settled Fulani in Nigeria, women are responsible for all milk processing and marketing and for deciding on the quantity of milk to be sold or consumed by the family. Marketing is seen as both an economic and social activity. The revenue the women generated from their dairy products contributed substantially to their household incomes (Waters-Bayer 1985).

Among the Fulani societies in Ferlo, Senegal, milk production is entirely controlled by women, who also have sole control over the sale of any surplus (Dieye et al. 2005). There are also mini-dairies run by women who source their milk through contract farmers (Corniaux 2003). These small processors or pasteurizers generally operate with the support of NGOs or development agencies.

A study of evolving pastoral markets in northeastern Somalia (Nori 2008) documents the crucial role that women play in the commoditization of pastoral camel milk. When pastoral women can sell milk, it enhances local food security (Dietz et al. 2001). Market exchanges and related terms of trade are of particular importance during the dry season, when food production does not always suffice to satisfy the energy requirements of pastoral households. This is supported by other case studies, such as that in the Ogaden in Ethiopia's Somali Region—a traditionally food-insecure area—which shows that women's participation in the sale of livestock milk products generates more than 80 % of the income needed to satisfy basic needs among pastoral households during the dry season (while it contributes about 40 % during the rainy season, when milk is in surplus) (Nori et al. 2006).

Physical, structural, and informational or organizational aspects of market access (Niamir-Fuller 1994) significantly affect women's ability to enter, engage in, and profit from livestock markets. Distances from villages to markets throughout Africa are often long, and milk is heavy to transport, particularly for women, who typically do not ride bicycles. The problem of long distances to markets is aggravated by structural problems—particularly inadequate roads and inefficient transport systems. Finally, lack of information can hinder women's access to and benefits from livestock marketing. In the Mandera triangle, at the intersection of the borders of Kenya, Ethiopia, and Somalia, Wabekbon Development Consultants (2009) cites lack of education and lack of access to accurate information and infrastructure as the most critical factors hindering women from selling milk and small ruminants.

In northern Kenya, Coppock et al. (2006) note that self-initiated groups convened and managed by women have managed to access livestock markets; they recommend that development initiatives facilitate more direct access by women to small local livestock markets or to cooperatives that could broker their livestock transactions so as to give women more control over the income generated.

Women as processors and retailers. Animal-source foods are among the most common street foods in most countries and often are derived from animals kept in cities (FAO/WHO 2005). In most African countries, most street-food processors and vendors are women (Canet and N'Diaye 1996). As much as 60 % of the milk sold in Dar es Salaam, Tanzania, is produced in and around the city (Canet and N'Diaye 1996).

In most cities in Pakistan, women provide the dairy needs from their urban and peri-urban plots. Similarly high levels of urban and peri-urban milk production are cited for Nairobi, Kenya, and Addis Ababa, Ethiopia. In South Africa, street food is probably the single largest informal sector employer (von Holy and Makhoane 2006). In Harare, around 9,000 people (81 % women) are involved in making and selling street food (Graffham et al. 2005).

A major concern about urban agriculture and informally marketed food is public health (Moy et al. 1997). The pathogens found in street food include *Escherichia coli*, *Staphylococcus aureus*, *Salmonella spp.*, and *Bacillus cereus*. Animal-source foods are the most common cause of diseases in urban areas. For example, in Zimbabwe, cooked meats posed the greatest health risk of all food sold on the street (Randolph et al. 2007). Zoonotic diseases, including most food-borne diseases, are both important and neglected in most developing countries (WHO 2006).

Authorities in many African countries have responded to this problem with weak and erratic implementation of legislation on street food and urban agriculture (Bryld 2003). As formal and informal standards grow, there is a real risk that the poor will be excluded from markets (Perry et al. 2005). Whereas food-safety/quality initiatives that have attempted to eliminate urban agriculture and informal food markets have been viewed as gender insensitive (Nduna 1990), the literature also provides examples of food-safety regulations that benefit women livestock keepers.

Impacts of commercialization of milk on women. Studies conducted among the Fulani in Nigeria (Waters-Bayer 1985, 1988) demonstrated how the commercialization of milk has eroded women's traditional control over milk products, thereby decreasing their power within the household. The men are most interested in ensuring that enough of the daily milk produced by the household cows is left for the suckling calves that the men are raising for the beef market. The women, who fully control the dairy earnings, are more interested in selling as much milk and dairy products as they can to obtain cash. Thus a change in the division of labor, with men taking over the milking role, reduced women's access to milk and thus to dairy income, diminishing their ability to control the welfare of their households.

Evidence from East Africa shows that where and which milk is sold can determine whether women manage the milk income or not. Women have greater control over the evening milk than the morning milk and manage more income from milk sold at local markets and to neighbors and mobile traders than they do from milk sold to collection centers or chilling plants (EADD 2008). A survey of dairy households in Kenya, Rwanda, and Uganda showed that women received dairy income in 35 % of the households that sold milk to individual traders but in only 16 % of households that sold milk to collection centers (EADD 2008). Formalizing milk markets through member-based collection centers and cooperatives can thus lead to women losing their income from milk. Njuki et al. (2011) found that the higher the income from livestock or livestock products, the less likely women were to manage the income.

In a review of literature on the impact of commercialization on the role of labor in African pastoral societies, Sikana and Kerven (1991) note that, where live animal marketing has come to dominate, women's labor in pastoral production is devalued, since dairying is no longer emphasized. Likewise, where marketing has led pastoralists to shift from large to small stock (which can have a higher market value), women's role in managing small stock is diminished.

Nevertheless, it may be too simplistic to conclude that commercialization only erodes women's power. Where a strong market value for milk and/or dairy products is established, women's roles in dairying may be enhanced and their labor refocused on marketing rather than production. This latter effect is described by Micheal (1987, cited in Sikana and Kerven 1991) for Baggara pastoralists of Kordofan, Sudan, where over the previous 30 years there had been a growth of seasonal cheese factories dependent on purchasing milk from Baggara women. These factories are the main suppliers of cheese to Sudan's urban areas, while cash income from milk sales is estimated to comprise about a quarter of pastoral family income for the Baggara. Micheal notes that, although it is men who sell cattle and men recognize that milk is important for herd growth, the increasing urban demand for milk and milk products has meant that women's traditional role in controlling milk output from the herd has evolved into their control of milk marketing (see, also, Nori (2008) on milk marketing by Somali women).

9.5.1 *What Kinds of Livestock Interventions Increase Women's Market Participation and Benefits?*

There appears to be more awareness of the importance of gender in market-related livestock projects than in projects focused on raising livestock productivity (although whether this awareness translates into effective livestock marketing strategies for women is unclear). A Heifer International report on activities in East Africa found that women provided more labor in dairy enterprises than did men, but the level of women's control of the dairy income did not usually match their contribution, and this despite Heifer's finding of a strong correlation between women's control of dairy income and the productivity and success of dairy projects.

Women's groups initiated by development projects are widely used to support women pursuing urban agriculture; these groups provide women with microcredit schemes and other forms of support for their dairying, poultry production, livestock marketing, and food transformation and sale (Niamir-Fuller 1994; de Haan 2001). Joining such groups may be the only way for many poor women to obtain sufficient resources to start up and profitably operate a livestock-related enterprise. Membership in such groups enables women to more effectively lobby government departments and other decisionmaking agencies affecting their livelihoods. Although the performance of such women's groups has been reported as variable, group membership gives many developing-country women the freedom to participate in livestock development activities, enabling them to protect their interests, to overcome legal hurdles facing them, and to access the training and equipment they need to increase their production and sale of safe livestock foods.

In Bangladesh, the Bangladesh Rural Advancement Committee (BRAC) poultry model is an interesting example of a market-oriented intervention because, in order to achieve its goal of increasing income and nutrition of poorest women through poultry production, the model also supports a range of supply (parent stock, feeds, vaccines) and service (training, credit, extension) activities and involves women in all these areas (Dolberg 2001). Women who provide supplies receive support so they can continue to do so, on a commercial basis, to program and nonprogram participants. By 1999, BRAC was reaching more than 1.4 million women with this model, and it has been scaled out to other NGOs and to several African countries.

9.6 Summary and Conclusions

This chapter reviewed the evidence for three main livestock-related pathways out of poverty for women—securing, building, and safeguarding assets; increasing and sustaining livestock productivity; and enhancing participation in and benefits from livestock markets.

Securing livestock assets: With respect to assets, while there is widespread evidence of women owning livestock, their circumstances and the kinds of livestock they keep vary considerably by region, culture, and even by household. Women's ownership can also change over time. The implication is that while it is important to be cognizant of existing ownership norms and patterns in the design of interventions, these should not be taken as given. They can change to the benefit or detriment to women, depending on how interventions are designed and implemented.

In many cases, women's ownership of stock does not correlate with their control over use of products or decisionmaking regarding management or sale. Although some women buy livestock in markets, many obtain animals through inheritance, gifts, and other informal mechanisms. The relatively informal means by which most developing-country women acquire livestock may help explain the limited rights women have over animals, if more informal means of acquisition are seen as conferring fewer rights to control than outright purchase. Interventions that secure women's rights to livestock—their own or those of their households in the event of dissolution—could be of great benefit to women. Other threats to livestock assets owned by women include their lack of access to complementary assets and to services for livestock health, production and marketing, and increased commercialization, particularly of milk and dairy enterprises. Reducing these threats will help make securing livestock assets a viable pathway out of poverty for women.

The review found relatively little information on the relative importance of livestock in women's current asset portfolios or on their preferences for livestock versus other assets. While animals are often among the few assets many developing-country women can own, the relative insecurity of their rights to these animals, coupled with the greater responsibility they may have for livestock-related tasks, could make them less desirable than other physical or financial assets. Addressing this gap should be a priority for research.

Increasing livestock productivity: When it comes to helping women increase the productivity of the livestock enterprises of themselves and their households, it is important to recognize the key roles women play in these enterprises. Women may have different production objectives than men. Interventions focused on areas for which women have responsibility (e.g., milking, tending young stock, poultry feeding) need to be targeted to women if they are to have an impact on how animals are managed, whether or not women are the "owners" of the animals in question. This implies that women need to be more involved in technology design and testing, and in dissemination processes.

Little information is available on the relative productivity of livestock enterprises managed by women versus men, although quite a lot is known about the constraints women face to accessing information, training, and improved technologies. Livestock-keeping women are disadvantaged by their lack of access to complementary assets, such as land for growing forages, and to livestock production inputs and services that could enhance their productivity. Greater access to livestock extension services seems to be especially important for women, with some examples of promising approaches targeting women being tried.

Through their close proximity to animals and their handling of animal products, women are in many cases differentially exposed to zoonotic diseases and other livestock-related health concerns. Addressing these issues could improve the productivity of livestock systems and improve the well-being of women and their families. Relatively little information is available on the relationships between gender and the negative environmental impacts of livestock production. Women's responsibility for gathering feed may contribute to degradation of forests and watersheds. At the same time, women are also likely to suffer the impacts of degradation; for example, when contamination by livestock of a water source requires them to get water from more costly (in time or money) sources. Addressing this gap will be important in order to reduce the environmental footprint of livestock in ways that help rather than hurt women.

Enhancing participation in livestock markets: The scarce literature that exists on women and livestock markets indicates that developing-country women participate in livestock value chains mainly as suppliers of dairy products and as producers and sellers of processed animal-source foods in informal markets. Although increasing the participation of women in livestock markets and value chains clearly has the potential to improve welfare, the increasing commercialization of livestock markets presents women with risks as well as rewards. The literature cites many cases where women's control over livestock enterprises and incomes is diminished rather than maintained or enhanced with increasing commercialization. Women stand to benefit substantially from improvements in food safety, especially in informal markets, but are often inadvertently hurt by the unintended consequences of inappropriate policies and regulations. The conditions leading to these different outcomes need to be much better understood. While market-oriented livestock projects, perhaps more than productivity-focused projects, are increasingly recognizing the need to pay attention to gender, the challenge remains to identify strategies that help women enter into and benefit more from livestock markets.

References

Aklilu HA, Almekinders CJM, Udo HMJ, Van der Zijpp AJ (2008) Village poultry consumption and marketing in relation to gender, religious festivals, and market access. Trop Anim Health Prod 39(3):165–177

Alene AD, Manyong VM, Omanya G, Mignouma HD, Bokanga M, Odhiambo G (2008) Smallholder market participation under transactions costs: maize supply and fertilizer demand in Kenya. Food Policy 33(4):318–328

Allport R, Mosha R, Bahari M, Swai E, Catley A (2005) The use of community-based animal health workers to strengthen disease surveillance systems in Tanzania. Office International des Epizooties Revue Scientifique et Technique 24(3):921–932

Ayoade JA, Ibrahim HI, Ibrahim HY (2009) Analysis of women involvement in livestock production in Lafia Area of Nasarawa State, Nigeria. Livest Res Rural Dev 21(220). http://www.lrrd.org/lrrd21/12/ayoa21220.htm

Bravo-Baumann H (2000) Gender and livestock: capitalisation of experiences on livestock projects and gender. Working document. Swiss Agency for Development and Cooperation, Bern. Available at: http://www.fao.org/wairdocs/LEAD/X6106E/x6106e00.HTM

Bryld E (2003) Potentials, problems, and policy implications for urban agriculture in developing countries. Agric Hum Values 20(1):79–86

Budaka DB, Darca N, Kantar M (2005) Women farmers and extension services in small ruminant production in mountain areas of Turkey. J Arid Environ 62(3):507–515

Canet C, N'Diaye C (1996) Street food in Africa. FAO/Food, Nutrition and Agriculture No. 17/18. http://www.fao.org/docrep/W3699T/W3699T00.htm

Carter MR, Barrett CB (2006) The economics of poverty traps and persistent poverty: an asset-based approach. J Dev Stud 42(2):178–199

Chawatama S, Mutisi C, Mupawaenda AC (2005) The socioeconomic status of smallholder livestock production in Zimbabwe: a diagnostic study. Livest Res Rural Dev 17(12). http://www.lrrd.org/lrrd17/12/chaw17143.htm

Coppock DL, Desta S, Wako A, Aden I, Gebru G, Tezera S, Tadecha C (2006) Collective action by women's groups to combat drought and poverty in Northern Kenya. Pastoral risk management project research brief 06–01. Global Livestock Collaborative Research Support Program. University of California–Davis, Davis, CA, USA. www.glcrsp.ucdavis.edu

Corniaux C (2003) La filière lait et produits laitiers dans la région de Saint-Louis. ISRA/CIRAD-PSI, Saint-Louis, Sénégal

de Haan N (2001) Of goats and groups: a study on social capital in development projects. Agric Hum Values 18(1):27–39

Delgado C, Rosegrant M, Steinfeld H, Ehui S, Courbois C (1999) Livestock to 2020: The next food revolution. Food, agriculture, and the environment discussion paper 28. International Food Policy Research Institute, Washington, DC

Deshingkar P, Farrington J, Rao L, Akter S, Sharma P, Freeman A, Reddy J (2008) Livestock and poverty reduction in India: findings from the ODI Livelihood Options Project. Discussion paper 8. International Livestock Research Institute, Nairobi (Kenya). http://mahider.ilri.org/bitstream/10568/281/1/LivesPovertReduc_DiscPaper8.pdf

Devendra C, Chantalakhana C (2002) Animals, poor people, and food insecurity: opportunities for improved livelihoods through efficient natural resource management. Outlook Agric 31(3):161–175

Dietz T, Abdirizak AN, Adano WR, Zaal F (2001) Pastoral commercialisation: on caloric terms of trade and related issues. In: Mohamed Salih M, Dietz T, Ahmed AGM (eds) African pastoralism, conflict, institutions, and government. Pluto Press, Amsterdam, pp 194–234, in association with Organization for Social Science Research in Eastern and Southern Africa, Amsterdam Research Institute for Global Issues and Development Studies (AGIDS), University of Amsterdam

Dieye PN, Ly C, Sane FCN (2005) Etude des services d'élevage dans la filière laitière au Sénégal. Pro-Poor Livestock Policy Initiative (PPLPI) internal report. Food and Agriculture Organization of the United Nations, Rome

Dolberg F (2001) A livestock development approach that contributes to poverty alleviation and widespread improvement of nutrition among the poor. Paper presented at the workshop on malnutrition in developing countries: generating capabilities for effective community action, international fund for agricultural development, 19–20 September. http://www.lrrd.org/lrrd13/5/dolb135.htm

Due JM, Magayane F, Temu AA (1997) Gender again: views of female agricultural extension officers by smallholder farmers in Tanzania. World Dev 25(5):713–725

EADD (East Africa Dairy Development) (2008) East Africa dairy development project Baseline report no 6. Gender, dairy production and marketing. EADD/International Livestock Research Institute, Nairobi

Engh I, Stloukal L, du Guerny J (2000) HIV/AIDS in Namibia: the impact on the livestock sector. Food and Agriculture Organization of the United Nations, Rome. http://www.fao.org/sd/wpdirect/Wpan0046.htm

FAO (Food and Agriculture Organization of the United Nations) (2003) Gender and development plan of action. Available at http://www.fao.org/docrep/005/y3969e/y3969e05.htm

FAO (Food and Agriculture Organization of the United Nations) (2009) Livestock in the balance: the state of food and agriculture. FAO (Food and Agriculture Organization of the United Nations), Rome

FAO/WHO (Food and Agriculture Organization of the United Nations/World Health Organization) (2005) Informal food distribution sector in Africa (street foods): importance and challenges. FAO/WHO Regional Conference on Food Safety for Africa, Harare, Zimbabwe, 3–6 October

Flintan F (2008) Women's empowerment in pastoral societies. World Initiative for Sustainable Pastoralism, Global Environment Facility, International Union for Conservation of Nature, United Nations Development Programme, Nairobi

Flintan F (2010) Sitting at the table: securing benefits for pastoral women from land tenure reform in Ethiopia. J East Afr Stud 4(1):153–178

George PS, Nair KN (1990) Livestock economy of Kerala. Centre for Development Studies, Trivandrum, India

Gladwin CH, Thomson AM, Peterson JS, Anderson AS (2001) Addressing food security in Africa via multiple livelihood strategies of women farmers. Food Policy 26(2):177–207

Goe MR, Mack S (2005) The linkages between HIV/AIDS and livestock in Eastern and Southern Africa: proceedings of a technical workshop. Animal Production and Health Unit, Food and Agriculture Organization of the United Nations, Rome. http://www.fao.org/docs/eims/upload/207140/hivlivestocksector.pdf

Graffham A, Zulu R, Chibanda D (2005) Improving the safety of street vended foods in Southern Africa. Final report, CPHP Project R8272. http://www.fao.org/docrep/meeting/010/a0215e/a0215e26.htm

Guèye EHF (2000) Women and family poultry production in rural Africa. Dev Pract 10(1):98–102

Guillet DW (1992) The impact of alfalfa introduction on common field agropastoral regimes: Quechua villagers in southwestern Peru. In: McCorkle CM (ed) Plants, animals, and people: agropastoral systems research. Westview Press, Boulder, pp 111–123

Haenlein GFW, Abdellatif MA (2004) Trends in small ruminant husbandry and nutrition and specific reference to Egypt. Small Rumin Res 51(2):185–200

Heffernan C, Misturelli F (2008) The delivery of veterinary services to the poor: preliminary findings from Kenya. Veterinary Epidemiology and Economics Research Unit, Department of Agriculture, University of Reading, Reading, UK. www.livestockdevelopment.org

Heffernan C, Misturelli F, Pilling D (2003) Livestock and the poor: findings from Kenya, India, and Bolivia. Animal Health Programme, Department for International Development, London

Heffernan C, Nielsen L, Misturelli F (2004) Restocking pastoralists: a manual of best practice and decision support tools. ITDG Publishing, Rugby

Heifer International (2008) 2007–2008 project profiles: where we work and what we do. Little Rock, AR, US. http://www.heiferproject.org

Herath S (2007) Women in livestock development in Asia. J Commonw Vet Assoc 24(1):29–37

Herrero M, Thornton PK, Notenbaert A, Wood S, Msangi S, Freeman HA, Bossio D et al (2010) Smart investments in sustainable food production: revisiting mixed crop-livestock systems. Science 327(5967):822–825

Hussain S, Zakaria M, Hassan Y, Mukhtar Y, Ali S (2004) Gender role in livestock rearing and effect of National Rural Support Programme (NRSP) in livestock training course in Kotli (Azad Kashmir). Int J Agric Biol 6(2):424–425. http://www.fspublishers.org/published_papers/57624_..pdf

IFAD (International Fund for Agricultural Development) (2005) Livestock services and the poor, Rome

ILRI (International Livestock Research Institute) (2002) Livestock—A pathway out of poverty. ILRI's strategy to 2010, Nairobi

Jaitner J, Sowe J, Secka-Njie E, Dempfle L (2001) Ownership pattern and management practices of small ruminants in the Gambia—implications for a breeding programme. Small Rumin Res 40(2):101–108

Kanyamurwa J, Ampek G (2007) Gender differentiation in community responses to AIDS in rural Uganda. AIDS Care 19(S1):64–72

Kimani TM, Ngethe EW (2007) Enhancing livestock management skills amongst Maasai Pastoral women of Magadi Division in Kenya: a CVA-funded Kenya Women Veterinary Association project. Fourth Pan Commonwealth Veterinary Conference, Barbados, West Indies, Nov 4–8

Kipuri N (1989) Maasai women in transition: class and gender in the transformation of a pastoral society. Unpublished PhD dissertation, Temple University, Philadelphia

Kristjanson P, Krishna A, Radeny M, Nindo W (2004) Pathways out of poverty in Western Kenya and the role of livestock. Pro-poor livestock policy initiative working paper. Food and Agriculture Organization of the United Nations, Rome. www.fao.org/ag/againfo/projects/en/pplpi/home.html

Kristjanson P, Krishna A, Radeny M, Kuan J, Quilca G, Sanchez-Urrelo A, Leon-Velarde C (2007) Poverty dynamics and the role of livestock in the Peruvian Andes. Agric Syst 94(2):294–308

LID (Livestock in Development) (2004) Livestock services and the poor. Veterinary Epidemiology and Economics Research Unit, Department of Agriculture, University of Reading, Reading, UK. www.livestockdevelopment.org

Little PD, McPeak J, Barrett C, Kristjanson P (2008) Challenging orthodoxies: understanding pastoral poverty in East Africa. Dev Chang 39(4):585–609

McPeak J, Doss C (2006) Are household production decisions cooperative? Evidence on migration and milk sales from Northern Kenya. Am J Agric Econ 88(3):525–541

Meinzen-Dick R, Pradhan R, di Gregorio M (2004) Collective action and property rights for sustainable development: understanding property rights. 2020 focus brief 11. International Food Policy Research Institute, Washington, DC

Micheal BJ (1987) Milk production and sales by the Hawazma (Baggara) of Sudan. Research in economic anthropology 9. JAI Press, Greenwich

Moy G, Hazzard A, Käferstein F (1997) Improving the safety of streetvended food. World Health Stat Q 50(1–2):124–131

Msoffe PLM, Bunn D, Muhairwa AP, Mtambo MMA, Mwamhehe H, Msago A, Mlozi MRS, Cardona CJ (2010) Implementing poultry vaccination and biosecurity at the village level in Tanzania: a social strategy to promote health in free-range poultry populations. Trop Anim Health Prod 42(2):253–263. http://ukpmc.ac.uk/classic/articlerender.cgi?tool=pubmed&pubmedid=19688307

Mullins GL, Wahome P, Tsangari A, Maarse L (1996) Impacts of intensive dairy production on smallholder farm women in coastal Kenya. Hum Ecol 24(2):231–253

Mutenje M, Mapiye D, Mavunganidze A, Mwale M, Muringai V, Katsinde C, Gavumende I (2008) Livestock as a buffer against HIV and AIDS income shocks in the rural households of Zimbabwe. Dev South Afr 25(1):75–82

Mwangi EN (2005) The transformation of property rights in Kenya's Maasailand: triggers and motivations. CAPRI working paper 35. International Food Policy Research Institute, Washington, DC. www.capri.cgiar.org/pdf/capriwp35.pdf

Nduna JN (1990) The struggles for survival of street traders in Umtata, Transkei, 1980–89. GeoJournal 22(3):1572–9893

Niamir-Fuller M (1994) Women livestock managers in the third world: focus on technical issues related to gender roles in livestock production. Staff working paper 18. International Fund for Agricultural Development, Rome

Njuki JM (2001) Gender roles in agroforestry: a socioeconomic analysis of Embu and Kirinyaga Districts, Kenya. PhD thesis, Sokoine University of Agriculture, Morogoro, Tanzania

Njuki J, Kihiyo M, O'ktingati A, Place F (2004) Male versus female labour in an agroforestry system in the Central Highlands of Kenya: correcting the misconception. Int J Agric Resour Gov Ecol 3(1–2):154–170

Njuki J, Kaaria S, Chamunorwa A, Chiuri W (2011) Linking smallholder farmers to markets, gender, and intrahousehold dynamics: does the choice of commodity matter? Eur J Dev Res 23(3):426–443

Nori M (2008) Milking drylands: gender networks, pastoral markets, and food security in stateless Somalia. Thesis, Wageningen University, Wageningen, The Netherlands. http://library.wur.nl/WebQuery/wda/lang/1929205

Nori M, Kenyanjui MB, Yusuf MA, Mohammed FH (2006) Milking drylands: the emergence of camel milk markets in stateless Somali areas. Nomadic People 10(1):9–28

Oboler RS (1996) Whose cows are they, anyway? Ideology and behaviour in Nandi Cattle "ownership" and control. Hum Ecol 24(2):255–272

Okali C, Sumberg JE (1985) Sheep and goats, men and women: household relations and small ruminant development in South-West Nigeria. Agric Syst 18(1):39–59

Oxby C (1987) Women unveiled: class and gender among Kel Ferwan Tuareg. Ethnos 52(1–2):119–136

Patel A (1998) Women and the White Revolution. Coop Dialogue 8(1):20–25

Perry B, Grace D (2009) The impacts of livestock diseases and their control on growth and development processes that are pro-poor. Philos Trans R Soc 364:2643–2655

Perry BD, Nin-Pratt A, Sones K, Stevens C (2005) An appropriate level of risk: Balancing the need for safe livestock products with fair market access for the poor. Pro-poor livestock policy initiative working paper 23. Food and Agriculture Organization of the United Nations/International Livestock Research Institute, Rome. http://www.fao.org/ag/againfo/programmes/en/pplpi/workingpapers.html

Place F, Kariuki G, Wangila J, Makauki A, Ndubi J, Kristjanson P (2004) Assessing the factors underlying differences in achievements of farmer groups: methodological issues and empirical findings from the Highlands of Central Kenya. Agric Syst 82(3):197–353

Quisumbing AR (ed) (2003) Household decisions, gender, and development: a synthesis of recent research. International Food Policy Research Institute, Washington, DC

Quisumbing AR, Pandolfelli L (2010) Promising approaches to address the needs of poor female farmers: resources, constraints, and interventions. World Dev 38(4):581–592

Randolph TF, Schelling E, Grace D, Nicholson CF, Leroy JL, Cole DC, Demment MW, Omore A, Zinsstag J, Ruel M (2007) Role of livestock in human nutrition and health for poverty reduction in developing countries. J Anim Sci 85(11):2788–2800

Rangnekar S (1998) Women in livestock production in developing countries. In: International conference on sustainable animal production, Hisar, India, 24–27 November

Rao SVN, Ramkumar S, Waldie K (2002) Dairy farming by landless women in the southern states of India. In: Morrenhof J, Ahuja V, Tripathy A (eds) Livestock services and the poor: papers, proceedings and presentations of the international workshop, Bhubaneswar, India, 28–29 October, pp 73–86

Riise JC, Permin A, Kryger KN (2008) Strategies for developing family poultry production at village level: experiences from West Africa and Asia. Network for Smallholder Poultry Development (NESPOD), Dyrlaegevej 2, 1870 Frederiksberg, Denmark

Rivière-Cinnamond A (2005) Animal health policy and practice: Scaling up community-based animal health systems, lessons from human health. Pro-poor livestock policy initiative working paper 22. Food and Agriculture Organization of the United Nations, Rome

Roth F, Zinsstag J, Orkhon D, Chimed-Ochir G, Nansalmaa M, Hutton G, Cosivi O, Carrin G, Otte J (2003) Human health benefits from livestock vaccination for Brucellosis: a case study. Bull World Health Organ 81(12):867–876

Roy S, Rangnekar DV (2007) Farmer Participatory Need-Based Extension (FPNE) approach: a sustainable model adopted by cooperative milk unions in Andhra Pradesh, India. Livest Res Rural Dev 19(144). http://www.lrrd.org/lrrd19/10/roy19144.htm

Rubin D, Tezera S, Caldwell L (2010) A calf, a house, a business of one's own: microcredit, asset accumulation, and economic empowerment in GL CRSP projects in Ethiopia and Ghana. Global Livestock Collaborative Research Support Program, U.S. Agency for International Development, Washington, DC

Salman MD, Morley PS, Ruch-Gallie R (eds) (1999) Proceedings of the 9th symposium of the international society for veterinary epidemiology and economics. Paper 333. International Society for Veterinary Epidemiology and Economics, Nairobi

Schelling E, Abdoulaye M, Wyss K, Randolph TF, Zinsstag J (2007) Human and animal vaccination delivery to remote nomadic families, Chad. Emerg Infect Dis 13(3):373–379. http://www.cdc.gov/ncidod/EID/index.htm

Sen AK (1997) Choice, welfare, and measurement. Harvard University Press, Cambridge, MA

Sikana PM, Kerven CK (1991) The impact of commercialisation on the role of labour in African pastoral societies. Pastoral Development Network, Overseas Development Institute, London

Smale M, Heisey PW (1994) Comment: gendered impacts of fertilizer subsidy removal programs in Malawi and Cameroon. Agric Econ 10(1):94–99

Smith LC, Ramakrishnan U, Ndiaye A, Haddad L, Martorell R (2003) The importance of women's status for child nutrition in developing countries. Research report 131. International Food Policy Research Institute, Washington, DC

Sparr P, Moser C (2007) International NGOs and poverty reduction strategies: The contribution to asset-based approach. Working paper no 08. Brookings Global Economy and Development, Brookings Institution, Washington, DC

Staal S, Poole J, Baltenweck I, Mwachero J, Notenbaert A, Thorpe W, Nzuma J, Herrero, M (2009) Targeting strategic investment in livestock as a vehicle for rural livelihoods. ILRI working document. International Livestock Research Institute, Nairobi, Kenya

Talle A (1988) Women at a loss: changes in Maasai pastoralism and their effects on gender relations. Thesis, Department of Social Anthropology, University of Stockholm, Sweden

Teufel N, Kuettner K, Gall C (1998) Contribution of goat husbandry to household income in the Punjab (Pakistan): a review. Small Rumin Res 28(2):101–107

Thornton PK, Kruska RL, Henninger N, Kristjanson PM, Reid RS, Robinson TP (2003) Locating poor livestock keepers at the global level for research and development targeting. Land Use Policy 20(4):311–322

Todd H (1998) Women climbing out of poverty through credit; or what do cows have to do with it? Livest Res Rural Dev. http://www.lrrd.org/lrrd10/3/todd103.htm

Torkelsson A, Tassew B (2008) Quantifying women's and men's rural resource portfolios—empirical evidence from Western Shoa in Ethiopia. Eur J Dev Res 20(3):462–481

Turner M (1999) Merging local and regional analyses of land-use change: the case of livestock in the Sahel. Ann Assoc Am Geogr 89(2):191–219

Upadhyay B (2005) Women and natural resource management: illustrations from India and Nepal. Nat Res Forum 29(3):224–232

Valdivia C (2001) Gender, livestock assets, resource management, and food security: lessons from the SR-CRSP. Agric Hum Values 18(1):27–39

von Holy A, Makhoane FM (2006) Improving street food vending in South Africa: achievements and lessons learned. Int J Food Microbiol 111(2):89–92

Wabekbon Development Consultants (2009) Women's milk and small ruminant marketing in Mandera Triangles (Kenya, Ethiopia, and Somalia). CARE-Ethiopia. http://edu.care.org/Documents/Womens%20Milk%20and%20Small%20Ruminant%20Marketing%20in%20Mandera%20Triangles.pdf

Waite W (2000) How is household vulnerability gendered? Female-headed households in the collectives of Suleimaniyah, Iraqi Kurdistan. Disasters 24(2):153–172

Wangui EE (2008) Development interventions, changing livelihoods, and the making of female Maasai pastoralists. Agric Hum Values 25(3):365–378

Waters-Bayer A (1985) Dairying by settled Fulani women in Central Nigeria and some implications for dairy development. ODI pastoral development network paper 20c. Overseas Development Institute, London

Waters-Bayer A (1988) Dairying by settled Fulani agropastoralists in Central Nigeria: the role of women and implications for dairy development. Wissenschaftsverlag, Vauk Kiel

Waters-Bayer A, Letty B (2010) Promoting gender equality and empowering women through livestock. In: Swanepole F, Stroebel A, Moyo S (eds) The role of livestock in developing communities: enhancing multifunctionality. University of the Free State/ILRI/CTA), Bloemfontein, pp 31–50

WHO (World Health Organization) (2006) The control of neglected zoonotic diseases: a route to poverty alleviation. Report of a Joint WHO/DFID-AHP Meeting with the participation of FAO and OIE, Geneva, 20–21 September 2005. http://www.who.int/zoonoses/Report_Sept06.pdf

World Bank (2001) Engendering development through gender equality in rights, resources, and voice. World bank policy research report. Management 1: report No. 36546-MW. World Bank, Washington, DC

Yisehak K (2008) Gender responsibility in smallholder mixed crop—Livestock production systems of Jimma Zone, South West Ethiopia. Livest Res Rural Dev 20(11). http://www.lrrd.org/lrrd20/1/yise20011.htm

Chapter 10
Gender and Social Capital for Agricultural Development

Ruth Meinzen-Dick, Julia A. Behrman, Lauren Pandolfelli, Amber Peterman, and Agnes R. Quisumbing

Abstract Social capital comprises the range of relationships, networks, and institutions that allow people to build trust and cooperation. This chapter documents gender differences in social capital related to agricultural development, defined as group membership and social networks, based on a critical literature review of key issues and a review of published and unpublished empirical studies

R. Meinzen-Dick (✉)
Environment and Production Technology Division, International Food Policy Research Institute (IFPRI), 2033 K Street, N.W, Washington, DC 20006, USA
e-mail: r.meinzen-dick@cgiar.org

J.A. Behrman
Department of Sociology, New York University, 295 Lafayette Street, 4th floor, New York, NY 10012, USA
e-mail: jab965@nyu.edu

L. Pandolfelli
Statistician, Evidence and Data for Gender Equality (EDGE), United Nations Statistics Division, Two United Nations Plaza, New York, NY 10017, USA

Stony Brook University, 100 Nicolls Rd, Stony Brook, NY 11790, USA
e-mail: pandolfelli@un.org

A. Peterman
Department of Public Policy, University of North Carolina, Chapel Hill, Chapel Hill, NC, USA
e-mail: amberpeterman@gmail.com

A.R. Quisumbing
Poverty, Health, and Nutrition Division, International Food Policy Research Institute (IFPRI), 2033 K Street, N.W, Washington, DC 20006, USA
e-mail: a.quisumbing@cgiar.org

A.R. Quisumbing et al. (eds.), *Gender in Agriculture: Closing the Knowledge Gap*, DOI 10.1007/978-94-017-8616-4_10,

conducted between 1999 and 2011. The authors focus on the types of groups and social networks that women and men join, the extent of their participation, as well as the gender-specific barriers that may affect women's full-scale participation. The analysis goes beyond simple dichotomies of men's and women's groups and networks to investigate whether, and under what circumstances, mixed-sex groups may be more effective than single-sex groups in achieving their development objectives. Following this, the authors examine the effects of women's participation on both group performance and extant gender relations and discuss what development actors can do to help realize gains in these areas. The chapter concludes with a summary of the evidence on whether women are disadvantaged in comparison to men in the accumulation of social capital, and if so, the extent to which programs are helping to overcome this gap.

Keywords Gender • Social capital • Social networks • Groups membership • Collective action

10.1 Introduction

While researchers and development practitioners have long recognized the importance of capital in agricultural development, including land, livestock, machinery, and other forms of tangible inputs, interest in social capital is relatively recent. Defined by Putnam (1995) as "features of social organization such as networks, norms, and social trust that facilitate coordination and cooperation," social capital has recently gained traction and attention from development practitioners, especially in grassroots participation and empowerment efforts. Social capital derived from formal and informal social networks and relationships that people draw upon in pursuit of their livelihood objectives plays an important role in agricultural production. For example, farmers with social networks can exchange information about farming practices, or those with informal social safety nets can use them in times of hardship to smooth consumption or acquire time-sensitive agricultural inputs. Access to social capital is particularly important for female farmers as it provides the formal and informal networks and groups in which they can gain valuable information, influence, and access to other resources such as seed, labor, or other assistance that they might otherwise be excluded from.

International organizations, governments, and nongovernmental organizations (NGOs) have enthusiastically embraced social capital as an alternative to government- or market-based approaches to development, with the World Bank hailing it as "the missing link" in development (Grootaert 1997). Working through groups or networks reduces the cost of delivering services to many individuals, making the outreach of programs more cost-effective. This is illustrated in well-known group-based microfinance such as the Grameen Bank's work, or Heifer International's group-based livestock acquisition programs, but also in a whole range of production and consumer cooperatives, and extension services such as Uganda's National Agricultural

Advisory Services (NAADS) (see Ragasa, Chap. 17). Decentralization programs for natural resource management usually depend on user groups such as water user associations, forest user groups, or pastoralist associations to take on responsibility for resource management, accompanied by varying degrees of rights and control over the resource (Meinzen-Dick et al. 2001; Pretty 2003). In addition, groups may be influential in creating political voice, such as India's Self-Employed Women's Association (SEWA), which, with over a million members, is a substantial trade union that organizes group activities to provide microfinance, improve water systems, and meet other basic needs (Chen 2008).

Aside from these externally-initiated formal programs, there are also many indigenous social groups and networks that provide important sources of social capital. Reciprocal labor groups enable people to mobilize enough labor for critical agricultural activities or build or repair housing. Other examples are funeral societies found across diverse cultures in developing countries that provide savings and mutual support in cases of death or even illness (Dercon et al. 2011). In addition to formal group membership, households and individuals may also invest in other, less formal, forms of social capital, such as social networks. Some studies (e.g., Katungi et al. 2008; Padmaja et al. 2006; Woolcock and Sweetser 2007) have examined the role of such networks for information exchange and technology adoption, or for accessing seed (e.g., Badstue et al. 2007). Other studies have emphasized the role of social networks in risk-smoothing (Fafchamps and Lund 2003; Hoddinot et al. 2005; Fafchamps and Gubert 2004).

A primary reason behind development practitioners' interest in social capital is the perception that it is relatively easier for the poor to acquire, unlike other assets such as land. Participation in groups is a commonly used indicator of social capital, although there are many alternative definitions (see the survey by Durlauf and Fafchamps 2005; Haddad and Maluccio 2003). Although not explicitly recognized, the poor may also face barriers to participation in groups. Participation in groups is not costless—networking takes time, especially when formal group meetings are required, and many groups require fees to participate. Social inequality and ethnic differences may also create barriers to social capital accumulation (Alesina and La Ferrara 2000).

Despite the growing importance of groups, their gender composition and the impact of membership in these groups on gender relations has been insufficiently explored within the social capital literature or the collective action literature on performance of groups and associations (Agarwal 2000; Molyneux 2002; Adkins 2005; Westermann et al. 2005). Are there gender-specific barriers to acquiring social capital? In societies where women are disadvantaged in acquiring assets, participation in groups, particularly credit groups, has been touted as a collateral substitute. However, women in poor households face particularly serious time constraints because of their various livelihood activities and childcare responsibilities. Membership fees may create a further barrier to participation by poor women, who have limited control over cash resources. Although both men and women with low levels of education may feel awkward about participating in groups, the fear that they will be perceived as "ignorant" or as having nothing to contribute may be more acute for women when cultural norms discourage women from speaking up in public or from socializing with men. Thus, women may decide that it is not worth their time and effort to participate in group meetings if they believe they will not be heard (Dikito-Wachtmeister 2001; Meinzen-Dick and Zwarteveen 1998). However, is it indeed the case that women have *less* social capital than men, or do they invest in *different* forms of social capital?

In this context, it is useful to consider differences between bonding, bridging, and linking social capital. Bonding creates social cohesion within peer groups, often based on ethnicity, location, religion, and shared values, but is often reinforced by working together, either formally in groups or through informal mechanisms (Njuki et al. 2008; Pretty 2003). Bridging creates structural relationships or networks that cross social groupings, involving coordination or collaboration, social support, or information sharing with people from different backgrounds. Linking social capital creates the ability to engage with external agencies, especially between poor groups and those in power and authority, to draw resources or influence policy.

This chapter documents gender differences in social capital related to agricultural development, defined as group membership and social networks, based on a critical literature review of key issues and a review of published and unpublished empirical studies conducted between 1999 and 2011. We focus both on the types of groups and social networks that women and men join, the extent of their participation, as well as the gender-specific barriers that may affect women's full-scale participation. We then examine the effects of women's participation on both group performance and extant gender relations and discuss what development actors can do to help realize gains in these areas. We conclude by summarizing the evidence on whether women are disadvantaged in comparison to men in the accumulation of social capital, and if so, assess the extent to which programs are helping to overcome this gap.

This review contributes to the literature in several ways. First, similar to the chapter on nonland inputs by Peterman et al. (Chap. 7), we focus on empirical household data from program evaluations and agricultural and socioeconomic research in order to summarize and bound parameters for estimates of the gender gap in a reasonable range. Second, our focus on studies published between 1999 and 2011 helps update the literature, given the rapidly evolving environmental, technological, and demographic trends in that period. Third, we treat both formal groups and informal networks as forms of social capital, and examine the evidence of their impact and effectiveness. We also attempt to go beyond simple dichotomies of men's and women's groups and networks to investigate whether, and under what circumstances, mixed-sex groups may be more effective than single-sex groups in achieving their development objectives.

10.2 Gender Differences in Social Capital: A Review

As previously mentioned, we consider two main pathways through which male and female farmers can gain access to social capital: (1) membership in groups, and (2) social networks. We define membership in groups as participation in local-level groups (such as agricultural co-ops, water user boards, and forest committees, as well as religious groups, savings, or funeral associations) that provide information, contacts, and collective action opportunities. We define social networks as the existence of ties not bounded by organized groups that facilitate informal exchange of information or other material, such as seeds or fertilizer. The latter may include kinship, friendship, or acquaintance networks. There is more empirical literature available on gender and group membership, in part because it is easier to measure; however, there are a growing number of studies analyzing social networks that account for and explore gender differences.

10.2.1 Group Membership

Group membership is an important means of accessing resources for agriculture and livelihoods. For example, microfinance groups provide for savings, credit, and in some cases, insurance; extension groups provide access to information; forest user groups provide access to firewood, timber, and other forest products; water users' associations improve access and control over water for productive and domestic uses; labor groups help overcome labor bottlenecks. Naturally, if groups restrict women's participation, it will affect women's individual ability to access the benefits of any given group.

In addition to these direct benefits of group participation, group membership (and especially leadership) creates ties with other members, which can raise one's social status, and may provide an empowerment effect, or offer other, less tangible but nonetheless important benefits. Thus, it is useful to look at the number as well as the type of groups to which men and women belong.

Table 10.1 summarizes quantitative studies that examine gender differences in access to social capital. The vast majority of empirical work that looks at gender-differentiated access to social capital does so by looking at group membership, typically examining the extent to which men and women are more (or less) likely to join groups, the types of groups they join, and what individual characteristics increase (or decrease) the probability of joining specific groups. Because specific sociocultural contexts influence the decision to join groups, the results presented here should not be taken as generalizable, but aim to present snapshots of gender differences in group membership in different countries and contexts. Of particular note is the World Bank/IFPRI (2010) research on gender and governance. In the study's survey of 966 households in India, researchers found that the sex of the household head does not play a significant role in determining the number of memberships in local community-based organizations (CBOs). However, the type of group joined varied along gender lines; women mainly joined self-help groups or women's groups, and men primarily joined forest groups, cooperative societies, and caste associations. The complementary studies in Ghana and Ethiopia also found that type of group membership varies by gender, with tending toward agriculture-oriented organizations, and women more likely to join religious groups. In Ghana,

Table 10.1 Gender differences in access to social capital

(1)	(2)	(3)	(4)	(5)	(6)	(7)	(8)	(9)	(10)	(11)	(12)
			Use of access to input								
Authors (year)	Country (crop)	Sample size	Input type	Women	Men	Unit	Gender indicator	Outcome measure	Effect size	Comments	Published
Agrawal and Chhatre (2006)	India (forest)	95 village forest management groups	Forest committees	–	–	–	Gender relations scale (1–3)	Forest condition scale (1–5)	0.235 (0.108)**	Gender relations scale (measuring whether women hold positions of power in village organizations) is significantly associated with better forest conditions using OLS regression. Qualitative evidence suggests women gain decisionmaking positions after local forests were viewed as deteriorating	x

Agrawal et al. (2006)	India (forest)	673 heads of forest protection committees	Forest committees	–	–	–	Female participants	Control of illicit grazing	0.241 (4.35)***	Probit analysis finds women's participation has substantial positive effects on regulating illicit grazing and felling, even after controlling for the effects of a range of independent variables	
				–	–	–		Control of illicit felling	0.275 (4.95)***		
Beard (2005)	Indonesia	10,098–11,000 women	Civil society organizations[a]	–	–	–	Female respondents	Knowledge	1.676 (0.179)***	Logistic regression model finds married women are significantly more likely than nonmarried women to know about and participate in civil society organizations	x
								Participation	2.884 (0.205)***		
Davis and Negash (2007)	Kenya	88 farmers	Dairy goat organization	52	48	%	Stratified by gender: female farmers	Participation	0.18 (0.67)	Descriptive analysis finds gender has a significant impact on type of group farmers participate in. Males dominate agricultural-oriented groups, while females dominate women/clan/village groups	
			Water group	9	33	%			8.09 (0.00)***		
			Farm group	2	12	%			3.24 (0.07)**		

(continued)

Table 10.1 (continued)

(1)	(2)	(3)	(4)	(5)	(6)	(7)	(8)	(9)	(10)	(11)	(12)
			Use of access to input								
Authors (year)	Country (crop)	Sample size	Input type	Women	Men	Unit	Gender indicator	Outcome measure	Effect size	Comments	Published
Fletschner and Carter (2008)	Paraguay	213 couples	Social network	–	–	–	Female respondents	Probability woman will demand capital	0.621 (nr)***	Probit analysis finds woman's demand for entrepreneurial capital is positively and significantly affected by the behavior of her reference group (social network). Women are more likely to demand entrepreneurial capital the larger the proportion of cooperative members in their reference group demand capital	x

Godquin and Quisumbing (2008)	The Philippines	304 households	Group membership (general)	63.2	58.7	%	Male and female household members	Participation	0.144 (0.97)	Simple model of participation in a group finds that gender does not impact group participation; however, there are gender differences in the types of groups to which men and women belong, and significantly more men are members of production-oriented groups	x
			Production group	14.5	22.4	%			−4.20 (−2.31)*		
Jagger and Pender (2006)	Uganda	451 households	Agriculture and environmental organizations	–	–	–	Female heads	Involvement in organization	0.128 (nr)	Probit regression finds that female head is not significantly associated with participation in agricultural/environmental NGOs and CBOs	

(continued)

Table 10.1 (continued)

(1)	(2)	(3)	(4)	(5)	(6)	(7)	(8)	(9)	(10)	(11)	(12)
			Use of access to input								
Authors (year)	Country (crop)	Sample size	Input type	Women	Men	Unit	Gender indicator	Outcome measure	Effect size	Comments	Published
Kariuki and Place (2005)	Kenya	442 households	Collective action via group membership[b]	(nr)	(nr)	–	–	–	–	Descriptive analysis suggests that men and women participate in similar groups but the motivation for joining groups and extent of participation are not the same. Women (subsistence farmers) join for social insurance and building assets; men join for commercialization and marketing	

Katungi et al. (2008)	Uganda (bananas)	351 heads	Agricultural information	–	–	–	Male heads	Informal exchange of information	0.079 (0.029)***	Multinomial logit model used to analyze multiple participation choices of information exchange. Findings demonstrate social capital is an important factor in information exchange, with men generally having better access to social capital than women	x

(continued)

Table 10.1 (continued)

(1)	(2)	(3)	(4)	(5)	(6)	(7)	(8)	(9)	(10)	(11)	(12)
			Use of access to input								
Authors (year)	Country (crop)	Sample size	Input type	Women	Men	Unit	Gender indicator	Outcome measure	Effect size	Comments	Published
Leino (2007)	Kenya	168 respondents	Water user committees	–	–	–	Female committee members	Number of women on commit-tees	1.060 (0.159)***	Study evaluated randomized intervention across 334 communities where 50 % of water user groups were given training designed to increase female participation. Analysis using instrumental variable approach finds number of females on committees increased, which did not, however, translate to changes in water source maintenance outcomes	
		805 respondents		–	–	–	Treatment group	Overall mainte-nance quality	0.023 (0.079)		

Maluccio et al. (2003)	South Africa	2,844 men, 3,433 women	Group membership	39	29	%	Male and female household members	Log per capita monthly expenditure, net of costs of group	0.07*** (same point estimate for men and women in 1998)	Regressions stratified by gender, while community fixed effects indicate a positive coefficient on men's and women's group membership in 1998, but not in 1993	x
Perdana et al. (2006)	Indonesia	7,200–10,000 households	Government	–	–	–	Female heads	Assistance received	0.039 (0.135)	Probit regression used to explore whether household head gender impacted access to assistance from a variety of groups in wake of the 1998 economic crisis	
			NGOs	–	–	–			0.126 (0.115)		
			Community	–	–	–			0.290 (0.130)***		

(continued)

Table 10.1 (continued)

(1)	(2)	(3)	(4)	(5)	(6)	(7)	(8)	(9)	(10)	(11)	(12)
			Use of access to input								
Authors (year)	Country (crop)	Sample size	Input type	Women	Men	Unit	Gender indicator	Outcome measure	Effect size	Comments	Published
Quisumbing (2009)	Bangladesh	745 men, 745 women (husband–wife pairs)	Any group	0.51	0.11	%	–	Probability of belonging to any group; number of groups	–	Regressions estimated separately for husbands and wives show that the probability of belonging to a group and the number of groups to which husbands belong increase with husband's years of schooling. Membership in women's groups is also progressive with respect to initial wealth, with both the probability of joining a group and the number of groups decreasing as the size of owned land increases. Having an older woman in the household increases both the probability of belonging to a group and the number of groups to which a wife belongs	
			Number of groups	0.69	0.14	Groups	–		–		

World Bank and IFPRI (2010)	India	966 households	Community-based organizations[c]	–	–	–	Female heads	Number of institutions household belongs to	0.033 (0.086)	OLS regression shows gender of head insignificant in determining number of institutional memberships per household; however, women participate mainly in self-help groups/women's groups, while men participate in forest groups, cooperative societies, and caste associations	x
	Ghana	1,168 heads	Farmer-based organizations	–	–	–	Male heads	Membership	0.079 (0.029)**	Probit regression shows that male heads are significantly more likely to belong to/participate in groups than are female heads (controlling for ecological zone, literacy of head, and household wealth proxy)	

(continued)

Table 10.1 (continued)

(1)	(2)	(3)	(4)	(5)	(6)	(7)	(8)	(9)	(10)	(11)	(12)
			Use of access to input								
Authors (year)	Country (crop)	Sample size	Input type	Women	Men	Unit	Gender indicator	Outcome measure	Effect size	Comments	Published
	Ethiopia	1,761 heads and spouses	Agricultural coopera-tives	4	24	%	Male heads	Involvement	S*** (nr)	Descriptive and bivariate analysis shows a significantly higher proportion of male than female respondents participate in agricultural cooperatives	

Notes: Articles listed in alphabetical order of first author's last name; *Effect size* refers to gender indicator coefficients with standard errors in parentheses unless otherwise noted; *S* significant, *NS* not significant, *nr* not reported; *indicates significant at 10%; **significant at 5 %; ***significant at 1 %. Outcome measures of *adoption* refer to rate of adoption of corresponding input type; *published* indicates that the study is published in a peer-reviewed journal

[a]Civil society organizations are defined as those that deliver public goods and services to territory-based communities. Men usually participate in civil society organizations related to community-level governance, physical infrastructure, environmental improvements, and neighborhood security, whereas women participate in organizations focusing on family welfare, economics, and health. As a result, the survey asked men and women about participation in different organizations

[b]Self-help groups (building household assets, social/economic support), water groups, dairy goat groups, and coffee groups were the four most common types of groups. Descriptive statistics reported in graphic form, but numbers were unassigned

[c]Including agricultural CBOs

probit regression showed that male household heads are significantly more likely to belong to a farmer-based organization than are female household heads, and in Ethiopia, a significantly higher proportion of male than female respondents are involved in agricultural cooperatives (24 % versus 4 %).

A number of other studies look at gender-based differences in group membership. Davis and Negash's (2007) study of 88 Kenyan farmers found that gender has a significant impact on the type of group that respondents participate in; males dominate agriculture-oriented groups, while females dominate women, clan, and village groups. Godquin and Quisumbing's (2008) study of 304 households in Bukidnon, the Philippines, found that men and women do not differ significantly in their probability of participating in groups or the number of groups they join. However, similar to previously mentioned studies, there are clear gender differences in the types of groups to which men and women belong; men are more likely to be members of production-oriented groups, while women belong to civic and religious groups. While it is unlikely that development initiatives would systematically target men rather than women, given the Philippines' relatively egalitarian society and women's higher educational attainment (Hausman et al. 2007), the specialization of men and women in different types of groups, and the possible delivery of services through those groups, may mean that women (and men) who are in groups where women (men) are less well represented run the risk of being marginalized. For example, using production groups as conduits for service delivery may not only be less likely to reach the poor, who participate less in those groups, but also women, who are less likely to participate in production groups. Conversely, using civic groups to reach men will mostly likely be ineffective, given the very low numbers of men in civic groups.

A noteworthy dimension of group membership in Bangladesh is that, despite cultural norms of women's seclusion, participation rates in NGO-supported groups are much higher among women than men (Quisumbing 2009). Membership in Bangladeshi NGOs tends to be progressive, with higher participation rates among the poor and those with smaller sizes of owned land, owing to the pro-poor orientation and membership criteria of these NGOs. While NGOs have been praised for empowering women in this context, it may be the case that those who do join NGOs already have higher bargaining power within their households: regression analysis of a sample of 957 households finds that women who bring more assets to marriage and who live closer to their natal villages are more likely to belong to a group.

Using panel data collected in 1994 and 1997 in Kwazulu-Natal, South Africa, Maluccio et al. (2003) examined the determinants of group membership separately for men and women. In both 1993 and 1998, both men and women were more likely to belong to more groups if they were the heads of their respective households. Better-educated individuals also belonged to more groups. Interestingly, in 1998, household heads belonged to twice the number of groups they belonged to in 1993—indicating that there may have been greater opportunities or returns to investing in social capital in the post-apartheid regime.

Kariuki and Place (2005) explored motivation for group membership in Uganda, finding that women, who are usually subsistence farmers, join groups for social insurance or household asset building, whereas men, who are more market oriented, join groups to enhance their marketing and commercialization ventures. Jagger and Pender (2006) found female-headed households in Uganda are equally as likely as male-headed households to be involved with local CBOs and NGOs focusing on agriculture and the environment, but are more likely to be involved as compared to male-headed households in other organizations focusing on a diverse set of topics such as poverty or credit. Beard (2005) found that married women are significantly more likely than nonmarried women to know about and participate in civil society organizations in rural Indonesia. Beard concluded that participatory community development organizations often restrict women's roles to that of "traditional" care-taking and topics such as family welfare and health.

Only one study explored differential access to resources and assistance from community groups, CBOs, and NGOs. Perdana et al. (2006) used a probit regression to explore whether the sex of the household head affected access to assistance from a variety of groups since the 1998 Indonesian economic crisis. This study found that female-headed household indicators are a significant determinant of assistance received with respect to CBOs, although not for government or NGO assistance.

Gender differences are also apparent in the leadership and management of local organizations. In Ethiopia, men are five times more likely than women to hold a leadership position within a cooperative; 3 % of female and 15 % of male cooperative members hold such roles (World Bank/IFPRI 2010). In Karnataka, India, women were underrepresented in the leadership of most organizations: no farmer cooperatives and only 10 % of dairy cooperatives studied had female chairpersons, and few had female secretaries (World Bank/IFPRI 2010). A survey of 73 farmer groups in Mozambique found that women do not enjoy the same chances as men to become president and represent the group, participate in meetings or seminars, and take final decisions: 12 % of the groups had female presidents, 27 % had female vice presidents, and 24 % had female secretaries. Women are more likely to hold treasurer positions than the other leadership positions (53 % of the groups; see Gotschi et al. 2009). This may relate to trust: Jemimah Njuki (personal communication, 2012) also found in Kenya that women are more likely to be treasurers in Kenyan watershed groups, because they are trusted with money.

10.2.2 Social Networks

Although there is a wide range of sociological literature on informal social networks and information exchange, there is relatively less empirical research that explores differential access to agriculture-related information exchange by gender. Using qualitative methods in Maharashtra, India, Padmaja et al. (2006) found that women in agriculture develop bonding social capital through kin networks and women-only groups, whereas men develop bridging social capital characterized by weaker, less dense but more cross-cutting ties such as with farmers, acquaintances, and friends from different ethnic groups. They suggest that gender-related bonding, where women talk to women and men talk to men, had some positive effects for empowerment, but also isolated women from outside opportunities (including new technologies), and made women more dependent on male relatives for various needs. A study by Katungi et al. (2008) examined the exchange of agricultural information in Uganda using multinomial logit modeling, and found social capital is an important factor in information exchange, with men generally having better access to social capital than women. Male and female networks may also perform different functions. In Bukidnon, the Philippines, networks related to price information and new technologies are smaller the larger the number of daughters living outside the village (Quisumbing et al. 2011). The network for new technologies, however, is positively associated with the number of sons living inside the village, but in separate households. This relationship may arise from the different roles of sons and daughters in Filipino society. For example, daughters are socialized to be responsible family members and often play the role of insurers, migrating to towns and cities and then sending remittances to their origin households (Lauby and Stark 1988). The number of daughters living outside the village negatively affects the combined number of persons in all networks, in addition to the number of people in price-information and technology-adoption networks. In contrast, sons who are living in separate households within the village are more likely to be engaged in agricultural production themselves, and are a local source of technology information for parents.

A study that explored households' informal social networks in Bangladesh found that husband's and wife's human and physical assets do not have the same influence on the strength of informal social relationships (Quisumbing 2009). Husband's years of schooling strengthen relationships with local officials, judges, lawyers, doctors, headmasters, big businessmen, and big landowners, while wife's years of schooling exert a positive influence on strong relationships with judges, lawyers, doctors, and NGO officials. This result probably reflects different spheres of influence of men and women.

10.3 Factors Affecting Participation in Groups and Networks

Most studies, including those reviewed here, define membership in groups as zero-one decisions, but it is not only participation in groups but also the quality of participation that affects the accumulation of social capital and the expected payoff. Several typologies of participation have been developed. Agarwal's classification developed in reference to community forestry programs in South Asia, in which participation ranges from nominal participation (membership in a group) to interactive participation in which a member has "voice and influence in the group's decisions" is particularly applicable to examination of gender differences (Agarwal 2001, 1624) (see Table 10.2).

Table 10.2 Typology of participation

Form/level of participation	Characteristic features
Nominal participation	Membership in the group
Passive participation	Being informed of decisions ex post facto; or attending meetings and listening in on decisionmaking, without speaking up
Consultative participation	Being asked an opinion in specific matters without guarantee of influencing decisions
Activity-specific participation	Being asked to (or volunteering to) undertake specific tasks
Active participation	Expressing opinions, whether or not solicited, or taking initiatives of other sorts
Interactive (empowering) participation	Having voice and influence in the group's decisions

Source: Agarwal (2001)

Here, the level of one's participation has a strong influence on the benefits that individuals experience, which can vary along gender lines. When membership in a group is limited to one member per household, women may not even get the chance to participate. In this scenario, women can be limited to the lower levels of nominal, passive, and consultative participation, which are reflected in lower benefits from participatory group action. Time-investment in group activities is likely to vary between men and women due to their different household roles and responsibilities. Yet, this "time poverty" is rarely considered in the way group meetings are scheduled. Although women may be interested in attending, they are often overburdened with childcare, food preparation, and agricultural activities, and therefore are unable to accommodate group meetings into their schedules. In addition, meetings may not be held in a place that is convenient or socially acceptable for women to attend. Despite these constraints, it is important to recognize that women are not a homogenous group, and may have greater or less ability to participate based on other socioeconomic factors, including income, ethnic group or caste, religion, and urban versus rural residence.

The gender composition of groups is also likely to influence the degree of women's participation. When there are very few women in a meeting, they may feel isolated or intimidated to speak. The literature on gender and governance discusses "threshold effects" created by a critical mass of women that encourages women to actively and vocally participate. "Although empirical verification of effective proportions is, to date, rather limited, among policymakers and practitioners globally it is the figure of one-third that has become widely accepted as *the* critical mass" (Agarwal 2010, 170–171). Agarwal's own studies in Nepal and India test this threshold primarily for executive committees, and find that the likelihood of women speaking up is significantly higher when one-third or more women serve on the committees. At the same time, all-female groups may be marginalized, or lack some of the skills and connections that men may have, especially for linking with influential individuals outside the group. In analyzing the role of social capital in adoption of soil conservation technologies in Zambia, Malawi, and Mozambique, Njuki and colleagues (2008) conducted factor analysis to aggregate indicators of social capital. The three factors that emerged related to bonding social capital, bridging/linking social capital, and gender relations at the community level. The latter factor had high positive loadings on women's ability to speak with confidence in pubic and men's respect and consideration of ideas given by women, conflict resolution, and abiding by norms and bylaws. This "gender factor" was positively associated with adoption of soil erosion structures, planting of agroforestry trees, use of cover crops, and fallows, suggesting that working with groups to ensure that women have the confidence to speak (and men to listen) can increase uptake of technologies that improve livelihood outcomes.

Despite evidence of gender differences in group dynamics, gender-blind approaches to group organizing, which do not explicitly consider sex of members, are still common. For example, many natural resource management programs, such as irrigators' associations or forest user groups simply define membership as one person per household or limiting membership to recognized owners of land or livestock. This approach frequently results in a predominance of male members, except from female-headed households, although even they may send a younger male to serve as a member rather than the female head. For example, in reviewing the evidence on water users' associations in South Asia, Meinzen-Dick and Zwarteveen (1998) found women's participation to be much lower than that of men, despite high involvement of women in irrigated agriculture and agricultural decisionmaking. The few documented cases of higher female involvement either stem from women-only organizations managing groundwater pumps or from areas where men were not interested or absent. This trend implies there is a significant role for policy, program design, and other mechanisms such as reform of group bylaws to ensure women participate and are given equal potential for voice in groups.

The literature points to several institutional mechanisms that enable women to join groups and remain active members, including allowing non-household heads and nonland owners to be group members; timing meetings to accommodate women's workloads; ensuring that poorer women have opportunities to voice their concerns in group meetings; and soliciting women's feedback in project monitoring and evaluation (Pandolfelli et al. 2008). A randomized evaluation of water infrastructure maintenance in rural Kenya found that speeches made by NGO facilitators about the importance of women's participation in the user committees, encouraging women to attend the community meetings at which committee members were selected, and holding the meetings at a convenient time for women all served to increase women's participation in the user committees (Leino 2007). Women are also more likely to participate when projects directly incorporate their concerns. In the Philippines, attempts to have women monitor lake water to determine if soil conservation techniques were reducing silting were unsuccessful until project staff realized that women were more interested in health issues than in soil loss. When the project began to raise awareness about how water quality affected the health of families and the program then expanded to include monitoring for *E. coli*, women's participation significantly increased (Diamond et al. 1997).

10.4 Gender, Participation, and Group Effectiveness

Women's participation in groups is not only important for their accumulation of social capital, but also for group performance. Theories of participatory management suggest that the lack of participation of a large number of the users of a resource would lead to performance weaknesses in the organization, because of weaknesses in communication, representation, democracy, and accountability, which may lead to free riding, rent seeking, and corruption (Ostrom 1992). Zwarteveen and Neupane (1996) found that the all-male organization for the Chhattis Mauja irrigation system in Nepal faced difficulties enforcing its rules on women. Female heads of farms in the head end of the system always took more water than their entitlements, while contributing less labor than they should, but it was difficult to solve the problem because women were not members of the organization and could thus not be punished. Women did not steal water or shirk from contributing labor to maintenance only because of opportunism. Water stealing by women occurred partly because women had an interest in applying more water to the paddy-field to reduce their labor requirements for weeding, and rules and prevailing gender norms made it difficult for women to make labor contributions.

Community forestry programs in India and Nepal have also found that the rules determined by men are too restrictive and increase the time burden on women who gather firewood for household use; in other instances, these regulations force women to violate community rules in order to meet their needs. Women who observe regulations are also not directly rewarded for their adherence: If cash or in-kind payments are made, they are typically distributed on a household basis and go to male household heads. For forest management in Nepal, Sarin (1995) found that noninvolvement of women made it easy for women (especially those from outside the village) to continue to gather firewood, in spite of strict regulatory rules set by the organization. In some communities, 90 % of the rule offenders were women. Male office-bearers found it difficult to stop these women, since they risked being accused of molesting them. As a result, the need for female participation in these organizations is now accepted, but not on grounds of equity, participation, or democracy, but because women are needed to help the organization enforce its rules, or to stop other women from taking firewood.

Women's participation in decisionmaking has also been found to significantly improve forest regeneration (Agarwal 2007, 2010), reducing the incidence of illegal harvesting and other unsanctioned activities (Agarwal 2009; Agrawal et al. 2006). Women's presence in forest user groups also enhances the group's capacity to manage and resolve conflicts (Westermann et al. 2005). Agrawal and colleagues' (2006) study of forest committees in Madhya Pradesh, India, found that women's participation has substantial positive effects on regulating illicit grazing and tree felling, even after controlling for the effects of a range of independent variables. Specifically, when women belong to forest protection committees, participate in committee meetings and patrol the forest, control of illicit grazing and felling increases by 24 % and 28 %, respectively, and the regeneration of allotted forest also increases by 28 % (Agrawal et al. 2006). However, it is essential not to idealize women as the "custodians" of resources (Jewitt 2000); other studies in forest management in India and Nepal suggest that women-only groups often lack information and are frequently allocated forest resources that are of marginal quality (Cornwall 2001; Pandey 1993; Rai and Buchy 2004). In comparative analysis of forest user groups in Kenya, Uganda, Mexico, and Bolivia, Mwangi et al. (2011) found that female dominant groups had lower rates of enforcement, and were less likely than mixed groups or male dominant groups to adopt regeneration practices and other forestry investments, perhaps because women had less access to information or resources to invest, or because of social norms. Mixed male and female user groups may encourage the use of their complementary strengths; men in mixed groups undertake monitoring when long distances are to be covered or due to dangers in forest patrolling (Watkins 2009; Westermann et al. 2005). Similarly, in Bangladesh, Sultana and Thompson (2008) found that compliance with rules limiting fishing in protected areas is higher when both men and women are actively involved in fishery management groups, because women, who control catches, exert pressure to ensure compliance with fishing rules, while men patrol the fish sanctuaries at night when it is unsafe for women to do so. In the highlands of central Kenya, where women are regarded as more trustworthy than men with money, Kariuki and Place (2005) report that men express more satisfaction with the group's financial management in mixed-sex groups than they do in all-male groups, because men are perceived as being more vulnerable to corruption.

Using group-based approaches that address women's needs may be better at achieving some development objectives compared to others, although trade-offs exist. An evaluation of the long-term impact of three programs disseminating agricultural technologies in Bangladesh found that, while gains in per capita expenditures and household assets were greatest in the site that targeted fishpond technologies to households, improvements in nutritional status (particularly of women and girls) and reduction of the gap between men and women in household assets were greatest in the site that targeted improved vegetable technologies through women's groups (Kumar and Quisumbing 2011; Quisumbing and Kumar 2011). Income and nutritional gains in the site that targeted fishpond technologies through women's groups were less, possibly because the proceeds from the pond had to be shared among many families.

As with all gender issues, the cultural context and local gender roles are crucial for understanding how the involvement of men and women will affect group outcomes. Where strong gender segregation exists, working with existing women's groups may help facilitate entry into communities and allow women to retain control of project benefits. Where women's and men's motivations for joining groups differ, projects that encourage mixed-sex groups may be less sustainable, particularly once external funding runs out.

Even many seemingly male (or female)-dominated institutions may, on closer examination, involve both men and women. Hambly Odame (2002) notes that in western Kenya, failure by an agroforestry extension project to understand the importance of men's role in the distribution of resources and benefits within women's groups led to a 67 % rate of collapse over a 12-year period, often resulting in a loss of labor, capital, and moral support for group members. Padmanabhan (2006) describes the tribal community of the Kurichyas in Kerala, India, which organizes collective action social networks around a *pittan* (headman) and his wife, who assume complementary roles in the monitoring, sanctioning, and exchange of seeds and their related knowledge. The *pittan* organizes official requests for seed from farmers outside of the community, while the *pittan*'s wife supervises the actual handling and storage of the seed. In her capacity as the guardian and custodian of women's knowledge of genetic wealth, she organizes other women within the household to weave storage baskets for the seeds, maintains a storage system to diversify risk, selects the quantity and the quality of seed to enter the exchange network, and cleans the seeds in preparation for exchange. Although this division of labor leads to the effective exchange of seeds, it is only the male members of a household who are allowed to formally represent the household's interest in acquiring seed.

Therefore, rather than gender-blind or single-sex organizations, a more nuanced, third approach, would be to try to develop mixed male and female organizations that allow for women's full participation, particularly where men and women share joint interests or are both users of a resource (e.g., water, communal fisheries). However, this is not easy. Although the evidence on this to date is fragmentary, we expect that establishing mixed sex groups has a higher transaction cost because of the need to overcome gender barriers. The corollary of this is that the larger the degree of gender inequality, the higher the transaction costs will be. Thus, establishing mixed organizations is likely to be easier in societies in which women already have education levels on a par with men's, and where women are used to autonomous movement and communication, and much more difficult in societies that practice female seclusion, with low levels of female education. Under such conditions of low female empowerment, working with all-women's groups to build capacity may be an important first step.

There are several examples of the equity-efficiency and other trade-offs in forming mixed-sex groups. For example, Sultana and Thompson's (2008) study of floodplain management in Bangladesh found that all-male community organizations took less time to establish than committees that included women (302 days vs. 340 days), but this was offset by a shorter time for the mixed organizations to start activities (179 days for all-male versus 106 days for mixed groups). All-male groups were able to obtain and disburse more credit, and undertook more fisheries management activities, but also had more conflicts and more rule-breaking, suggesting that involvement of women in decisionmaking was instrumental for compliance and conflict resolution.[1] In addition, Acharya and Gentle's (2006) study of the SAMARPAN (Strengthening the Role of Women and Civil Society in Democracy and Governance) in Nepal illustrates some of the

[1] Interestingly, in detailed case studies, Sultana and Thompson found that an all-women's group formed a men's advisory committee, whereas the all-male group did not have a separate advisory committee.

complexities involved in building gender-balanced organizations. The program provided advocacy literacy training to women, but also engaged with local community group leaders. Community forestry user groups participating in this program showed an increase in women in leadership positions, who became active in auditing group funds, expanding women's membership, and influencing the group activities to include fodder and multiple use activities, biogas, and a range of activities to help very poor households. The integration of a critical mass of women into the regular user groups led to better outcomes than all-women's user groups, which lacked the support of men, and had smaller overall forest areas and less land per household than the mixed groups.

Despite the higher transaction costs of establishing them, the examples above indicate that mixed groups can also have higher payoffs because they can tap into the differential strengths of men and women. Even when women's participation in user groups does not lead to greater group effectiveness, their increased participation may have indirect benefits for the community. Leino's (2007) study of water infrastructure maintenance in rural Kenya found that although the intervention designed to increase female participation in water user committees was successful, the increased levels of female participation did not have a significant impact on water source maintenance outcomes, possibly because elected women were not given technical training on water infrastructure maintenance. However, according to Leino (2007), the potentially restrictive effects of the lower average education and experience of female committee members may have been offset by other factors in which women may have a comparative advantage, such as lower monitoring costs of water infrastructure or better knowledge of the provisioning and safeguarding of water. Thus, the effectiveness of all-male, all-female, or mixed groups may depend on whether effectiveness assessed in the long or short term and whether direct or indirect indicators are used.

10.5 Impact of Social Capital on Gender Relations

Whereas the *effectiveness* of social capital refers to the ability of groups to meet their immediate purposes (e.g., the management of a natural resource), *impact* of social capital refers to changes (in this case, changes in gender relations) that go beyond the initial objective. For example, a microcredit scheme designed to raise the income of its members would measure its effectiveness in terms of income earned, while measurements of impact on gender relations would include the ability of women to control that income within the household.

Often the impacts of social capital include improvements in women's empowerment, and therefore gender equity. Although definitions of women's empowerment are debated, in essence empowerment is the individual or group capacity to make self-informed and effective choices (Alsop and Heinsohn 2005). The concept of women's empowerment can be viewed a continuum, ranging from emergence from isolation on one end of the continuum to participation in the public sphere on the other.[2] Thus, the criteria selected to measure the impact of social capital on gender relations, including women's empowerment or gender equity, will vary according to where along the continuum individuals are situated at the initial state. However, because of the importance of public participation as a dimension of empowerment, the Women's Empowerment in Agriculture Index includes group membership and feeling comfortable speaking in public as the indicators for the leadership dimension (USAID et al. 2012).

Impacts on gender equity can be evaluated by several indicators, including (1) level and distribution of income, taking into account the fact that women may make trade-offs, or tactical choices, between different material, psychological, and symbolic aspects of poverty (Chant 2003); (2) the ability to secure basic needs; (3) the degree of social and political inclusion; (4) security against violence (including specifically violence against women); (5) vulnerability to shocks; and more broadly, (6) the opportunity set for livelihood improvement.

Strengthening social capital can affect gender relations at four levels: relations within the household, relations within the collective action group itself, relations of the group vis-à-vis the community, and relations of the community vis-à-vis the outside.

Analysis of the impact of social capital on gender equity cannot be divorced from analysis of the household because activities undertaken as a collective feed back into women and men's social bargaining within the household. For example, income-generating social capital schemes may increase a woman's fallback or exit options if she is able to strengthen her asset endowments (e.g., financial capital) and draw upon them as action resources to increase her bargaining power within the household. The aforementioned group-based fishpond or vegetable technology programs in Bangladesh resulted in significantly higher empowerment levels on such criteria as keeping control over money, and reduced domestic violence among program participants (Hallman et al. 2007). A follow-up study of the same site also found that group-based approaches did better than household targeting in closing the household gender asset gap (Quisumbing and Kumar 2011).

Fletschner and Carter (2008) found that for women in rural Paraguay, demand for entrepreneurial capital is positively driven by the behavior of members of their reference group. Thus the larger the membership of a co-op (which they used as in indicator of an entrepreneurial mentality), the more likely the woman is to demand entrepreneurial capital herself. In Kwazulu-Natal, South Africa, Maluccio et al. (2003) found that in 1993, neither men's nor women's membership in groups was a significant determinant of per capita expenditures, but in 1998 both men's and women's group membership positively affect household expenditures, and to the same degree. However, because women's participation is greater, the elasticity is greater for women's social capital—that is, the percentage increase in per capita household expenditure is greater for a percentage increase in women's group membership compared to men's.

[2] James-Sebro (2005) defines gender equality in four stages: (1) engagement of women to come out of isolation; (2) empowerment through acquired ideas, knowledge, skills, and resources; (3) enhancement of lives in households and communities; and (4) emergence into the public sphere.

At the community level, groups, particularly mixed-sex groups, may alter perceptions of women's socioeconomic contributions, thereby increasing their status within the community. In the same Bangladesh programs previously mentioned, the group-based fishponds changed the gendered division of work because, although men were involved at various stages, negotiation over the activities and output took place above the household—men had to negotiate with groups of women backed by an NGO, rather than with their wives individually (Naved 2000).

Groups may also mobilize enough social and political capital to contest the state. A well-known example is the Chipko movement in India, which began as a group of women literally embracing trees to prevent against deforestation in their community, spread across the state, and resulted in a major victory in 1980 with a 15-year ban on green felling in the Himalayan forests of Uttar Pradesh. The Green Belt Movement in Kenya similarly grew into a significant political force. In Argentina, Andujar (2005) found that women's ability to secure clean water for Villa Jardin rendered them "indisputable interlocutors" with institutions outside of their neighborhood. Collective lobbying efforts have also been influential in strengthening women's legal rights and share of state expenditure at the national level, for example in Uganda, Tanzania, and South Africa. Even at the transnational level, the global women's movement may be seen as a web of social networks that has had an impact on development discourse and policy, such as through the Convention on the Elimination of All Forms of Discrimination against Women (CEDAW) and the Beijing Platform for Action.

Changes in gender relations may also occur at multiple levels simultaneously. Panda (2006) observes, for example, that women who participated in SEWA's water campaign grew more confident to participate in the public domain as a collective and thus challenged (male) alcohol consumption at both the village and household levels, resulting in both increased confidence and decreased alcoholism in some villages.

However, group membership may have negative impacts on women's empowerment if group-based programs are designed "gender-blind" or with false assumptions regarding women's motivations for joining a given group. For example, Arganosa-Matienzo (2005) notes that women who engaged in collective soap production faced additional time constraints due to the high labor inputs required for soap making while also earning less than they did as paid farm laborers. Thus, development actors need to be aware that social capital can be used as a vehicle for women's empowerment, but it can also contribute to women's disempowerment.

Indeed, given the complexity of gendered norms and roles and their variances across cultures, there is no "one-size-fits-all" strategy for fostering gender equity via social capital groups. In some instances, particularly where there are deeply entrenched levels of gender inequity, women-only groups may be more effective strategies for bringing women out of isolation, fostering their self-confidence, and building their capacity to bargain within the household. In other instances, mixed-sex groups may be more effective vehicles for enabling women to build their asset base and negotiate in the public arena.

Therefore, if development actors conceive of women's empowerment along a continuum, as suggested above, mixed-sex groups that respond to both women's and men's needs may be more appropriate the further along the continuum women are situated. Such groups may also affect more transformative change in gender roles if, through repeated interactions, women receive greater recognition by men for both their paid and unpaid contributions to the community, although clearly this hypothesis warrants further investigation.

Another factor for external organizations to consider is what mechanisms work best to integrate gender concerns into groups. In some contexts, groups that explicitly address gender equity as an end-goal may have a greater impact on women's empowerment, whereas in other contexts, groups that address gender issues only in terms of the obstacles and constraints they present for realizing the group's (non-gender-related) objective may have a greater impact. The Bangladeshi NGO vegetable program discussed earlier was successful because it worked within the confines of existing gender norms that discourage women from working outside of their homesteads. Yet, the project did not address another byproduct of those gender norms—women's inability to market their produce directly. In another example from Bangladesh, Sultana and Thompson (2008) find that an NGO's insistence that it would work only with women to create aquatic resource management committees failed to involve women in the long run because it openly challenged local norms discouraging women's public participation, and men within the communities refused to allow their wives to participate. Thus, strategies for group formation that challenge gender norms must be weighed against other project objectives, such as increased food security and better management of natural resources, which, over time, may transform gender norms. Encouraging women to define their needs and preferences prior to the design of projects may help ensure balance between challenging and respecting local norms.

10.6 Conclusion

There is little doubt that social capital—as developed through formal group membership and informal social networks—plays a critical role in agricultural production and poverty reduction. Indigenous institutions in almost every rural society provide access to mutual support, labor, seeds, and information. Government and NGO programs rely on more formally organized groups to deliver extension, technologies, credit, and other services. In the context of decentralization of natural resources, local organizations are playing an increasing role in managing land, water, and forests.

Since the 1990s, the attention to social capital in development has inspired enthusiasm that social capital can substitute for other, often tangible assets. For example, in group-based microfinance programs, social capital—members vouching for and insuring each other—provides an alternative to other forms of collateral. The thousands of women's microfinance and "self-help" groups have similarly prompted enthusiasm that social capital could help women make up for lack of other assets.

On closer examination, however, the evidence on social capital through formal group membership and informal social networks indicates that a gender gap remains. Although cultural context clearly matters, studies in many societies have found that, while men and women both join groups, men are more likely to be in

production-oriented groups like farmer cooperatives, while women a
involved in civic or religious groups. This has important implications for th
ery of agricultural services, especially extension: to reach women as well as
may be better to work with the types of groups where women are already participating, rather than to target only seemingly production-related local organizations.

There are many potential mechanisms that group-based programs can implement to increase women's participation. The first step is to remove structural barriers to women's participation, for example, addressing membership criteria to ensure that not only "heads of households" or landowners can be members, and finding times and locations for meetings that allow women to attend. Attention to the process of meetings and group activities is also needed to make women feel that it is worthwhile to participate, given their time constraints. This may involve helping them build their confidence to speak, as well as facilitating meetings in a manner that they will be listened to.

In societies with strong segregation of the sexes, as in many parts of South Asia and many areas of West Asia and Northern Africa, working through women's organizations may be the best way to strengthen women's social capital and build confidence for women to participate actively in groups. There is some evidence that mixed-sex groups may be more effective in some areas, especially for managing resources like forests or water, where both men and women are important users of the resource. This is because the mixed groups can draw upon the skills of both men and women. However, this is not automatic, and often requires particular efforts to ensure that women's voices are heard, such as by ensuring a critical mass of women's participation in groups, and structuring meetings or activities so that they are relevant to women's interests and draw upon their expertise.

As with any other asset, building social capital requires investments. Further research and program evaluation is needed to assess the extent of the gender asset gap in social capital, as well as to identify effective ways to close this asset gap to enable women, men, their families, and rural communities to benefit.

References

Acharya KP, Gentle P (2006) Improving the effectiveness of collective action: sharing experiences from community forestry in Nepal. CAPRi working paper 54. International Food Policy Research Institute, Washington, DC

Adkins L (2005) Social capital: the anatomy of a troubled concept. Feminist Theory 6(2):195–211

Agarwal B (2000) Conceptualizing environmental collective action: why gender matters. Camb J Econ 24(3):283–310

Agarwal B (2001) Participatory exclusions, community forestry, and gender: an analysis for South Asia and a conceptual framework. World Dev 29(10):1623–1648

Agarwal B (2007) Gender inequality, cooperation, and environmental sustainability. In: Baland JM, Bardhan PK, Bowles S (eds) Inequality, cooperation, and environmental sustainability. Russell Sage Foundation/Princeton University Press, New York/Princeton, pp 274–313

Agarwal B (2009) Gender and forest conservation: the impact of women's participation in community forest governance. Ecol Econ 68(11):2785–2799

Agarwal B (2010) Gender and green governance: the political economy of women's presence within and beyond community forestry. Oxford University Press, Oxford

Agrawal A, Chhatre A (2006) Explaining success on the commons: community forest governance in the Indian Himalaya. World Dev 34(1):149–166

Agrawal A, Yadama G, Andrade R, Bhattacharya A (2006) Decentralization and environmental conservation: gender effects from participation in joint forest management. CAPRi working paper 53. Consultative Group on International Agricultural Research, Washington, DC

Alesina A, La Ferrara E (2000) Participation in heterogeneous communities. Q J Econ 115(3):847–904

Alsop R, Heinsohn N (2005) Measuring empowerment in practice: Structuring analysis and framing indicators. World Bank policy research working paper no. 3510. World Bank, Washington, DC

Andujar A (2005) When women say enough is enough: the struggle for drinking water in Villa Jardin, Argentina. Paper presented at the CAPRi international research workshop on gender and collective action, Chiang Mai, Thailand, 17–21 October

Arganosa-Matienzo EL (2005) When collective action leads to overburdening of rural women. Paper presented at the CAPRi international research workshop on gender and collective action, Chiang Mai, Thailand, 17–21 October

Badstue LB, Bellon MR, Berthaud J, Ramírez A, Flores D, Juárez X (2007) The dynamics of farmers' maize seed supply practices in the Central Valleys of Oaxaca, Mexico. World Dev 35(9):1579–1593

Beard VA (2005) Individual determinants of participation in community development in Indonesia. Environ Plann C Gov Policy 23(1):21–39

Chant S (2003) New contributions to the analysis of poverty: methodological and conceptual challenges to understanding poverty from a gender perspective. United Nations Economic Commission for Latin America and the Caribbean, New York, http://www.eclac.cl/publicaciones/UnidadMujer/5/LCL1955/lcl1955i.pdf

Chen M (2008) A spreading banyan tree: the self-employed women's association, India. In: Mathie A, Cunningham G (eds) From clients to citizens: communities changing the course of their own development. Practical Action, Rugby

Cornwall A (2001) Making a difference? Gender and participatory development. IDS discussion paper no. 378. Institute of Development Studies, Sussex

Davis K, Negash M (2007) Gender, wealth, and participation in community groups in Meru Central District, Kenya. CAPRi working paper series 65. Consultative Group on International Agricultural Research, Washington, DC

Dercon S, Hoddinott J, Krishnan P, Woldehannna T (2011) Burial societies in rural Ethiopia. In: Mwangi E, Markelova H, Meinzen-Dick R (eds) Collective action and property rights for poverty reduction: insights from Africa and Asia. Penn Press for the International Food Policy Research Institute, Philadelphia, pp 51–78

Diamond N, Sims Feldstein H, Gambill D, Hill Rojas M (1997) A working session on communities, institutions, and policies: moving from environmental research to results. WIDTECH, Washington, DC, photocopy

Dikito-Wachtmeister M (2001) Social capital. In: Quisumbing AR, Meinzen-Dick RS (eds) Empowering women to achieve food security. 2020 Focus 6. International Food Policy Research Institute, Washington, DC

Durlauf SN, Fafchamps M (2005) Social capital. In: Aghion P, Durlauf SN (eds) Handbook of economic growth, vol 1. Elsevier, Amsterdam, pp 1639–1699

Fafchamps M, Gubert F (2004) The formation of risk sharing networks. University of Oxford, UK. Photocopy

Fafchamps M, Lund S (2003) Risk-sharing networks in rural Philippines. J Develop Econ 71(2):261–287

Fletschner D, Carter M (2008) Constructing and reconstructing gender: reference group effects and women's demand for entrepreneurial capital. J Socio-Econ 37(2):672–693

Godquin M, Quisumbing AR (2008) "Separate but equal? The gendered nature of social capital in rural Philippine communities. J Int Develop 20(1):13–33

Gotschi E, Njuki J, Delve R (2009) Equal numbers, equal chances? A case study of gender differences in the distribution of social capital in smallholder farmer groups in Búzi District, Mozambique. Eur J Develop Res 21(2):264–282

Grootaert C (1997) Social capital: the missing link? In: Indicators and Environmental Valuation Unit (ed) Expanding the measure of wealth: indicators of environmentally sustainable development. ESD studies and monographs series no. 17. World Bank, Washington, DC, pp 77–93

Haddad L, Maluccio J (2003) Trust, membership in groups, and household welfare: evidence from KwaZulu-Natal, South Africa. Econ Develop Cult Change 51(3):573–601

Hallman K, Lewis D, Begum S (2007) Assessing the impact of vegetable and fishpond technologies on poverty in rural Bangladesh. In: Adato M, Meinzen-Dick R (eds) Agricultural research, livelihoods, and poverty: studies of economic and social impacts in six countries. International Food Policy Research Institute, Washington, DC, pp 103–148

Hambly Odame H (2002) Men in women's groups: a gender and agency analysis of local institutions. In: Cleaver F (ed) Masculinity matters! Men, gender, and development. Zed Press, London, pp 138–165

Hausmann R, Tyson LD, Zahidi S (2007) The global gender gap report 2006. World Economic Forum, Geneva

Hoddinot J, Dercon S, Krishnan P (2005) Networks and informal mutual support in fifteen Ethiopian villages: a description. International Food Policy Research Institute, Washington, DC

Jagger P, Pender J (2006) Influences of programs and organizations on the adoption of sustainable land management technologies in Uganda. In: Pender J, Place F, Ehui S (eds) Strategies for sustainable land management in the East African Highlands. International Food Policy Research Institute, Washington, DC, pp 277–308

James-Sebro M (2005) Revealing the power of gender mainstreaming: enhancing development effectiveness of nongovernmental organizations in Africa. InterAction, Washington, DC

Jewitt S (2000) Unequal knowledge in Jharkhand, India: de-romanticizing women's agroecological expertise. Develop Change 31(5):961–985

Kariuki G, Place F (2005) Initiatives for rural development through collective action: the case of household participation in group activities in the Highlands of Central Kenya. CAPRi working paper 43. International Food Policy Research Institute, Washington, DC

Katungi E, Edmeades S, Smale M (2008) Gender, social capital, and information exchange in rural Uganda. J Int Develop 20(1):35–52

Kumar N, Quisumbing AR (2011) Access, adoption, and diffusion: understanding the long-term impacts of improved vegetable and fish technologies in Bangladesh. J Develop Eff 3(2):193–219

Lauby J, Stark O (1988) Individual migration as a family strategy: young women in the Philippines. Popul Stud 42(3):476–486

Leino J (2007) Ladies first? Gender and the community management of water infrastructure in Kenya. University of California, Berkeley, CA, US. Electronic copy

Maluccio J, Haddad L, May J (2003) Social capital and gender in South Africa, 1993–1998. In: Quisumbing AR (ed) Household decisions, gender, and development: a synthesis of recent research. International Food Policy Research Institute, Washington, DC, pp 145–152

Meinzen-Dick RS, Zwarteveen M (1998) Gendered participation in water management: issues and illustrations from water users associations in South Asia. Agric Hum Values 15(4):337–345

Meinzen-Dick RS, Knox A, Di Gregorio M (eds) (2001) Collective action, property rights, and devolution of natural resource management: exchange of knowledge and implications for policy. Feldafing, Germany: Zentralstelle für Ernährung und Landwirtschaft. Available online at http://www.capri.cgiar.org/workshop_devolution.asp

Molyneux M (2002) Gender and the silences of social capital: lessons from Latin America. Dev Chang 33:167–188

Mwangi E, Meinzen-Dick R, Sun Y (2011) Gender and sustainable forest management in East Africa and Latin America. Ecol Soc 16(1):17. http://www.ecologyandsociety.org/vol16/iss1/art17/

Naved RT (2000) Intrahousehold impact of transfer of modern agricultural technology: a gender perspective. Food consumption and nutrition division discussion paper 85. International Food Policy Research Institute, Washington, DC

Njuki JM, Mapila MT, Zingore S, Delve R (2008) The dynamics of social capital in influencing use of soil management options in the Chinyanja Triangle of Southern Africa. Ecol Soc 13(2):9. http://www.ecologyandsociety.org/vol13/iss2/art9/

Ostrom E (1992) Crafting institutions for self-governing irrigation systems. Institute for Contemporary Studies Press, San Francisco

Padmaja R, Bantilan MCS, Parthasarathy D, Gandhi BVJ (2006) Gender and social capital mediated technology adoption. Impact series no. 12. Patancheru 502 324, International Crops Research Institute for the Sem-Arid Tropics, Andhra Pradesh, India

Padmanabhan M (2006) Collective action in agrobiodiversity management: gendered rules of reputation, trust, and reciprocity in Kerala, India. J Int Develop 20(1):83–97

Panda SM (2006) Women's collective action and sustainable water management: case of SEWA's water campaign in Gujarat, India. CAPRi working paper 61. International Food Policy Research Institute, Washington, DC

Pandey D (1993) Empowerment of women for environmentally sustainable development through participatory action research. In: Ahmed S (ed) Gendering the rural environment: concepts and issues for practice. Workshop report no. I. Institute of Rural Management Workshop proceedings, Anand, India, 23–24 April

Pandolfelli L, Meinzen-Dick R, Dohrn S (2008) Gender and collective action: motivations, effectiveness, and impact. J Int Develop 20(1):1–11

Perdana A, Matakos K, Radin E (2006) Does it pay to participate? CSIS working paper series: WPI 095. Centre for Strategic and International Studies, Washington, DC

Pretty J (2003) Social capital and the collective management of resources. Science 302(5652):1912–1914

Putnam R (1995) Bowling alone: America's declining social capital. J Democr 6(1):65–78

Quisumbing AR (2009) Beyond the Bari: gender, groups, and social relations in rural Bangladesh. CAPRi working paper 96. International Food Policy Research Institute, Washington, DC

Quisumbing AR, Kumar N (2011) Does social capital build women's assets? The long-term impacts of group-based and individual dissemination of agricultural technology in Bangladesh. J Develop Eff 3(2):220–242

Quisumbing AR, McNiven S, Godquin M (2011) Shocks, groups, and networks in Bukidnon, the Philippines. In: Mwangi E, Markelova H, Meinzen-Dick R (eds) Collective action and property rights for poverty reduction: insights from Africa and Asia. Penn Press for the International Food Policy Research Institute, Philadelphia, pp 79–109

Rai B, Buchy M (2004) Institutional exclusion of women in community forestry: is women-only strategy a right answer? In: Kanel K, Mathema P, Kandel BR, Niraula DR, Sharma AR, Gautam M (eds) Twenty-five years of community forestry: contribution in Millennium Development goal, pp 399–408. Proceedings of the fourth national conference of community forestry international, Kathmandu, Nepal, 4–6 August

Sarin M (1995) Regenerating India's forests: reconciling gender equity with joint forest management. IDS Bull 26(1):83–91

Sultana P, Thompson P (2008) Gender and local floodplain management institutions: a case study from Bangladesh. J Int Develop 20(1):53–68

USAID/IFPRI/OPHI (United States Agency for International Development/International Food Policy Research Institute/Oxford Poverty and Human Development Initiative) (2012) Women's empowerment in agriculture index. International Food Policy Research Institute, Washington, DC

Watkins CA (2009) Natural resource use strategies in a forest-adjacent Ugandan village. Hum Ecol 37(6):723–731

Westermann OJ, Ashby J, Pretty J (2005) Gender and social capital: the importance of gender differences for the maturity and effectiveness of natural resource management groups. World Dev 33(11):1783–1799

Woolcock M, Sweetser AT (2007) Social capital: the bonds that connect. Asian Development Bank, Manila

World Bank/IFPRI (International Food Policy Research Institute) (2010) Gender and governance in rural services: insights from India, Ghana, and Ethiopia. Gender and Governance Author Team. International Bank for Reconstruction and Development/World Bank, Washington, DC

Zwarteveen M, Neupane N (1996) Free-riders or victims: women's nonparticipation in irrigation management in Nepal's Chhattis Mauja Irrigation Scheme. Research report no. 7. International Irrigation Management Institute, Colombo

Chapter 11
Gender Implications of Poor Nutrition and Health in Agricultural Households

Jody Harris

Abstract For most of the rural poor, their most important asset is their own physical capacity for work. This depends crucially on individual nutritional and health status. This chapter summarizes the evidence on gender differences in vulnerabilities to poor nutrition and health, and their potential effects on the productivity of men and women in farming households. Adopting a life-cycle perspective, the chapter examines the implications of four key health and nutritional disorders—undernutrition, iron-deficiency anemia, HIV, and malaria—for the productivity and well-being of men and women in agriculture. These disorders have both direct and interacting impacts, with the nature of the disorder and the context in which it is found determining its exact impacts and the strategies required to cope with nutrition and health shocks. In each case, the impact on the productivity of women is different from that on the productivity of men for biological, social, and cultural reasons. The author discusses several promising policies and interventions to prevent and mitigate some of the negative impacts of specific disorders discussed above on women's agricultural productivity and production. The chapter concludes by proposing further research on understanding the complexities of women's time use and trade-offs in coping with ill health and poor nutrition in agriculture, and on evaluating the most promising policies and programs to protect poor women and enhance their productivity in agriculture and income-generating activities.

Keywords Gender • Agricultural households • Health • Human capital • Undernutrition

J. Harris (✉)
Poverty, Health, and Nutrition Division, International Food Policy Research Institute, 2033 K Street, N.W, Washington, DC 20006, USA
e-mail: j.harris@cgiar.org

A.R. Quisumbing et al. (eds.), *Gender in Agriculture: Closing the Knowledge Gap*, DOI 10.1007/978-94-017-8616-4_11,

11.1 Introduction

Nutrition and health status are intricately linked with agriculture. In agricultural households, agriculture affects food security and diets and dictates the balance of daily activities contributing to the nutrition and health of different household members; these, in turn, contribute to the household's human capital inputs for agricultural work, and so the cycle continues. Understanding and responding to these complex linkages and their gender dimensions is both a challenge and an opportunity for development.

The effects of poor health and undernutrition are likely to affect human capital inputs into agriculture differently for men and women: the contributions of women to agriculture and the impacts of agriculture on women vary enormously with context, as a result of differences in intrahousehold decisionmaking, labor allocations, access to resources, family structure, traditional gender roles in agriculture and in the home, and the level of agricultural commercialization, among other factors. Women are often more vulnerable than men to diseases found in developing countries, due to a complex mixture of biological and social factors, including changes in immunity during pregnancy, lowering disease resistance; cultural norms reducing a woman's control over sex and increasing her risk of sexually-transmitted diseases such as HIV/AIDS; working patterns increasing exposure to disease vectors; and lack of knowledge and information about disease and care-seeking (women's educational levels tending to be lower than men's). Women play the major role in family healthcare, providing the majority of caring time within the household and often having responsibility for seeking care and treatment outside of the home, but not always having the power within the household to decide when and how to seek care, nor the control over resources with which to purchase treatment, either for themselves or their children (Okwa 2007; Magadi 2011). Furthermore excessive nutrient losses and increased requirements during the reproductive period, combined with high energy expenditure from physically demanding agricultural work and domestic chores, and poor nutrient intakes where diets are suboptimal, is likely to also place women at higher nutritional risk than their male counterparts (UNSCN 1990).

This chapter aims to summarize the evidence on gender differences in vulnerabilities to poor nutrition and health, and potential effects on the agricultural productivity of men and women in farming households. The chapter provides an overview of nutrition and health as human capital in agriculture, and of specific intergenerational aspects experienced by women and girls. It then concentrates on four major health and nutrition conditions that interact with agriculture and for which men and women tend to have different susceptibilities and outcomes: (1) undernutrition; (2) anemia; (3) HIV, as a chronic, often terminal illness; and (4) malaria, as an acute but seasonally recurring illness. The chapter concludes with a discussion on the interactions between gender, agriculture, nutrition, and health.

11.2 Nutrition and Health as Human Capital in Agriculture

Human capital in agriculture is often measured using indicators such as attained education, farming knowledge, or ability to provide labor (FAO 2011), but nutrition and health underpins all of these measures. Good health and nutrition status are valued by individuals and societies in and of themselves, as well as for the value they can add to productivity in agriculture and other sectors. Good nutrition and health are particularly important in agriculture precisely because the poorest and most marginal groups in society are also those working in agriculture with the most arduous physical requirements, and those most vulnerable to nutritional deficiencies and poor health. The marginal impact of improved health and nutrition is therefore likely to be greater in these populations (Strauss and Thomas 1998).

Figure 11.1 provides a conceptual framework of the potential consequences of poor nutrition and health for agricultural productivity, household income and food security (see, also, the framework designed by Hawkes and Ruel (2006) illustrating the multidirectional linkages between agriculture, nutrition, and health). Poor nutrition and health in a household affect agriculture through loss of work time to absenteeism and caregiving, loss of farming investments due to reduced savings and

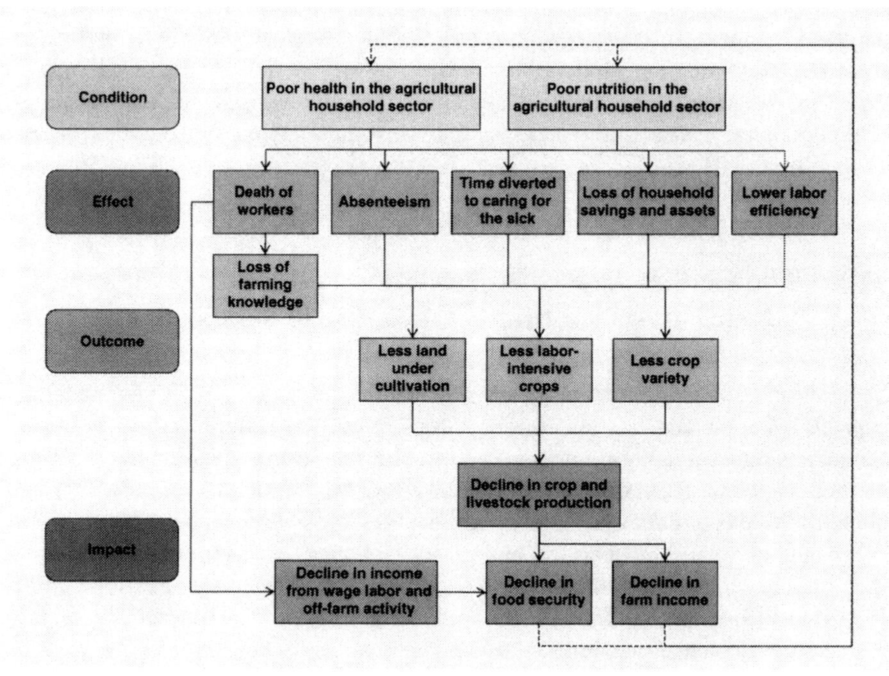

Fig. 11.1 Conceptual framework for the impact of poor health and nutrition on agriculture (Source: Adapted from Asenso-Okyere et al. 2011)

assets, and ultimately death of agriculture workers. Poor nutrition and health can also have direct effects on labor efficiency, referred to as "impaired presenteeism," whereby an individual attends work but accomplishes less than optimal/usual performance. Major outcomes and impacts include loss of farming knowledge due to premature death; loss of cultivated land or converting to less favorable crops if plots revert to the control of other family members or if there is insufficient labor available; and reduced crop variety. These, in turn, result in an overall reduction in crop and livestock productivity and production, and reductions in household income, food security, and diet quality. The broken line linking impacts back to conditions in Fig. 11.1 indicates that the process is a feedback loop, often resulting in a vicious cycle of low productivity, poverty, poor nutrition, and ill health.

Within this framework, the particular route by which poor health and nutrition affect agriculture may be short-run or long-run. In many cases, the most important time for specific agricultural tasks (planting, harvesting) will coincide with the season in which illness or lack of food is most likely (Doss 1999), and absenteeism or reduced labor capacity due to malnutrition, illness, or caring will have a short-run effect on productivity. But if these nutrition and health shocks occur every year, or if labor and knowledge are permanently affected through death of a household member or loss of productive assets due to health costs, then, as well as lowering productivity within certain agricultural enterprises, illness and malnutrition may lock people into lower return enterprises altogether, for the long term, and lead to chronic poverty (Yajima et al. 2005). The potential effects on productivity therefore are many and varied, and coping with different shocks will require different strategies; ultimately, a household's ability to cope with these shocks will depend on its asset base and social, human, and physical resources (Asenso-Okyere et al. 2011).

11.3 The Gender Aspects of Nutrition and Health as Human Capital in Agriculture—A Life-Cycle Perspective

In addition to the different nutrition and health needs, vulnerabilities, and input capacity of men and women in agriculture related to differences in biological, social, and cultural factors, a woman's particular role in child-bearing has an impact on her agricultural productivity, both directly and indirectly: *directly*, through changes in labor productivity due to reduced ability to perform heavy tasks in late pregnancy and early postpartum, and through changes in labor availability due to the time needed for breastfeeding and childcare in the first few years of the child's life (Doss 1999); and *indirectly*, through early, repeated, or frequent pregnancy, which may deplete a woman's nutrient levels and consequently reduce physical capacity, endurance, and labor productivity and raise susceptibility to illness (Smith et al. 2003). Childcare responsibilities generally fall to women, who are left with the choice of undertaking agricultural tasks compatible with this care (usually home gardening or subsistence farming), or leaving infants with other family members or young siblings, which may have negative effects on child welfare and further implications for their future productivity.

The reproductive role of women has significant implications not only for agricultural production during her lifetime, but also for the intergenerational impact of her nutrition and health status on future agricultural productivity through her children. Frequent pregnancy and lactation may deplete a mother's nutrient reserves, which, in turn, can reduce the child's access to nutrients during gestation and through breast milk (King 2003). This increases the risks that children will be born small, will continue to experience growth faltering during early childhood, will have impaired cognitive development and lower schooling performance, and will become smaller, less healthy, and less economically productive adults (Martorell et al. 2010). In the many areas of the developing world where societal norms discriminate against girls, these effects will disproportionately affect girls and women, and perpetuate the transmission of poverty, poor health, and undernutrition into the next generation.

Further, there are occupational health hazards in agriculture that can have an impact on the life cycle: the use of agrochemicals, particularly pesticides, is a common and expanding practice in developing countries, and the exposure of women to these through agricultural work can affect their children, either in utero or through breast milk, with negative outcomes ranging from intrauterine growth retardation to neurological effects (Garry 2004; Sanborn et al. 2004) and potential implications for later health and productivity. Similarly, hookworm parasitic infections are common in men and women working in agriculture and exposed to contaminated soil. The blood loss associated with hookworms often results in anemia (Stoltzfus et al. 1997), which, again, can have potential implications for future productivity through in utero iron deficiency if women are affected during pregnancy.

The good nutrition and health of women, then, is key not only to their own well-being and productivity, but also to that of future generations; while the impact of poor nutrition and health for men is felt in their current productivity, for women the burden affects their own productivity and potentially that of their children, through intergenerational effects (Alderman et al. 2005).

11.4 Undernutrition

Undernutrition is a broad term describing deficiency in energy or other nutrients in the diet. In adults, a commonly used indicator of undernutrition is low weight-for-height measured using body mass index (BMI), which is computed as weight (kg)/height (m) squared. A BMI of less than 18.5 in adults is considered a sign of energy deficiency (World Health Organization 1995).

Early studies examining undernutrition-agriculture linkages typically assessed energy expenditure in a range of agricultural activities, paving the way for later work on productivity (Bleiberg et al. 1980, 1981; Berio 1984; Singh et al. 1989). Although there was debate surrounding the measurement of energy intake or expenditure in these studies, they broadly agreed on two main findings: (1) that agricultural activities tend to make up a major share of rural women's energy expenditure, often at high levels of effort and in addition to normal domestic duties; and (2) that women are more likely than men to suffer from seasonal energy deficiency (negative balance of energy used for physical activity to energy consumed from food, particularly at harvest time, when food stocks are low) and seasonal weight loss.

Several studies have looked at caloric intake and productivity in agriculture, measured as wages or output, controlling for the impact of increased wages on access to calories (Strauss 1986; Sahn and Alderman 1988). These studies documented a significant positive relationship between nutritional status and productivity, although others have not found this association (Deolalikar 1988), suggesting a role for context. Subsequent reviews have found mixed but generally positive evidence for an influence of calorie intake on productivity, even in intervention trials, although these stress that measurement errors are still a problem in this field (Strauss and Thomas 1998; Thomas and Frankenberg 2002; Behrman et al. 2004).

To get around the problem of error in energy intake assessment, studies have taken anthropometric measurements as a proxy for current nutritional status in looking at productivity, particularly weight-for-height and BMI. Regional statistics indicate that where anthropometric figures for men and women differ, the majority show higher prevalence of low BMI among women (although this varies by region, with women in South America and Asia, and especially Southeast Asia, particularly disadvantaged) (FAO 2011). Weight-for-height has been found to be a significant predictor of farm wages and production, even in the absence of a calorie effect (Deolalikar 1988). A study looking at several factors for an impact on productivity found that weight was causally related to women's earnings from all sources at low and normal weights (Thomas and Frankenberg 2000). It has been suggested that the mechanism for increased productivity as measured by these indicators is likely to be increased strength, an asset in agricultural work (Haddad and Bouis 1991).

Several studies have posited height as a useful and measurable indicator of early-life (rather than current) human capital investment, and found subsequent correlations with productivity. Height has been found to be a significant predictor of income and productivity in several studies, with taller individuals being more productive with higher earnings both generally (Strauss and Thomas 1998; Schultz 2002; Hoddinott et al. 2008) and in agriculture specifically (Haddad and Bouis 1991); this last study, looking specifically at gender differences in agricultural productivity, found that weight-for-height indicators (such as BMI) were a stronger predictor of productivity for women than height alone, however (Haddad and Bouis 1991). The literature broadly shows a 2–2.4 % increase in earnings for a 1 % increase in height (Behrman et al. 2004); undernutrition in its broad sense is economically costly, reducing lifetime earnings of affected individuals by up to 10 %, and reducing gross domestic product (GDP) of affected countries by up to 3 % (World Bank et al. 2009).

In summary, while methodological issues persist and evidence is somewhat mixed, there is both biological plausibility and experimental evidence that undernutrition at different points in the life course, and energy imbalance in particular, has an impact on agricultural productivity in both men and women, but that in many contexts, women are more likely to suffer from insufficient energy intake and undernutrition.

11.5 Iron-Deficiency Anemia

Anemia (a deficiency of red blood cells or a reduction in their hemoglobin content) affects over 1.5 billion people, or a quarter of humanity, with prevalence among women significantly higher than among men (13 % of men; 30 % of nonpregnant women; and 42 % of pregnant women) (de Benoist et al. 2008). A major cause of anemia is iron deficiency (leading to iron-deficiency anemia, IDA), although other contributing factors include malaria, parasitic infections, and deficiencies of other micronutrients (WHO 2001). A woman's biologically-determined loss of iron cyclically each month starting in adolescence, and her related reproductive role, takes a toll on her nutritional status, predisposing her to IDA (Smith et al. 2003). The biological plausibility of both direct and indirect effects of iron deficiency or IDA on productivity has caused it to be much studied. Iron deficiency is known to cause fatigue and affect cognitive performance, which may affect not only a woman's agricultural productivity but also her energy and performance in other tasks such as childcare (Smith et al. 2003). Iron deficiency and anemia also affect aerobic capacity, and therefore the capacity for strenuous work and endurance (Haas and Brownlie 2001; Thomas and Frankenberg 2002). Iron deficiency in early life also leads to reduced or delayed cognitive development, which in turn has been linked to reduced adult wages (Horton and Ross 2003).

Several studies have sought to assess the association between IDA and agricultural productivity, or to measure the impacts of iron supplementation on IDA and productivity. Many of these studies are dated, but they document an association between IDA and work capacity (strength and stamina). For example, work performance capacity was found to be reduced in anemic female tea estate workers in Sri Lanka (Gardner et al. 1977) and in anemic male rubber tappers in West Java, Indonesia (Basta et al. 1979). Productivity (measured as quantity picked) was found to increase with subsequent iron supplementation in the first study, with a dose-response relationship between tea picked and increased hemoglobin concentrations, particularly in participants who were more severely anemic at baseline (Edgerton et al. 1979). Supplementation with either iron or iron with vitamin C (a nutrient that enhances iron absorption) has also been associated with higher wages from agricultural labor in India (Weinberger 2003). Subsequent reviews have concluded that iron deficiency negatively affects agricultural productivity and can be corrected through iron supplementation (Haas and Brownlie 2001; Thomas and Frankenberg 2002; Behrman et al. 2004). Iron supplementation has also been shown to increase nonagricultural productivity (such as factory work) in both men and women (Li et al. 1994; Thomas et al. 2006).

Estimates of productivity losses have attempted to quantify both the direct (physical) and indirect (cognitive) economic impacts of iron deficiency and iron replacement. Median annual direct losses, calculated for ten developing countries, have been estimated at around $2.32 per capita, or 0.57 % of GDP, and combined direct and indirect effects at around $3.64 per capita, 0.81 % of GDP (Horton and Ross 2003). This study estimates that iron replacement in the diet could lead to a 17 % increase in productivity in heavy manual labor such as in agriculture, and that the benefit-cost of fortification in this context is 6:1 for direct effects, rising to almost 9:1 when future indirect effects are factored in. Nationally, it has been estimated that a 10 % rise in iron intake could produce a 1–3.4 % rise in productivity in India (Weinberger 2003).

In summary, iron deficiency significantly reduces productivity in men and women, evidenced by several strong intervention trials in which iron supplementation increased work output in both sexes. Although none of the studies cited looked specifically at gender differences in the effects of iron deficiency or iron replacement, women's increased susceptibility to IDA, especially during the reproductive period, is likely to expose this group disproportionately to the productivity losses it induces.

11.6 HIV

Human Immunodeficiency Virus (HIV) is the infection that can lead to Acquired Immunodeficiency Syndrome (AIDS). It attacks the immune system, decreasing immunity and increasing susceptibility to opportunistic infections that are the major cause of morbidity and mortality. An estimated 31 million adults were living with HIV in 2010, with an estimated 52 % of all HIV infections in women (UNAIDS 2010). Gender inequality is a key driver of the epidemic, with women more culturally and biologically vulnerable to the disease; women are at an estimated 60–70 % higher risk of HIV infection than their male counterparts in Sub-Saharan Africa, where the epidemic is at its worst (WHO/UNAIDS 2004; Magadi 2011).

The majority of people infected and affected by HIV in the developing world are supported predominantly by agricultural livelihoods, to the extent that in some regions agriculture is being severely affected (Gillespie 2006a). As with other illnesses, HIV has an impact on agriculture through reduced labor, both directly and through increased caring burden, and increased healthcare costs. But unlike many other illnesses, HIV is chronic, so far incurable, and debilitates over the long term, particularly affecting prime-age adults who would normally be the main agricultural workers and income earners (Gillespie and Kadiyala 2005). This has a major and prolonged impact particularly on women's time diverted to caring, even if they themselves are not sick. Ultimately, the death of male household members leaves unusually large numbers of female-headed households in some areas, and these households are often more vulnerable to poverty, with less labor available, and less land and fewer assets with which to buffer future shocks (FAO et al. 2010).

Changes in household productivity are therefore common in households affected by HIV, as livelihood strategies are adjusted to respond to the shock. It is unknown to what extent households may abandon agriculture for nonfarm activities due to HIV-related illness or death; however, common coping strategies include changes in cropping patterns to farming that demands less labor and fewer resources, and an abandoning of cash crops when labor is short or when women cannot take on previously "male" tasks as well as household duties. Conversely, men may be unwilling or unable to take on culturally "female" agricultural work when female household members are sick, so this work may pass to other women or children, increasing their work burden. Death of a male household member has implications for agricultural knowledge transfer, HIV being a prime example of circumstances under which the remaining family members (often women) resort to lower-return agricultural practices altogether. Death due to HIV may also affect the asset and landholdings of women left behind; widows are often not recognized in inheritance practices in high-HIV-prevalence areas, thereby undermining their independence as farmers (Gillespie 2006b; FAO et al. 2010; Peterman 2010).

Individuals living with HIV may also have reduced work capacity, leading to a vicious cycle of reduced productivity and income, and reduced ability to purchase the drugs or food necessary to maintain strength for work (Gillespie 2006b). Although not disaggregated by gender, a study looking at both productivity and work attendance, this time in tea-estate workers in Kenya, found that those who terminated their employment due to HIV-related illness or death were absent more often, and had significantly impaired presenteeism, picking significantly less tea per day and earning 16–18 % less than non-affected colleagues in the 2 years before termination (Fox et al. 2004). A subsequent study looking at the impact of antiretroviral drug therapy (ART) on HIV-positive tea pickers here found that while the drugs allowed male workers to maintain similar working patterns to their non-infected counterparts, women on ART worked 30 % fewer days, and 100 % more days on non-picking duties, than their peers. No explanation for this apparent gender difference, nor an evaluation of the possible longer-term consequences to women in agriculture, was given in this research (Larson et al. 2009).

The increased caring burden in HIV deserves particular mention, as it is predominantly women on whom this burden falls, and it is a particularly long-term burden in HIV. Additional caring is required not only for the long-term sick, but also for additional children who may join the household as orphans of relatives who have died as a result of HIV. This burden is added to a woman's existing care, household, agricultural and other activities; in several studies, time allocated by women to agricultural activities is lower in HIV-affected households as a result of these multiple time obligations (FAO et al. 2010; FAO 2011).

It is not only at household level that HIV has an impact: both government services and national output are also disrupted in the worst-affected regions. Workers within government departments, including ministries of agriculture and agricultural extension, are just as likely to be infected or affected as the rest of the population, with consequences for the agricultural sector as a whole. Even when government workers are not themselves living with HIV, the sheer volume of absent days due to funeral attendance can have an impact in some areas (Gillespie and Kadiyala 2005; Gillespie 2006b). The sum of effects on individual, household, and sectoral productivity, as well as increased public healthcare costs, means that HIV can exert a significant negative impact on GDP, reducing it by up to 1 % through lost productivity and probably more than that as much economic activity is not captured in national statistics in high-burden countries (Gillespie and Kadiyala 2005).

In summary, the household impacts of HIV are disproportionately felt by women as they are biologically and socially more vulnerable to the disease, they are the main caregivers in most households, and they are disadvantaged socially by the death of a husband or male relative whose agricultural knowledge, labor, and often land is no longer available to them.

11.7 Malaria

Malaria is caused by the Plasmodium parasite, transmitted via the bites of infected mosquitoes to infect red blood cells causing fever, headache, and vomiting and, if not treated, becomes life-threatening, particularly for young children. Malaria-carrying mosquitoes breed in standing water, and therefore malaria is seen to be more prevalent in areas in which agriculture is a primary livelihood (as agriculture thrives particularly in areas of high rainfall or irrigation). Malaria surveillance is not standardized across countries, with different methods and indicators used, so there are no global prevalence data available. It has been noted, however, that the introduction of irrigation for agriculture tends to raise malaria infection rates in areas of sporadic malaria transmission, although in endemic areas natural immunity keeps rates stable, even after the introduction of irrigation, the so-called "paddies paradox" (Ijumba and Lindsay 2001).

Women's vulnerability to malaria is increased in particular, as women tend to have more contact with water than men in general household tasks (with potentially increased contact with the malaria vector) and due to general barriers to accessing treatment and care (Okwa 2007). Women are also more susceptible to malaria in pregnancy, probably due to a combination of hormonal, immunological, and physiological effects (Rogerson et al. 2007). Nonsymptomatic placental malaria infection in pregnancy is often undetected and untreated, increasing the risk of anemia, low birth weight in infants, and obstetric complications and maternal mortality (Desai et al. 2007; Uneke 2008).

As with other disorders reviewed here, malaria can affect household and individual productivity either directly, through days off work or reduced work capacity, or indirectly through effects on caring, cognitive development and education levels. The number of productive days lost to malaria has been found to vary significantly according to patterns of transmission, with endemic areas having higher transmission than epidemic, seasonal areas, but with a shorter duration of each bout due to partial immunity induced by such high bite rates (Cropper et al. 2004). Studies have found between 4 and 11 days per bout of illness lost to malaria at household level (no data available disaggregated by sex) (Cropper et al. 2004; Bloom et al. 2006; Canning 2008). The value of this lost time will vary according to wage rates or market prices for produce, and has been calculated at between US$6 and US$31 per bout in different African countries (Ettling and Shepard 1991; Sauerborn et al. 1991; Shepard et al. 1991; Cropper et al. 2004). Impact on productivity may also be affected by the season in which malaria transmission occurs, with time off work having a larger impact at key points in the agricultural cycle, such as planting and harvesting (Cropper et al. 2004).

Indirect impacts on agricultural productivity are many and varied: Productive time lost by those other than the infected household member, including substitute labor by other household members and time devoted to caring, will affect household productivity. This care will usually be provided by women, as will care for the children who are the prime targets of malaria, disproportionately taking women out of agricultural work even when they are not themselves sick. Exposure to malaria in childhood can impede both physical and cognitive development (Cutler et al. 2010), and has been found to negatively impact female literacy, all of which can depress later productivity. Where malaria recurs seasonally or is endemic, repeated bouts can cause a sustained decline in productivity and affect the long-term nature of agriculture in farming households (Cropper et al. 2004; Canning 2008). It has been noted that many studies examining the impact of malaria on productivity fail to consider the effect of household coping strategies, such as labor substitution, in their calculations (Bloom et al. 2006). Finally, resources that might otherwise have been invested in agriculture may need to be redirected to healthcare costs where malaria is common, with the cost of treatment estimated at US$0.41 to US$3.88 per person per month, or up to 28 % of household income, in various Sub-Saharan countries (excluding peripheral costs such as travel to a clinic), and the cost of prevention (mainly bednets and repellents) at slightly less (US$0.05 to US$2.10 per person) (Chima et al. 2003). Malaria eradication programs in India have been found to raise per capita household expenditure for men but not for women, possibly reflecting the higher labor force participation of men (Cutler et al. 2010).

At the national and regional level there have been several estimates of impacts on GDP. Malaria-endemic countries are also some of the poorest countries in the world; correlations show far higher mortality rates from infectious diseases such as malaria in developing economies, and countries with severe malaria have income rates at a third of non-malarial countries (Gallup and Sachs 1998; Strauss and Thomas 1998). It has been estimated that GDP growth in Sub-Saharan Africa was reduced by 1–2 percentage points annually before 1990 due to malaria, and that countries with endemic malaria grew 1.3 % less per person per year than non-malarial countries, with a 10 % reduction in malaria associated with 0.3 % higher growth (Gallup and Sachs 2001). Other studies in Africa have shown country-specific reductions in GDP of up to 6 %, with the proportion of women working in each sector significantly affecting the impact on that sector; in Kenya the agricultural sector was found to be most affected, at 13 % reduction in productivity, due to the preponderance of women in that sector (Leighton and Foster 1993).

In summary, in addition to an increased biological and social susceptibility to malaria, women again disproportionately absorb the additional burden of care related to malaria and other tropical diseases, and yet women often do not have the social or financial resources to access either preventative or curative measures. Direct and indirect impacts on agriculture, as well as the recurring, seasonal nature of the disease, have major implications for agricultural productivity of both men and women.

11.8 Discussion

This chapter summarizes evidence of the gender-specific effects on agriculture of the nutrition and health status of women and their families, illustrated through four key nutrition and health disorders. While nutrition is seen to impact predominantly on work capacity, it also indirectly affects health through reduced disease resistance. Ill health, on the other hand, carries additional indirect effects through demands for caring time and resources for accessing treatment and healthcare. It should be noted that the four disorders mentioned above can and do interact: micronutrient deficiencies have been found to contribute significantly to malaria morbidity and mortality (Caulfield et al. 2004); HIV increases malaria risk and the ill effects of malaria (Desai et al. 2007); undernutrition is both a contributing factor to and a consequence of deteriorating HIV status (Gillespie et al. 2001); and malaria is known to cause anemia in pregnant women (Desai et al. 2007), even while supplementation with iron has been seen to increase morbidity and mortality in malaria-endemic areas (Sazawal et al. 2006), for instance.

Although linkages between nutrition, health, and agricultural productivity have been explored, there are several methodological issues that still need to be addressed in measuring diet, nutrition, health, and productivity. Both energy and micronutrient intakes have been hypothesized to have effects on the ability to perform labor-intensive activity and therefore on agricultural productivity; however, teasing out precisely which nutrients have which effects, and how, has not been conclusive. Studies of nutrition and productivity fall broadly into two groups: those that assess diet, calorie, or nutrient intakes, through measurement of food consumed by individuals or extrapolated from food available in the household; and those that assess physical status, through anthropometry or biomarkers. The former face all the usual challenges of measuring diet, such as recall bias, measurement error, and difficulties in assessing the individual intake of different household members, especially if some meals are taken outside the home. The latter face the problem that anthropometry and biomarkers are far from ideal proxies for dietary intake because of the multiple factors, including health status, that affect how food and nutrients are used by the body and how they affect body composition and overall nutritional status. Other methodological issues surround the feedback loops between income and nutrition: as wages rise, more money is likely to be spent on food, blurring the direction of causality between nutrition, health, and agricultural productivity.

In terms of measuring productivity, using wages as a proxy is not a perfect reflection of productivity and may mask household coping strategies such as work-sharing. It should also be noted that many studies looking at national-level productivity are based on males, especially if they are conducted in countries with very low female labor force participation; estimates of losses or gains in national-level productivity in previous sections are therefore provided only for background information, as it is not possible to separate out male and female contributions in most cases. The potential for GDP bias in not counting women in these estimates, as much of their work is undertaken in or around the home and unpaid, is well-established.

Further, there is no single agreed indicator for measuring "health," and measurement is systematically biased toward those who do seek care, with those who do not visit health facilities often lost to data collection. Measuring women's time use and decisionmaking power is also complex, and therefore rarely undertaken. To fully understand the complexities of the gender aspects of nutrition and health in agricultural livelihoods, a vast array of quantitative and qualitative data are needed, which is rarely possible in practice (Strauss and Thomas 1998; FAO et al. 2010). Field studies on this topic must choose carefully their indicators across productivity, income, time allocation, coping strategies, nutrition, and health in order to fully understand the interactions and their implications (Horton and Levin 2001).

The cases of the disorders reviewed above serve as illustrations of the interactions and impacts summarized in Fig. 11.1; the effects of each disorder on agriculture fall under the headings of workdays lost to illness and caring for family members, reduced labor efficiency, and losses of household savings, assets, and resources, but it is the nature of the disorder and the context in which it is found that will determine its exact impacts and the strategies required to cope with the shock. In each case, the impact on the productivity of women is different to that on the productivity of men for biological, social, and cultural reasons; it is a combination of a woman's distinct reproductive role and different social and cultural contexts that exposes women disproportionately to these impacts.

Several promising policies and interventions have been suggested to prevent and mitigate some of the negative impacts of specific disorders discussed above on women's agriculture productivity and production (Behrman et al. 2004; Gillespie and Kadiyala 2005; Quisumbing and Pandolfelli 2010). Promising approaches include the consideration of local gender norms in the design of agricultural interventions; recognizing and managing trade-offs in women's time and resource allocation; ensuring public-health and education policies that explicitly support girls and women, including providing nutritional supplements and reproductive health services to reduce the intergenerational transmission of poor nutrition and health, and bed nets for malaria reduction; and securing women's access to preventive and curative healthcare, particularly antiretroviral drugs for HIV and intermittent sulfadioxine-pyrimethamine in pregnancy for malaria prevention. Further research is needed particularly on understanding the complexities of women's time use and trade-offs in coping with ill health and poor nutrition in agriculture, and on evaluating the most promising agricultural policies and programs to protect poor women and enhance their productivity in agriculture and income generating capacity.

Broadly, investment in nutrition and health as human capital does appear to have the potential to increase the productivity of women in agriculture, but the social barriers to that investment, such as women's time allocation and traditional roles, must be recognized and addressed in order for that to happen; without this attention to broader gender issues, any gains from nutrition and health interventions for women in agriculture are unlikely to achieve their potential.

Acknowledgments I am grateful for comments on concepts and early drafts from Amber Peterman, Ruth Meinzen-Dick, and Agnes Quisumbing; for input to the malaria section from Andrew Dillon; and for substantial input and edits from Marie Ruel. Any errors are the responsibility of the author.

References

Alderman H, Behrman JR, Hoddinott J (2005) Nutrition, malnutrition, and economic growth. In: López-Casasnovas G, Rivera B, Currais L (eds) Health and economic growth: findings and policy implications. Massachusetts Institute of Technology Press, Cambridge, MA, pp 169–194

Asenso-Okyere K, Chiang C, Thangata P, Andam KS (2011) Interactions between health and farm-labor productivity. Food policy report. International Food Policy Research Institute, Washington, DC

Basta S, Soekirman S, Karyadi D, Schrimshaw NW (1979) Iron deficiency anemia and the productivity of adult males in Indonesia. Am J Clin Nutr 32(4):916–925

Behrman JR, Alderman H, Hoddinott J (2004) Malnutrition and hunger. In: Lomborg B (ed) Global crises, global solutions. Cambridge University Press, Cambridge, pp 363–442

Berio A-J (1984) The analysis of time allocation and activity patterns in nutrition and rural development planning. Food Nutr Bull 6(1):53–68

Bleiberg FM, Brun TA, Goihman S, Gouba E (1980) Duration of activities and energy expenditure of female farmers in dry and rainy seasons in Upper Volta. Br J Nutr 43(1):71–82

Bleiberg F, Brun TA, Goihman S, Lippmann D (1981) Food intake and energy expenditure of male and female farmers from Upper-Volta. Br J Nutr 45(3):505–515

Bloom D, Bloom L, Weston M (2006) Business and malaria: a neglected threat? World Economic Forum, Geneva

Canning D (2008) Diseases. A commentary on "Disease control" by Dean T Jamison, Prabhat Jha, and David E Bloom. Copenhagen consensus 2008 perspective paper. Copenhagen

Caulfield LE, Richard SA, Black RE (2004) Undernutrition as an underlying cause of malaria morbidity and mortality in children less than five years old. Am J Trop Med Hyg 71(Supp. 2):55–63

Chima RI, Goodman CA, Mills A (2003) The economic impact of malaria in Africa: a critical review of the evidence. Health Policy 63(1):17–36

Cropper ML, Haile M, Lampietti J, Poulos C, Whittington D (2004) The demand for a malaria vaccine: evidence from Ethiopia. J Develop Econ 75(1):303–318

Cutler D, Fung W, Kremer M, Singhal M, Vogl T (2010) Early-life malaria exposure and adult outcomes: evidence from malaria eradication in India. Am Econ J Appl Econ 2(2):72–94

de Benoist B, McLean E, Egli I, Cogswell M (eds) (2008) Worldwide prevalence of anaemia 1993–2005: WHO global database on anaemia. World Health Organization, Geneva

Deolalikar AB (1988) Nutrition and labour productivity in agriculture: estimates for rural South India. Rev Econ Stat 70(3):406–413

Desai M, ter Kuile FO, Nosten F, McGready R, Asamoa K, Brabin B, Newman RD (2007) Epidemiology and burden of malaria in pregnancy. Lancet Infect Dis 7(2):93–104
Doss CR (1999) Twenty-five years of research on women farmers in Africa: lessons and implications for agricultural research institutions; with an annotated bibliography. Economics program paper 99-02. Centro Internacional de Mejoramiento de Maiz y Trigo, Mexico City
Edgerton VR, Gardner GW, Ohira Y, Gunawardena KA, Senewiratne B (1979) Iron-deficiency anaemia and its effect on worker productivity and activity patterns. Br Med J 2(6204): 1546–1549
Ettling MB, Shepard DS (1991) Economic cost of malaria in Rwanda. Trop Med Parasitol 42(3): 214–218
FAO (Food and Agriculture Organization of the United Nations) (2011) The state of food and agriculture 2011—women in agriculture: closing the gender gap for development. FAO, Rome
FAO/IFAD/ILO (Food and Agriculture Organization of the United Nations/International Fund for Agricultural Development/International Labour Office) (2010) Gender dimensions of agricultural and rural employment: differentiated pathways out of poverty—status, trends, and gaps. Food and Agriculture Organization of the United Nations, Rome
Fox MP, Rosen S, MacLeod WB, Wasunna M, Bill M, Foglia G, Simon JL (2004) The impact of HIV/AIDS on labour productivity in Kenya. Trop Med Int Health 9(3):318–324
Gallup J, Sachs J (1998) The economic burden of malaria. Center for International Development at Harvard University, Cambridge, MA
Gallup J, Sachs J (2001) The economic burden of malaria. Am J Trop Med Hyg 64(Supplement 1): 85–96
Gardner GW, Edgerton VR, Senewiratne B, Barnard RJ, Ohira Y (1977) Physical work capacity and metabolic stress in subjects with iron deficiency anemia. Am J Clin Nutr 30(6):910–917
Garry V (2004) Pesticides and children. Toxicol Appl Pharmacol 198(2):152–163
Gillespie S (2006a) HIV and agriculture. Understanding the links between agriculture and health. 2020 vision for food and agriculture. International Food Policy Research Institute, Washington, DC
Gillespie S (ed) (2006b) AIDS, poverty, and hunger: challenges and responses. International Food Policy Research Institute, Washington, DC
Gillespie S, Kadiyala S (2005) HIV/AIDS and food and nutrition security: from evidence to action. Food policy review 7. International Food Policy Research Institute, Washington, DC
Gillespie S, Haddad L, Jackson R (2001) HIV/AIDS, food, and nutrition security: impacts and actions. In: United Nations Administrative Committee on Coordination. Sub-committee on Nutrition (ACC/SCN) (eds) Nutrition policy paper No 20 (Report of the 28th session symposium, 3–4 April 2001, Nairobi, Kenya). ACC/SCN, Geneva, Oct 2001
Haas JD, Brownlie T (2001) Iron deficiency and reduced work capacity: a critical review of the research to determine a causal relationship. J Nutr 131(2):676S–690S
Haddad LJ, Bouis HE (1991) The impact of nutritional status on agricultural productivity: wage evidence from the Philippines. Oxf Bull Econ Stat 53(1):45–68
Hawkes C, Ruel M (2006) The links between agriculture and health: an intersectoral opportunity to improve the health and livelihoods of the poor. Bull World Health Organ 84(12):984–990
Hoddinott J, Maluccio JA, Behrman JR, Flores R, Martorell R (2008) Effect of a nutrition intervention during early childhood on economic productivity in Guatemalan adults. Lancet 371(9610): 411–416
Horton S, Levin C (2001) Commentary on "evidence that iron deficiency anemia causes reduced work capacity". J Nutr 131(2):691S–696S
Horton S, Ross J (2003) The economics of iron deficiency. Food Policy 28(1):51–75
Ijumba JN, Lindsay SW (2001) Impact of irrigation on malaria in Africa: paddies paradox. Med Vet Entomol 15(1):1–11
King JC (2003) The risk of maternal nutritional depletion and poor outcomes increases in early or closely spaced pregnancies. J Nutr 133(5):1732S–1736S
Larson B, Fox M, Rosen S, Bii M, Sigei C, Shaffer D, Sawe F, McCoy K, Wasunna W, Simon JL (2009) Do the socioeconomic impacts of antiretroviral therapy vary by gender? A longitudinal study of Kenyan agricultural worker employment outcomes. BMC Public Health 9(1):240

Leighton C, Foster R (1993) Economic impacts of malaria in Kenya and Nigeria. Abt Associates, Bethesda

Li R, Chen X, Yan H, Deurenberg P, Garby L, Hautvast JG (1994) Functional consequences of iron supplementation in iron-deficient female cotton mill workers in Beijing, China. Am J Clin Nutr 59(4):908–913

Magadi MA (2011) Understanding the gender disparity in HIV infection across countries in Sub-Saharan Africa: evidence from the Demographic and Health Surveys. Sociol Health Illn 33(4):522–539

Martorell R, Melgar P, Maluccio JA, Stein AD, Rivera JA (2010) The nutrition intervention improved adult human capital and economic productivity. J Nutr 140(2):411–414

Okwa O (2007) Tropical parasitic diseases and women. Ann Afr Med 6(4):157–163

Peterman A (2010) Widowhood and asset inheritance in Sub-Saharan Africa: empirical evidence from 15 countries. Presented at the Chronic Poverty Research Centre/Overseas Development Institute Roundtable on Inheritance and the Intergenerational Transmission of Poverty, ODI, London, 11 October 2010)

Quisumbing AR, Pandolfelli L (2010) Promising approaches to address the needs of poor female farmers: resources, constraints, and interventions. World Dev 38(4):581–592

Rogerson SJ, Hviid L, Duffy PE, Leke RF, Taylor DW (2007) Malaria in pregnancy: pathogenesis and immunity. Lancet Infect Dis 7(2):105–117

Sahn DE, Alderman H (1988) The effects of human capital on wages, and the determinants of labor supply in a developing country. J Develop Econ 29(2):157–183

Sanborn M, Cole D, Sanin LH, Bassil K (2004) Systematic review of pesticide human health effects. Ontario College of Family Physicians, Ontario

Sauerborn R, Shepard DS, Ettling MB, Brinkmann U, Nougtara A, Diesfeld HJ (1991) Estimating the direct and indirect economic costs of malaria in a rural district of Burkina Faso. Trop Med Parasitol 42(3):219–223

Sazawal S, Black RE, Ramsan M, Chwaya HM, Stoltzfus RJ, Dutta A, Dhingra U, Kabole I, Deb S, Othman MK, Kabole FM (2006) Effects of routine prophylactic supplementation with iron and folic acid on admission to hospital and mortality in preschool children in a high malaria transmission setting: community-based, randomised, placebo-controlled trial. Lancet 367(9505):133–143

Schultz TP (2002) Wage gains associated with height as a form of health human capital. Am Econ Rev 92(2):349–353

Shepard DS, Ettling MB, Brinkmann U, Sauerborn R (1991) The economic cost of malaria in Africa. Trop Med Parasitol 42(3):199–203

Singh J, Prentice AM, Diaz E, Coward WA, Ashford J, Sawyer M, Whitehead RG (1989) Energy expenditure of Gambian women during peak agricultural activity measured by the doubly-labelled water method. Br J Nutr 62(2):315–329

Smith LC, Ramakrishnan U, Ndiaye A, Haddad L, Martorell R (2003) The importance of women's status for child nutrition in developing countries. Research report 131. International Food Policy Research Institute, Washington, DC

Stoltzfus RJ, Dreyfuss ML, Chwaya HM, Albonico M (1997) Hookworm control as a strategy to prevent iron deficiency. Nutr Rev 55(6):223–232

Strauss J (1986) Does better nutrition raise farm productivity? J Polit Econ 94(2):297–320

Strauss J, Thomas D (1998) Health, nutrition, and economic development. J Econ Lit 36(2):766–817

Thomas D, Frankenberg E (2000) Links between women's health and labor market outcomes in Indonesia. The POLICY project working papers 2. Futures Group International, Inc, Washington, DC

Thomas D, Frankenberg E (2002) Health, nutrition, and prosperity: a micro-economic perspective. Bull World Health Organ 80(2):106–113

Thomas D, Frankenberg E, Friedman J, Habicht J-P, Hakimi M, Ingwersen N, Jaswadi, Jones N, McKelvey C, Pelto G, Sikoki B, Seeman T, Smith JP, Sumantri C, Suriastini W, Wilopo S (2006) Causal effect of health on labor market outcomes: experimental evidence. California Center for Population Research (CCPR) 070-66. University of California-LA, Los Angeles, CA

UNAIDS (2010) UNAIDS report on the Global AIDS Epidemic 2010. Geneva

Uneke CJ (2008) Impact of placental Plasmodium falciparum malaria on pregnancy and perinatal outcome in Sub-Saharan Africa: part III: placental malaria, maternal health, and public health. Yale J Biol Med 81(1):1–7

UNSCN (United Nations Standing Committee on Nutrition) (1990) Women and nutrition. Nutrition policy discussion paper no. 6. United Nations, Administrative Committee on Coordination/Subcommittee on Nutrition, Geneva

Weinberger K (2003) The impact of micronutrients on labor productivity: evidence from rural India. Presented at the 25th International Conference of Agricultural Economists (IAAE), Durban, South Africa, 16–22 August 2003

WHO (World Health Organization) (1995) Physical status: the use and interpretation of anthropometry. Report of a WHO expert committee. World Health Organization technical report series 854. Geneva

WHO (World Health Organization) (2001) Iron deficiency anaemia assessment, prevention, and control: a guide for programme managers. WHO, Geneva

WHO (World Health Organization)/UNAIDS (2004) Ensuring equitable access to antiretroviral treatment for women. WHO/UNAIDS policy statement. Geneva

World Bank/FAO/IFAD (Food and Agriculture Organization of the United Nations/International Fund for Agricultural Development) (2009) Gender in agriculture: sourcebook. World Bank/FAO/IFAD, Washington, DC

Yajima M, Chiwona-Karltun L, Kambewa P, van Huis A, Jiggens J (2005) Learning about cassava in an HIV/AIDS-affected area of Malawi. Presented at the eighth Oxford conference on learning and development: learning and sustainable rural livelihoods in the times of AIDS—challenges and prospects. Oxford, UK, 13–15 September 2005)

Part IV
Gender and Markets: Moving Beyond the Farm

Chapter 12
Promoting Gender-Equitable Agricultural Value Chains: Issues, Opportunities, and Next Steps

Deborah Rubin and Cristina Manfre

Abstract This chapter reviews the growing body of work on reducing gender-based barriers to value chain development. It highlights key questions that are emerging within the gender and value chain community related to methodologies for promoting both greater gender equity and efficiency. The authors lay out the rationale and evidence for promoting gender equitable value chains focusing on business, social justice, and development goals. The chapter then reviews the terms and assumptions used in value chain approaches and provides evidence and examples of different gender and value chain approaches. The authors also look at gender issues in value chain performance and gender issues benefitting from value chain production, including employment and income and social capital and networking. This is followed by a review of current debates in the field of gender and value chain studies. The concluding section identifies new questions and challenges facing researchers and practitioners, for example, on chain selection, targeting of women, and achieving food security and improved nutrition in value chain development.

Keywords Gender • Agriculture • Value chains • Markets • Employment

12.1 Introduction

Fears of new food crises have recently raised the profile and funding of agricultural activities for the developing world, from donor programs to foreign private investors seeking to stabilize food supplies at home (World Bank 2007; von Braun and Meinzen-Dick 2009). This has created a corresponding growth in efforts to

---

D. Rubin (✉) • C. Manfre
Cultural Practice, LLC, 4300 Montgomery Avenue,
Suite 305, Bethesda, MD 20814, USA
e-mail: drubin@culturalpractice.com; cmanfre@culturalpractice.com

A.R. Quisumbing et al. (eds.), *Gender in Agriculture: Closing the Knowledge Gap*, DOI 10.1007/978-94-017-8616-4_12,

understand how the globalizing food system is shaping (or thwarting) opportunities for women to participate more fully in agricultural value chains, the organizational links that structure how products move and are improved from the farm to the consumer. Globally, women—as unpaid family workers, as farm managers, processors, wage workers, managers, and entrepreneurs—contribute time, energy, creativity, and knowledge to the production, processing, and marketing of crops and livestock in amounts that are, on average, nearly equivalent and, in some instances, surmount the contributions of men (FAO 2011), but they typically remain small farmers and laborers, reaping low returns from production, packing, and processing jobs. Furthermore, only a small minority are entrepreneurs in transportation, marketing, and exporting, where more value is added and returns are higher (see Hill and Vigneri, Chap. 13). This imbalance reinforces the importance for continued attention to engendering the analyses that often determine which crops to commercialize and what form that commercialization should take.

Value chain approaches are popular because they clarify how market relationships are organized among different stakeholders, but there is little consensus on the most useful methodologies for either analysis or chain development. Most approaches do not clearly address how to organize markets in gender equitable ways, e.g., how to best increase women's participation in agricultural enterprises or how to effectively reduce gender inequalities in accessing inputs or services. The gender and development community has been quick to point to the need to engender value chain approaches or risk exacerbating gender inequalities and marginalizing women. Only recently has the question of how to promote more gender-equitable agricultural development become an explicit component of value chain development efforts[1] (Chan 2010; Mayoux and Mackie 2007; Rubin et al. 2009).

This chapter looks at this growing body of work to highlight key questions that are emerging within the gender and value chain community. It reviews the terms and assumptions used and provides evidence and examples of different gender and value chain approaches. It concludes with noting some new questions and challenges facing researchers and practitioners. Because the gender and value chain literature is still limited in scope, this review also draws on a wider range of studies related to gender and agriculture, food security, and cash cropping.

12.2 Value Chains: Theory and Practice

A core element of pro-poor growth approaches in agriculture is the emphasis on a more diversified agricultural sector to generate employment and offer more inclusive participation of farmers and enterprises. A value chain charts the sequence of

[1] Even the comprehensive *Gender in Agriculture Sourcebook* published jointly by the World Bank, IFAD, and FAO in 2009 (World Bank et al. 2009) did not include a chapter on gender and value chains as a separate reference topic, and other recent contributions to setting research priorities for value chains continue to downplay or ignore the gender dimensions of the topic (see Gómez et al. 2011).

actions and the organizational links that move a product or service from conception, through a series of steps, including production, processing, marketing, and delivery to final consumers, through to its consumption and disposal. The "farm to fork" connections are the subject of value chain analysis, a focused process of data collection and analysis that grows out of earlier studies of markets and their operations across different geographical and political terrains: national, regional, and international. With the intensification of globalization, scholars have recognized the emergence of new forms of connectivity between producers, buyers, and consumers. Sales operations have become better organized and more controlled, with procurement processes shifting from wholesale markets, where multiple sellers competed with each other to gain access to multiple buyers, to a well-coordinated "chain" of known suppliers selling to a single purchaser (Gereffi 1994; Kaplinsky and Morris 2000; Reardon and Berdegué 2002).

Value chain analysis involves collecting information about these market connections to help firms identify their strengths (or weaknesses) vis-à-vis other members of the chain. In this context, the firm can be a farming household, a business, or an association engaged in producing, processing, transporting, or marketing agricultural goods—or supporting those activities. When the results of the analysis are applied, firms should be able to become more competitive by reducing their costs or enhancing the distinctiveness of their products or services (or both), a process known as "upgrading." At the same time, the analysis should also be able to point to ways that the chain as a whole can become more efficient. In a development context, value chain analysis is used to promote value chain development, a directed effort to build both the competencies of the firm and the improved functioning of the chain, often with a pro-poor focus (Mayoux and Mackie 2007). It may also involve a collaborative visioning to help the different chain actors assess their own roles in the chain (USAID MicroLinks Wiki n.d.). A value chain analysis oriented toward achieving the goal of poverty reduction should also assess how well the different options for chain development are able to both encourage broad-based growth and achieve greater gender equality.

Engendering value chain analysis involves consistently making explicit the different levels and categories of men's and women's participation in value chain activities. It requires the use of a gender-sensitive methodology for value chain selection. It includes seeking out women-owned and women-managed firms and analyzing in which subsectors of the economy they function. It considers how these firms choose suppliers, reach customers, and develop business plans, and whether aspects of these tasks are distinguishable from firms owned by men. Both women and men, as workers and entrepreneurs, should be represented when mapping the chain and convening stakeholders to discuss the results of the analysis (Gammage et al. 2009; Rubin et al. 2009).

Whether working with donors, community development organizations, or the private sector, the work of engendering agricultural value chains is based on the premise that it can be a win-win opportunity, where the results of developing a more inclusive work force and entrepreneurial base will benefit workers (both employees and smallholders) and firm owners alike. This premise is based on three assumptions[2]:

- First, value chains are embedded in a social context that reflects the operation of gender relations from the household to the firm. This assumption is rooted in the concept of the *gendered economy*, that economic systems express the consequences of gender relations, for example, shaping which jobs are open to men or women (Elson 1999). Understanding the functions and operations of value chain actors cannot be isolated from an examination of how gender roles and relations shape and have an impact on particular behaviors within value chains;
- Second, value chain development affects gender roles and relations. A large body of qualitative research has documented shifts in allocation of responsibilities in the household, often increasing demands on women's labor. There are examples of how increasing women's participation in market-oriented production can either increase or decrease their access to and control over income, depending upon the character of their involvement and the specific characteristics of the chain (Hamilton et al. 2002; Dolan and Sorby 2003; Coles and Mitchell 2009).
- Third, gender equity and value chain competitiveness are mutually supportive goals. Large-scale comparative studies have demonstrated that greater gender equality and economic growth can go hand-in-hand and that gender inequalities are costly and inefficient (World Bank 2001; World Bank et al. 2009).

In practice, achieving both gender equality and efficient value chains is often elusive. A recent DANIDA review of gender and value chains finds few evaluations of value chain interventions that address gender dimensions (Riisgaard et al. 2010). Among the evaluation studies analyzed, the results of women's participation in value chains are mixed, with the authors concluding that there is no automatic association between increasing women's participation in value chains and increases in women's decisionmaking power in the household. There is need for much more serious attention to careful measurement of the changes in women's time allocation patterns, access to and control over income, and decisionmaking opportunities in the household and in the community resulting from increased participation in value chains.

Clearly, achieving this "win-win" of economic growth and increasing gender equality will require intentional efforts to create positive synergies between gender relations and value chain development. The reasons and methods for doing so are described in the next Section.

12.3 What Is the Rationale and Evidence for Promoting Gender-Equitable Value Chains?

The renewed focus on agriculture since the global food price crises of 2007–2008 has brought a growing recognition that past efforts failed to dynamize agricultural markets in sustainable, equitable, and commercially viable ways. Moreover, they

[2] These assumptions are discussed in greater detail in Rubin et al. (2009).

have not engaged the full range of actors, from smallholder farmers to multinational food corporations. Increasingly, arguments are being made that achieving an end to hunger and poverty will require addressing gender inequalities in agricultural value chain development, framed in terms of the business, social justice, or development cases.

12.3.1 The Business Case

Gender equality and women's economic empowerment outcomes have not always translated well into compelling arguments for engaging private-sector actors. Yet being able to articulate the business case for addressing gender inequalities and supporting women's economic empowerment is critical to winning the support of local, regional, and multinational firms and leveraging their power in the chain. The business case rests on the positive impact to firm-level and economy-wide performance that can be achieved by proactively promoting women's participation in value chains and removing gender-based impediments limiting their productivity.

- Ensure the flow of quality goods

Many women are involved in producing and handling crops. However, as low wage and unpaid workers, women have few incentives to invest their time and energy into improving production and processing practices. While women often have control over the small amounts of income they generate through local sale of food stuffs and other microenterprises, larger sums of money are often controlled or can be appropriated by men. Evidence from Kenya on the chili pepper value chain reveals that under these circumstances women may withdraw their labor, particularly if others, such as spouses, reap the economic benefits from their work, thus endangering the constant supply of materials necessary for a functioning value chain (Rubin et al. 2009). Addressing women's lack of incentives to participate in the value chain can go a long way toward ensuring the long-term supply of quality products to the value chain.

- Improve the efficiency of business

On the farm, men are still often assumed by business, government, and development representatives alike to be the "real" farmers and thus receive a greater proportion of available technical assistance and extension services, even for tasks and crops that women manage (see Ragasa, Chap. 17). As a result, information about new techniques and upgrading does not flow to the appropriate individuals, costing firms through decreased volume and quality of goods. Sex-segmentation and discrimination in hiring practices also create inefficiencies in human capital and productivity. Adopting business practices that reduce these inefficiencies, for example by hiring women extension officers and by targeting both men and women for technical assistance, will improve product quality, and create a more efficient chain by raising productivity and reducing waste.

- Take advantage of new market opportunities

Women are often invisible and underserved buyers and suppliers in agricultural value chains. Firms may be able to increase their client base by sweeping aside preconceived notions of who participates in agriculture, targeting women through the design and delivery of business development services (both financial and non-financial) and reaching out to them directly. In the mobile phone sector, it has been estimated that revenue opportunities of US$13 billion could be achieved by closing the mobile gender gap through the addition of 300 million women subscribers in low and middle-income countries (GSMA 2010, 7).

- Target niche markets and corporate social responsibility opportunities

In industrialized countries, consumers increasingly ask where and how goods are produced and who produces them, creating opportunities for actors along specific value chains to market their socially responsible actions to discerning consumers. Examples include marketing products as fair trade and certified organic. Women-only coffee cooperatives like Café Femenino in Peru and Las Hermanas in Honduras are supplying large coffee retailers eager to meet consumer interest in social responsibility.

12.3.2 The Social Justice Case

Gender equality has a sound footing in international agreements and is widely accepted by the development community. As a basic human right, gender equality is recognized as an end in and of itself. This is affirmed in numerous global and regional agreements from the Convention on the Elimination of All Forms of Discrimination against Women or CEDAW (1979) to the Millennium Development Goals (2000). In value chain development, social justice principles support efforts to:

- Ensure the dignity of work and economic equity for all

Despite women's significant participation in agricultural value chains, the terms and conditions of their work are often unjust and unfavorable, with limited and unequal rewards for their inputs. Whether in the fields or in packing and processing plants, they are often hired as temporary workers and are not given benefits. Value chain development needs to uphold the principles of dignified work, and fair and equal pay (Raworth 2004; Dolan 2005).

- Remove discriminatory beliefs and practices

Gender inequalities often result from discriminatory beliefs and practices that restrict women's (or men's) full participation in value chains and the terms and conditions of their participation, inhibiting economic efficiency and social development (ILO 1999). Discrimination in hiring and firing based on age, pregnancy status, or union affiliation are illegal practices, for example, that continue to occur in garment-sector value chains (Maquila Solidarity Network n.d.). As humans, both men and women have a right to live free from discrimination that reduces their access to education, skills, and employment opportunities for which they are qualified.

12.3.3 The Development Case

The development case builds on the mutually supportive links between the business case argument for improving economic performance and the social justice argument for greater gender equality. Studies now exist to show that gender equality is positively correlated with economic growth. They establish a strong relationship between women's increased earnings and greater investments in children's health and education (see Quisumbing 2003). Gender inequalities exact a high cost on economic and human development in the long run and affect competitiveness by creating rigid labor markets and restraining productivity and growth. When roughly half of a country's potential labor force is not used efficiently, competitiveness with other countries is negatively affected (Lopez-Carlos and Zahidi 2005). Other evidence suggests that women's participation is good for firm performance (Catalyst 2007). Value chains that address gender inequalities can create the conditions for the greatest number of men and women to participate in and access the benefits from increased economic growth.

The goal of integrating gender into agricultural value chains should be to identify how actors with different capacities and interests can work together to build broad-based economic growth with poverty-reducing impacts. When different arguments are seen as complementary instead of competing, the benefits to the men and women involved in value chains can be maximized. Identifying the points of mutual interest, understanding where unavoidable trade-offs exist, and equipping all stakeholders with the tools to make sound judgments offers a greater likelihood of achieving both greater gender equality and increased competitiveness.

12.4 Identifying and Addressing Approaches to Integrating Gender Issues into Value Chain Development

The advent of value chain analysis in development programs has given rise to a proliferation of methodologies for correcting market failures and improving coordination and cooperation among actors bringing agricultural products from field to fork. In spite of its popularity—or perhaps as a result of it—scrutiny over the potential and ability of value chain development to address equity concerns and reduce poverty has emerged in recent years. This is understandable, given the importance of women to the agriculture sector and women's mixed experience in other global supply chains (e.g., garment and textiles). Attention is increasingly drawn to how the development of the agricultural value chains can be executed in equitable, sustainable, and commercially viable ways.

Two considerations are central to developing gender equitable agricultural value chains:

1. How gender relationships shape men's and women's participation in value chains; and,
2. How gender relationships influence access to the benefits of participation in value chains.

The gender issues that mediate participation in chains and access to benefits from them include differences in men's and women's access to and control over productive resources, agricultural practices and responsibilities, the beliefs and perceptions over appropriate types of work and division of responsibility, and the differential impact of laws, policies, and institutions.

12.4.1 Gender Issues in Value Chain Participation

The heterogeneity of women's participation in agricultural value chains already noted is the result of a number of factors, most important of which is access to and control over factors of production, which vary both by location, product (e.g., sweet potatoes or chickens), and over time. What is important for this discussion is how this interaction influences the way in which men and women enter agricultural value chains.

12.4.1.1 Wage Work

The last several decades have seen an expansion in wage work for women in export-oriented agriculture. Supermarket retailing, both in developed and developing countries, has induced these changes through the development of global agricultural supply chains that source food and exert control over agricultural production and processing in developing countries. Up to 80 % of food retailing in the United Kingdom is controlled by supermarkets; in South Africa, it is between 60 and 80 % (Barrientos 2001, 3).

This process has expanded wage employment opportunities for both men and women. Although women's participation in traditional agricultural exports varies by crop, the growth of high value agriculture (fruits, vegetables, and cut flowers) has paralleled a rise in women's employment in the sector. Women are employed in the production and processing functions of these agricultural value chains. In cut flowers, women make up 80 % of workers in Colombia. In Zimbabwe, they constitute 91 % of horticultural employees (cited in Randriamaro 2006, 22). In Chile, women's employment in the fruit sector quadrupled between 1982 and 1992 (Raworth 2004, 76).

Despite the rising numbers, women hold fewer permanent positions compared to men, with higher levels of temporary or seasonal work at lower wages and with few or no benefits. Women become "permanently temporary" in jobs that are short-term but which roll over for long periods (Raworth 2004, 19). In the South African fruit industry in the late 1990s, women formed 69 % of all temporary and seasonal workers, but only 26 % of long-term employees. The gap was even more pronounced in Chile, with the same categories at 52 % and 5 % women, respectively (Barrientos 2001; Dolan and Sorby 2003; Raworth 2004.)

12.4.1.2 Smallholder Farming

Smallholder farmers are increasingly drawn into agricultural value chains as a result of rising food demands and donor imperatives to ensure that agricultural development reaches the rural poor. Women's participation in smallholder agriculture is mediated by their access to productive assets, and their often lower access to land, credit, social networks, and information relative to men can limit their opportunities to enter higher value and more competitive value chains. As a result, women are less able than men to engage as independent farmers in more profitable chains. Porter and Philips-Howard (1997) observed few women contract farmers participating in barley and sugar value chains in South Africa. Dolan (2001) noted that only 10 % of contracts in tea and horticulture in Kenya were with women. Low participation of women farmers was observed in rice, sorghum, and sunflower schemes in Uganda (Elepu and Nalukenge 2009 cited in Schneider and Gugerty 2010), in a sugar authority scheme in Malawi (cited in Porter and Philips Howard 1997), and in French bean exports in Senegal (Maertens and Swinnen 2009).

Landownership in particular is a key criterion for participation in the chain for independent producers, for members in producer associations, or for outgrowers in contract farming schemes. Women's low landownership rates, their often less secure usufruct rights, and their typically smaller land parcels of lesser quality are all disadvantages when seeking to enter into independent agreements with buyers. Recruitment into these schemes is often biased toward men and their greater access to land. Von Bulow and Sorensen (1993) found that the Kenya Tea Development Authority contracted exclusively with men household heads because of their access to land titles.

Women can reduce these gender inequalities by participating in associations, groups, and horizontal organization. Small firms, whether led by men or women, collaborate to overcome constraints they face individually. Moreover, these small, often local businesses play an important role in facilitating change in the value chain. In Tanzania, for example, horizontal linkages among smallholder producers were consistently associated with upgrading (Bloom et al. 2008). Gender and value chain guides note that to successfully build horizontal linkages, associations need to have gender-equitable governance systems that promote inclusive membership criteria and leadership opportunities for both women and men in associations. In addition, access to value chain participation is enhanced when women as well as men are able to actively engage in group discussions and activities (Van Ingen et al. 2002). Analyzing sociocultural norms to determine the appropriateness of forming either mixed or single-sex associations is an important aspect of success.

12.4.1.3 Entrepreneurship

Value chain interventions can draw in local businesses, especially input suppliers (e.g., retailers, distributors, and wholesalers), and other specialized business service providers (e.g., pruning, grafting, or artificial inseminator specialists). This process

Box 12.1 Women and Men in Food Processing and Petty Trading in Africa

In Uganda, few women sell food or cash crops, approximately 30 % and 9 %, respectively.

Tanzanian men dominate as urban food traders and wholesalers, representing up to 75 % of traders in both activities at the national level. In Dar es Salaam, 60 % of women are mainly self-employed street vendors, selling fruits, vegetables, and cakes.

Around Lake Victoria in Kenya, women make up 75 % of the artisanal fishing sector, as processors and traders.

Nigerian women make up 68 % of urban and 78 % of rural informal-sector cowpea processors and vendors across 12 states. Men's involvement increases as the business grows.

Sources: S. White (1999); GATE (2008).

generates rural off-farm employment opportunities and supports backward and forward linkages with the local economy. The process of formalizing chain activities may also bypass or eliminate intermediary actors such as traders and middlemen. Unfortunately, often little consideration is paid to the gendered composition of these enterprises or whether men- and women-owned enterprises are equally well-placed to join the value chain.

The focus on women's roles as wage workers and household laborers comes at the expense of considering how women entrepreneurs participate in or are excluded from other parts of the agricultural value chain. Women operate a variety of businesses, most often in the services sector. Like men with few assets, they tend to concentrate in the informal sector, where they require little more than their labor to operate and where opportunities for accessing credit, technical assistance, and market opportunities are limited.

Although the number of studies on women's engagement with value chains through women's groups is growing, data on women's participation in the agriculture sector as independent, off-farm entrepreneurs are scant, except in small-scale food processing and trading activities where some data exist (Box 12.1). Informally, women-owned businesses support national and regional trade in raw and processed agricultural products. In parts of Sub-Saharan Africa, women play an important role in cross-border trading of foodstuffs. In the Southern African Development Community (SADC) region, where informal cross-border trade contributes between 30 and 40 % of intra-SADC trade, women make up 70 % of the traders. Estimates for Benin, Mali, and Chad calculate the contributions of women informal traders to national gross domestic product (GDP) at 64 %, 46 %, and 41 % respectively. Often these traders are bypassed or excluded as more formal value chain activities are developed around them (United Nations Economic Commission for Africa/African Union Commission/African Development Bank 2010; UN Women 2010).

12.4.2 Gender Issues in Value Chain Performance

For smallholders, the gender inequalities that lead to differences in productivity (see Peterman et al., Chap. 7) also shape men and women farmers' performance in value chains. Differences in access to labor, inputs, information, and training influence how well farmers are able to upgrade their practices and maintain their participation in value chains.

There are four types of economic upgrading involving efforts to gain productivity and to build in value-added aspects to the chain: making "better products … more efficiently, or [moving] into more skilled activities" (Pietrobelli and Rabelloti 2006, 1). Each type might involve different sets of constraints and opportunities for women or men in different types of chains:

- *Process upgrading*, which aims to increase the efficiency of production processes, resulting in reduced unit costs. Process upgrading can involve improved organization of the production process or improved technology.
- *Product upgrading*, which improves the quality of a product or variety that increases its value to consumers.
- *Functional upgrading*, which refers to entry into a new, higher value-added function in the value chain that moves the value chain actors and/or the overall value chain closer to the final consumer and positions it to receive a higher unit price for the product.
- *Channel upgrading*, which refers to entry into a marketing channel that leads to a new end market in the value chain, for example, from the domestic to the export market for the same product (Humphrey and Schmitz 2001; Bolwig et al. 2008, 17).

In addition, there is an emerging literature on *social upgrading*, defined as improvements in living standards, not only as measurable by increases in wages and work conditions, but also in consideration of greater gender equality and resistance to shocks (see Milberg and Winkler 2010).

There have been specific and localized studies that have investigated some of the gender dimensions of upgrading, much of it related to the impact on women workers in different global value chains (see, for example, Barrientos 2001; Dolan 2005; Harilal et al. 2006). In women-owned enterprises, process and product upgrading can be partially addressed by training in new skills or by providing new avenues for accessing credit, allowing businesses to expand and prosper. The USAID Lulu Livelihoods Program, for example, supported the creation and development of the shea nut value chain in southern Sudan, providing women with technical skills, equipment, and assistance in forming marketing linkages to buyers in Sudan and the region (Armstrong et al. 2008). Functional upgrading or channel upgrading by women-owned firms can also be supported through training and credit, but successfully negotiating these more complex forms of upgrading may depend on more sophisticated business networking, higher levels of education, and/or technical capacity (or the ability to hire it) in the information technology and communications arena.

The degree to which women employees are benefitting as social upgrading occurs within the value chain is addressed in the next Section.

12.4.3 Gender Issues in Benefitting from Value Chain Participation

Gender relationships also mediate how participation translates into benefits for the individual, the household, and the community. The benefits of value chain participation include employment, wages or other income, and empowerment, all of which can accrue to an individual or a household. Additionally, participation can build skills and capacity, increase knowledge and bargaining power, and promote networking that allow actors–individuals, associations, and firms–to enhance their ability to improve the terms of their participation within the chain. Gender dynamics and power relations at multiple levels of the value chain determine who gains, and how these benefits are accessed and distributed. As Coles and Mitchell (2009) highlight, gendered patterns of benefit distribution are such that participation in the value chain does not always translate into gains, such as in the case in Kenya, where women provided 72 % of the labor but obtained only 38 % of the income from their work (Dolan 2001). At the same time, nonparticipation does not equate to a lack of benefit. What matters is not simply the level of income derived from value chain activities, but a combination of factors related to the perception of ownership or management of a particular commodity, the scheduling of payment, and the point of entry into the chain.

12.4.3.1 Employment and Incomes

The increase of women in export-oriented value chains has had mixed results. In conditions where low-skilled labor is needed, women have often been the preferred employee for export-oriented industries, both as a result of gender stereotypes, "a number of stylized assumptions that equate production imperatives of quality, consistency, and speed with ostensibly 'feminine' traits of dexterity and conscientiousness" (Dolan 2004, 107) and because of their typically lower cost. Coupled with women's lack of access to on-the-job training opportunities to upgrade their skills, women are often locked into these low-wage and low-skilled occupations, unable to move into senior positions in the chain. This vertical segregation repeats women's experiences in global value chains in other sectors, such as manufacturing, for example, where despite women's high level of participation in Southeast Asian manufacturing, the last 20 years have seen a "defeminization" of the sector as it upgraded to maintain competitiveness and replaced jobs that women held with technology (IANGWE 2011, 5). These patterns appear to vary nationally and regionally, as well as by the type of commodity.

Concerns about discriminatory hiring practices, sexual harassment, and unequal wages exist in agricultural value chains as in other industries. Emerging research efforts address these inequalities by analyzing the impact of economic upgrading on social upgrading (Milberg and Winkler 2010). Other initiatives, like the Ethical Trade Initiative, focus on building multipartite alliances between companies, trade unions, and voluntary organizations to improve working conditions through the adoption of codes of labor practices.[3]

Despite these concerns, access to employment and earnings has had a positive impact for many women and their families. Wages from off-farm employment are often higher than from other opportunities available. Research on women's participation in the garment sector shows that from Honduras to Bangladesh, access to paid employment has had an effect on women's autonomy, independence, and bargaining power within the household (Fontana et al. 1998; Ver Beek 2001; Raworth 2004).

There are very few studies, however, that are able to measure the trade-offs for women between working on their own or family farms linked to domestic, regional, or export markets and those who work in agricultural packhouses and processing plants. A study of Kenyan horticulture looked at the total income of households engaged in horticultural farming and compared them to households who had members, many of whom were women, working in horticultural packhouses (McCulloch and Ota 2002). The data conclusively found that households of packhouse workers had higher incomes than those relying on only smallholder production, but left open a number of questions about causality of the relationship as well as the specific impact on women and their control of those higher levels of household income. In an earlier study from Kenya comparing the control over income by women in sugarcane producing households with that of women in other agricultural households and nonagricultural households found that the degree of control by women was highest among other agricultural households, but that their levels of income were lower than among sugarcane producers (Kennedy and Cogill 1987). This research points to the need for other comparative studies across a wider number of locations and in different value chains, so that arguments about the benefits of women's participation in agricultural value chains can be properly assessed against other employment options, when they exist.

The benefits women derive are not always commensurate with their labor and participation in agricultural supply chains. In household farming enterprises, women's unpaid labor is harnessed for the production of cash crops but they do not reap the rewards, since gender norms often set expectations about who controls income and the decisionmaking over how it will be spent. While women often have control over the small amounts of income they generate through local sale of food stuffs and other microenterprises, larger sums of money are often controlled or can be appropriated by men (Kennedy and Cogill 1987, 58). On smallholder farms, married women and daughters work as unpaid family laborers with the expectation that income derived from the sale of crops will return to the household. In the Kenyan tea sector, women supplied the labor, but when the income did not return to the household, it gave rise to increased marital conflicts (Von Bulow and Sorensen 1993 cited in Schneider and Gugerty 2010). Some reports describe situations where

[3] See Ethical Trade Initiative (ETI) at http://www.ethicaltrade.org. ETI works with both large- and small-scale sourcing operations in multiple sectors including horticulture. Their codes concentrate on wages, hours of work, health and safety, right to association, and rights of workers or small-scale suppliers.

women have lost control over the production and income from crops as their commercial value increases,[4] but more research is needed on this issue. Women's lack of secure property rights means they are often vulnerable to displacement or encroachment by others, reducing their incentives to participate or improve their performance in value chains.

In another case, Koczberski (2007) found that the lack of an individual economic incentive for women in an oil palm project in Papua New Guinea reduced their participation and the potential income increases to the household. To address women's unremunerated labor, the processing company began to pay women directly into their own bank accounts and hired more women extension workers to provide technical assistance. Subsequently, 26 % of smallholder income was paid directly to women and overall household income increased by 5 %. The project also found a reduction in domestic violence and a shift in the perception of women from "household helpers" to producers.

12.4.3.2 Social Capital and Networking

Women's participation in value chains can bring them further benefits by expanding their networks and building social capital that allows them to advocate and challenge social norms and inequalities (see chapter on social capital by Meinzen-Dick et al., Chap. 5). Women's access to horizontal linkages can improve their performance in the value chain by facilitating their access to credit, information, and marketing opportunities. It can also facilitate their ability to improve social and economic conditions, for example, by joining trade unions to advocate for improved wages and labor conditions.

Value chain participation allows women to expand their sometimes limited business networks and increases their interaction with institutions and individuals from a diversity of backgrounds and with a range of skills, information, and attributes. Women are able to build both bonding and bridging social capital to strengthen their role in the value chain and beyond. *Bonding* social capital is present in many associations and is a strong feature of horizontal cooperation where individuals with common backgrounds or of similar socioeconomic groups link together for different purposes. For women, bonding capital has been shown to build confidence and leadership skills in single-sex groups in Mozambique (Gotschi et al. 2009). *Bridging* social capital fosters collaboration across and among smaller groups or networks of firms and individuals, such as through producers' associations and trade unions, thus broadening the base of resources to draw upon for information, inputs, services, and markets. *Linking* social capital connects unlike groups, such as between producers and market agents.

Women often mobilize high levels of bonding capital, developed through savings groups, church groups, and other networks close to home. However, their wealth of bonding capital may reinforce household roles, inhibiting their links to other actors

[4] See, for example, Wold (1997) and von Braun et al. (1994).

in the value chain. Social norms that restrict their interactions outside of socially acceptable networks limit their ability to gain access to information about inputs and markets, and constrain women's ability to develop bridging capital. In rural Uganda, for example, women are restricted from entering drinking clubs where men network and conduct business (Katungi et al. 2006); business women in Central Asia similarly reported that many business arrangements are made by men through connections established and maintained in social settings, such as bars and bathhouses, to which women have no access.

By comparison, men's networks often are larger and more extensive. Greater mobility allows them to interact with a broader range of individuals including buyers, input dealers, and public officials. Social norms place fewer restrictions on whom they may publicly interact with, facilitating their ability to seek out new relationships. This creates greater opportunities for men to develop linking capital. In a study of the ability of producer groups in Tanzania to improve their crop marketing, Barham and Chitemi (2009, 54) found that "groups with a greater ratio of male to female leaders are more likely to improve their market situation." In this situation, women-only groups often faced more challenges.

Research in Mozambique confirms that although mixed-sex groups present their own challenges for women to articulate their needs and become leaders, they may provide a more conducive environment for women to access benefits in the form of bridging and linking capital (Gotschi et al. 2009). Building bridging capital through access to men's networks offers women more opportunity to access inputs, networks, and information than they might on their own.

12.4.4 Approaches to Addressing Gender Issues in Agricultural Value Chains

The last several years have witnessed a proliferation in approaches to integrating gender into value chain development interventions. They include research methodologies, tool kits for practitioners, and reports monitoring or evaluating the impact of value chain interventions on men and women. Within these categories there are differences in both what constitutes value chain development and what the gender-related goals and outcomes should be. Some of these differences reflect donor priorities, the level of the intervention (e.g., production, processing, etc.), and the partners involved in the activity. In many ways the diversity of approaches is evidence of the infancy of the intersection of these two technical areas and the steep learning curve facing value chain and gender practitioners as they attempt to integrate their activities, goals, and objectives. It also reflects the evolving context as issues like climate change and food security emerge as key priorities on the development agenda.

The impetus for analyzing gender in value chains stems in part from the concept of the gendered economy (Elson 1999). This counters traditional economic theory that the economy is gender-neutral, and instead insists that labor markets and the economy are gendered institutions revealing constructions of gender norms and inequalities. It integrates reproductive activities into the economy because these underpin productive, market-oriented activities. In value chain analysis, it refers to the sex-segmentation of men and women in different activities along the value chain, and within firms and production systems.

The first analytical studies on gender and value chain analysis were built on these concepts to understand the factors contributing to the intensity of women's employment and the flexibility (and informality) in buyer-driven export sectors (Barrientos 2001).[5] These analyses highlight the sex-segmentation of men and women along the chain and across occupational categories, drawing attention to the concentration of women in low-wage, low-skilled positions. They link sex-segmentation to governance of the value chains, revealing how power in chains governs not only upgrading, but also the terms and conditions of participation in the chain. Evidence from South Africa, Chile, Egypt, and Kenya illustrates the clustering of women in labor-intensive, low-return activities such as weeding, pruning, grading, and sorting (Barrientos 2001; Dolan and Sorby 2003; El Messiri 2001). Moreover, these studies revealed how the perceptions about the femininity of particular jobs translated into vertical segregation and lower wages for women.

More recently, several tools and guides have emerged on addressing gender issues in projects using a value chain approach. These are attempts to translate the analytical approaches and learning into action-oriented interventions, providing field practitioners with more specific direction on how to address gender issues along the chain. Although some cast a wider net, a large proportion of these tools are oriented toward agriculture. The majority have been tailored to organizations, NGOs, and implementing agencies that play a coordinating or facilitating role in value chain programs (Rubin et al. 2009; Dulón 2009; RUTA n.d.; Mayoux and Mackie 2007). Their aim is to ensure that the organizations and their field staff increase their understanding of whether gender-based constraints are inhibiting the achievement of women's economic empowerment in their program activities. The guides build staff capacity to undertake more gender-sensitive programming. Many of these methods map a gender analysis process over the project cycle to illustrate how project staff can address gender issues at each stage of the process, providing tips and actions to practitioners at each step of the process (see Fig. 12.1). The number of guides and tools directed at this audience is proliferating rapidly.

A small number of guides address gender in value chain development at other levels of action. Mayoux and Mackie (2007, 33) emphasize the need for "push-up" approaches that involve men and women who are "near the bottom of the power hierarchies in the chain" and include a community-based methodology to empower these men and women. Building on the same idea, the Women's Empowerment Mainstreaming and Networking for Gender Justice in Economic Development

[5] For a methodological summary, see Gammage et al. (2009).

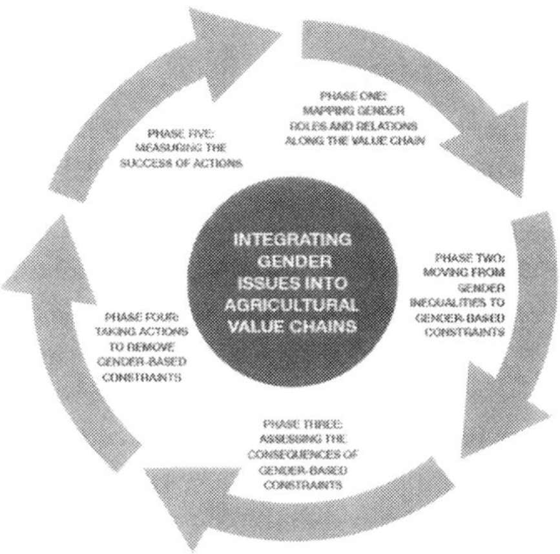

Fig. 12.1 Integrating gender into agricultural value chains (INGIA-VC)
Phase One helps researchers/practitioners/businesses collect data on the factors that shape outcomes for men and women in value chains, collect and organize the data on gender roles and responsibilities using the Gender Dimensions Framework, and understand the sex-segmented character of the value chain. Phase Two assists in identifying areas of gender inequalities as a guide to identifying gender-based constraints. Phase Three guides in thinking through the consequences of the constraint for value chain development. Phase Four develops appropriate actions to reduce or remove the most critical constraints. Phase Five develops indicators to measure success of actions to remove gender-based constraints and progress toward achieving gender equality outcomes (Source: Rubin et al. 2009, 61–62)

(WEMAN n.d.) developed the Gender Action Learning System (GALS), a participatory, community-driven approach aimed at empowering men and women as economic, social, and political actors (Oxfam NOVIB 2004). Currently being piloted in Uganda, the GALs approach works with men and women producers to identify household- and community-level constraints that may be reducing productivity. Household behaviors, such as alcoholism and domestic violence, are discussed alongside production constraints, with the hope that behavior change in the household will improve economic empowerment and well-being.

At the other end of the value chain, efforts are building on the early analytical work on gender and value chains by targeting the private sector, particularly the firms at the helm of buyer-driven chains. Directed at large food companies, these guides make the business case that women's critical role in the production and processing of raw materials into different food items makes them important stakeholders in their supply chain (Chan 2010). Arguing that women's access to services, inputs, and technology is less than men's, this approach guides companies that wish to improve the quality and quantity of their raw materials by making greater efforts to target women when they engage with smallholders, and use their leadership in the chain to influence the business of other stakeholders in ways that can support women.

12.5 Current Debates in the Field of Gender and Value Chain Studies

- Does a focus on women-dominated value chains achieve gender equality?

The question of whether one can best address gender disparities by targeting the development of value chains where women are already active at the node of production or processing or by lowering barriers to women's participation in all types of value chains is highly contested. Proponents of the former argue that in this way, the activity supports women and will expand their income-earning opportunities. Often for this group, there is a lack of understanding of the larger social context that is segregating women into the targeted crop or enterprise. Others maintain that the goals of engendering value chains should be the reduction of gender disparities and enhancement of economic competitiveness. These goals can be met more effectively when the chain development not only includes, but is both profitable for and equalizes, opportunities between men and women.

A project in Senegal, for example, was promoting the development of a value chain for hibiscus flowers, a crop known to be cultivated by women on the perimeters of their spouses' fields. The initial economic projections for hibiscus production and marketing did not reflect the real-world environment in which women operate. The farm budgets assumed that land, labor, and equipment had no or little cost; in fact, the women in some groups had to pay for labor and could not access their husband's equipment in a timely manner, causing delays in cultivation that reduced the quality of the product and, as a result, the prices they received. In these situations, it is not uncommon to find that as these women become more successful producers and become more tightly linked to the market, their rights to the factors of production are contested and men seek to gain access or control of their land or profits. More sophisticated farm-level economic analyses of the crops cultivated by women is needed to determine whether it is economically feasible to encourage women to pursue expanded production in them, or to take up other activities at different points in the value chains of other crops, or even to choose nonagricultural options (Rubin 2010).

It can be rewarding over the short term to work with women in value chains where they are already active, and examples of women benefitting from increased participation are well-known, such as the case of the shea butter value chain in West Africa or women's production in the cut flower industry. Unless these efforts are able, however, to move women into positions of greater control along the chain, over the long term this tactic may be unable to sustainably reduce gender inequalities. The high levels of women's participation in these chains are often the result of gender biases in the economy that restrict their entry into other more profitable ventures or occupations. Failing to address differences in men's and women's capabilities may only exacerbate inequalities.

A more inclusive approach starts from a gendered analysis of a range of value chains and identifying how men and women participate in those chains, while seeking to understand what factors channel men one way and women another. The resulting analysis should clarify steps that could be taken to bring more women into potentially profitable nodes in the chain where they have been underrepresented. In Mali, where sorghum production and marketing (although not processing) is dominated by men, discussions with sorghum researchers led to the suggestion that women's groups could become certified sorghum seed producers. In this way, the women could enter the chain as input suppliers, creating a new niche in an existing chain, with scope to expand as the chain expanded (Rubin with Me-Nsope 2010).

- Can value chain methodologies move us beyond a focus only on farm-level participation for women?

Until now, much of the work to improve women's participation in agricultural value chains has started from where women are: on the farm and in the packhouses and processing plants. While a reasonable initial strategy, this approach may be self-limiting over the longer term, as more smallholders' farms are consolidated and global value chains increasingly seek efficiencies of scale to minimize supply disruptions. It is also constrained by the very real problem of women's limited access to land in many countries, and the cultural sensitivities to making wholesale shifts in landownership. It is thus important to look forward to a range of possible agricultural scenarios, including consideration of climate change on smallholder systems and to identify alternative avenues for women's economic advancement.

The value chain focus on multiple actors allows for a parallel discussion of gender issues at different levels of the chain. It permits a discussion of the capabilities and opportunities for a range of different women and men beyond the farm and different entry points in the chain for addressing gender issues. Support and technical assistance can be channeled toward building horizontal and vertical connections between larger value chain actors and women's enterprises. For example, more creative approaches can be taken to support women's participation in business development and transportation services (see Box 12.2).

Successfully moving women into nodes of the value chain where they have been historically underrepresented will require dedicated resources, careful analysis, and support to change behavior. Gender-equitable opportunities can be enhanced in business development services, in processing, packaging, transport, exporting, and in financing. Support to women to enter and become leaders in agriculture, such as through training and mentoring fellowships provided by the African Women in Agricultural Research and Development (AWARD) program or the Borlaug Leadership Program, are exciting examples of what can be achieved in this arena. The 10,000-women initiative is another model for building the capacity of women entrepreneurs.[6]

[6] For more information about these programs, see AWARD (http://awardfellowships.org/), USDA Borlaug Fellowship Programs (http://www.fas.usda.gov/programs/borlaug-fellowship-program), the USAID Borlaug Leadership Enhancement in Agriculture Program (http://borlaugleap.org/), and 10,000 Women initiative (http://www.goldmansachs.com/citizenship/10000women/index.html).

Box 12.2 Innovative Opportunities for Expanding Women's Participation in Agricultural Value Chains

- The USAID-funded Kenya Dairy Sector Competitiveness Program implemented by Land O'Lakes links up providers of artificial insemination (AI) services to local farmers. One of them is a young mother who received a certificate in AI after completing secondary school. She finds that providing AI services is a good way to earn an income even part-time. Her biggest constraint is the travel to and from the farms, since she has to be ready to go whenever the farmer calls to report that his/her cows are ovulating, even late at night. Assistance to women like her could take the form of building up networks or associations of such service providers that could collectively purchase and share a vehicle or link members to credit options for individual vehicle purchases.
- Another innovative woman entrepreneur in Kenya, Mary Mwangi, started the Double M bus company, so that women could get to work comfortably and safely. Her bright purple buses now employ over 200 people, and are relied on by workers in Nairobi. She remains one of a select group of women in African transport, yet her success can be an inspiration to others. Interviews conducted in Kenya and Tanzania, suggested that safety concerns are a limiting factor for women in transport, whether real or perceived, but women's reluctance to become drivers or automotive mechanics should not hinder their involvement in other aspects of the work, from dispatchers to fleet owners.

Sources: Authors' interview notes.

- Value chains and food security: Are these mutually reinforcing or mutually exclusive?

The argument for supporting a value chain approach to address food security is that the potential increases in efficiency in the food chain, e.g., from higher productivity to decreased transportation costs, will lower food prices, benefitting both urban and rural consumers. It also has the advantage of providing a conceptual framework for linking producers to consumers, establishing the logic for the scope and direction of production.

Even if or when the development of agricultural value chains enhances food security at a local, national, or regional level, there is as yet little evidence to determine whether increasing the extent and scope of women's participation in such chains enhances food security in the household. Historically, rural producers' strategies to reduce risk and to maximize food security involved diversifying household production, growing a mix of varieties to cope with uncertain rainfall patterns, and only selling surplus produce. This strategy initially appears to contradict value chain approaches that focus on the production of a single crop, often in isolation of broader household livelihood strategies. In aligning production more closely to market needs, new risk reduction methods are needed, many of which will rely on extra-household support services, such as crop insurance schemes.

At this time, there is not a wealth of evidence to support or refute the assertions that increasing women's participation in agricultural value chains will necessarily improve household food security, nor is there careful analysis of the pathways that this improvement could follow. In one of the few existing studies, Bolwig and Odeke (2007) look at the consequences of the conversion to organic export-oriented production of coffee and pineapples in Uganda. They found that, overall, the benefits of increased income to the household from the marketed production outweighed any decrease in on-farm food production. From a gendered perspective, the different crop and farm characteristics, however, did have different consequences for women: coffee production increased women's labor as the shift was on existing farmland; pineapple cultivation was expanded through acquisition of new farmlands and used hired labor, with limited impact on household labor. Although the women in these households did not gain control over the added household income, they reported their involvement in the cash crop production as beneficial to their households and worth their added labor input. In both the pineapple and coffee growing sites, increases in income added to food security (Bolwig and Odeke 2007). Earlier studies from Kenya discussed previously (Kennedy and Cogill 1987; McCulloch and Ota 2002) also report positive impacts at the household level. Hamilton et al. (2002) also found that household production of snowpeas and other horticultural crops appeared to increase women's ownership of land and other economic resources in Guatemala and highland Ecuador. A wider comparative study, however, could offer a more robust conclusion than these isolated cases, and provide a clearer description of the pathways that support women's likelihood of benefiting from market involvement. Such research could inform gender-equitable program design.

12.6 The Challenges Ahead

As the field of gender integration into value chain development matures, several topics emerge for further research and investment:

- Developing value chains in crops with added nutritional and health benefits

Attention to the role of nutrition in agricultural development, while long-standing, has recently received new attention in the programs of key donors. Much of the focus has been on improving the nutritional status of women and children through direct (supplementation) and indirect (income-related) interventions, building on findings that income controlled by women has positive outcomes for children's nutritional status (Quisumbing 2003). Until now, the emphasis for many donor programs and commercial buyer-driven export chains has been on high value crops, but, more recently, there is growing interest in developing value chains for staple foods at national and regional levels, including grains such as millet and sorghum, roots and tubers such as orange-fleshed sweet potato, and indigenous vegetables, all of which have nutritional benefits. There are other ways to think about enhancing nutritional benefits alongside value chain development as well. In addition to looking for new crops, efforts can be taken to decrease the loss of nutrient value in the transport and processing nodes of the value chain and to build demand from consumers for more nutritious products (Hawkes and Ruel 2011).

- Improving the database on gender-related value chain impacts on women's income and food security

This brief review has identified a number of topics that could benefit from new research to test hypotheses about the benefits of women's participation in agricultural value chains in comparison to other employment options for their ability to enhance asset ownership, increase income, and boost food security. There is also a need for greater attention to the measurement of changes in women's time allocation patterns, access to and control over income, and decisionmaking opportunities in the household and in the community that result from market involvement and agriculturally-related employment. Findings from these studies could refine the design of donor- or government-led interventions that support value chain development and provide guidance to private investors.

A recent article (Gómez et al. 2011) articulated six research principles for developing value chains and strategies for a future research agenda. We would suggest an additional one that includes more explicit attention to the gender dimensions of value chain development (Box 12.3). They also noted that we need to recognize "how little we know about complex food value chains and their effects on poverty and the environment and [be] cautious in our policy prescriptions" (2011, 1155). They also proposed the development of "a transdisciplinary, multidimensional conceptual framework" (2011, 1155) for studying value chains. In our view, the work of the gender and value chain community[7] has already been developing such frameworks.

- Developing gender-equitable, climate-smart value chains

Adaption to climate variability and climate change will be an increasingly critical aspect of smallholder farming. It will involve investments in agricultural research to test different crop varieties that are more resistant to drought or have increased tolerance of moisture or higher temperatures, depending on the location. Research that offers new crops or crop mixes and sequences with greater climate resilience that do not add to women's labor and time burdens should be

[7] One of the more active communities of practice on gender and value chains is that of the Agri Pro Focus Learning Network (http://genderinvaluechains.ning.com/), which links practitioners, researchers, and others in a global network. The website provides resources on many aspect of integrating better attention to gender into agricultural value chains.

Box 12.3 Engendering the Future Value Chain Development and Research Agenda

1. Focus on opportunities available in domestic markets.
2. Pay attention to indirect effects, not only to increased sales from smallholders, *including ways to enhance the equitable distribution of benefits to all participants in the household.*
3. Enhance marketing channel efficiency.
4. Pay attention to postharvest losses, both in volume and quality, *providing extension information on improved technologies and practices to men and women within the household, and supporting women's enterprises to address these losses.*
5. On-farm natural resources conservation can enable, and benefit from, smallholder food value chain participation.
6. Certification appears necessary but not sufficient.
7. *Conduct gender analyses of value chain options and activities that will contribute to interventions and investments designed to both achieve greater economic gain AND gender equality.*

Source: Gómez et al. (2011).
Note: Authors' additions on gender in italics.

the goal. Introducing new crops can be advantageous to women when they are either integrated into existing gendered responsibilities or create new opportunities (Olson et al. 2010). Expanding the value chain for more climate resistant crops, such as sweet potatoes and other tubers, which tolerate poorer soils and drought conditions, has the potential to increase the incomes of the women who are its primary producers, as well as the nutrition of their household members under conditions of climate variability.[8]

In sum, the value chain approach helps us to consider a wide range of alternatives for women's engagement with the global food system. It forces us both to look not only at increasing women's participation, but also to consider the quality and prospects of that participation. It makes us look beyond the farm to both acknowledge and build on the potential of off-farm agricultural employment (including urban employment) to contribute to total household income for the benefit of all household members. Finally, an engendered value chain model offers a way to compile a diverse set of strategies that encourage women to benefit from expanding economic opportunities.

[8] See International Potato Center (n.d.), "Sweetpotato for profit and health initiative" http://sweetpotatoknowledge.org/sweetpotato-introduction/overviewsweetpotato-for-profit-and-health-initiative.

Acknowledgments The chapter draws heavily on research conducted under the USAID-funded project, "Greater Access to Trade Expansion," implemented by Development and Training Services, Inc., under the Women in Development IQC Contract No. GEW-I-00-02-00018-00, Task Order No. 02, but it does not necessarily express the views of USAID. The authors thank several organizations for the opportunity to present and get useful feedback from participants at workshops that they sponsored on gender and value chain development: the Bill and Melinda Gates Foundation (2010), the International Center for Research on Women with the International Livestock Research Institute (2011), and Oxfam International (2011).

References

Armstrong S, Wagner G, Belknap K, Viezee C (2008). Value chain framework and the Lulu Livelihoods Program. MicroReport #102. USAID AMAP project. US Agency for International Development, Washington, DC

Barham J, Chitemi C (2009) Collective action initiatives to improve marketing performance: lessons from farmer groups in Tanzania. Food Policy 34(1):53–59

Barrientos S (2001) Gender, flexibility, and global value chains. IDS Bull 32(1):83–93

Bloom D, Dunn E, Atcha S, Fan V, Chaoudhary V, Salyer P, Evans S, Akin B, Church P (2008) Analyzing the integration of micro and small enterprises into value chains: evidence from Tanzanian horticulture. USAID Accelerated Microenterprise Advancement Project, Washington, DC

Bolwig S, Odeke M (2007) Food security effects of certified organic export production in tropical Africa: case studies from Uganda. SAFE Policy Brief No. 7. Danish Institute for International Studies, Copenhagen

Bolwig S, Ponte S, du Toit A, Rüsgaard L, Halberg N (2008) Integrating poverty, gender, and environmental concerns into value chain analysis: a conceptual framework and lessons for action research. DIIS working paper 2008/16. Danish Institute for International Studies, Copenhagen

Catalyst (2007) The bottom line: connecting corporate performance and gender diversity. BMO Financial Group, Toronto

Chan M (2010) Improving opportunities for women in smallholder-based supply chains: business case and practical guidance for international food companies. Prepared for the Bill and Melinda Gates Foundation. http://www.gatesfoundation.org/learning/Documents/gender-value-chain-guide.pdf

Coles C, Mitchell J (2009) Gender and agricultural value chains: a review of current knowledge and practice and their policy implications. Paper prepared for UN FAO SOFA 2010. Overseas Development Institute, London. Unpublished

Dolan C (2001) The good 'wife': struggles over resources in the Kenyan horticulture sector. J Dev Stud 37(3):39–70

Dolan C (2004) On farm and packhouse: employment at the bottom of a global value chain. Rural Sociol 69(1):99–126

Dolan C (2005) Benevolent intent? The development encounter in Kenya's horticultural industry. J Asian Afr Stud 40(6):411–437

Dolan C, Sorby K (2003) Gender and employment in high-value agriculture industries. Agriculture and rural development working paper 7. World Bank, Washington, DC

Dulón RF (2009) Género en Cadenas de Valor: Manual para la Transervalización. Catholic Organization for Relief and Development Aid, Den Haag

El Messiri S (2001) Women laborers in horticulture: 2001 gender report. Agricultural Technology Utilization and Transfer (ATUT) and the RONCO Consulting Corporation, Giza

Elepu G, Nalukenge I (2009) Contract farming, smallholders, and commercialization of agriculture in Uganda: the case of sorghum, sunflower, and rice contract farming schemes. CEGA working paper series no. AfD-0907. Center of Evaluation for Global Action, University of California, Berkeley

Elson D (1999) Labor markets as gendered institutions: equality, efficiency, and empowerment issues. World Dev 27(3):611–627

FAO (Food and Agriculture Organization of the United Nations) (2011) The state of food and agriculture 2011—women in agriculture: closing the gender gap for development. FAO, Rome

Fontana M, Joekes S, Masika R (1998) Global trade expansion and liberalisation: gender issues and impacts. BRIDGE Report 42. Institute of Development Studies, University of Sussex, Sussex

Gammage S, with Manfre C, Cook K (2009) Gender and pro-poor value chain analysis: insights from the GATE Project Methodology and Case Studies. USAID Greater Access to Trade Expansion (GATE) Project. dTS, Arlington

GATE (Greater Access to Trade Expansion) (2008) A study of the cowpea value chain in Kano State, Nigeria, from a pro-poor and gender perspective. USAID GATE Project. Development & Training Services, Inc., Arlington http://www.usaidwidworkshops.org/GATE_CD/images/pdf's_Nov_17/GATE%20Publications%20List.doc

Gereffi G (1994) The organisation of buyer driven global commodity chains: how US retailers shape overseas production networks. In: Gereffi G, Korzeniewicz M (eds) Commodity chains and global capitalism. Praeger, Westport, pp 95–122

Gómez MI, Barrett CB, Buck LE, De Groote H, Ferris S, Gao HO, McCullough E et al (2011) Research principles for developing country value chains. Science 332(6034):1154–1155

Gotschi E, Njuki J, Delve R (2009) Equal numbers, equal chances? A case study of gender differences in the distribution of social capital in smallholder farmer groups in Búzi District, Mozambique. Eur J Dev Res 21(2):264–282

GSMA Development Fund and Cherie Blair Foundation for Women (2010) Women and mobile: a global opportunity. http://www.gsma.com/mobilefordevelopment/wp-content/uploads/2013/01/GSMA_Women_and_Mobile-A_Global_Opportunity.pdf

Hamilton S, Asturias de Barrios L, Sullivan G (2002) Nontraditional agricultural export production on small farms in highland Guatemala: long-term socioeconomic and environmental impacts. IPM working paper 02-1. Virginia Polytechnic Institute and State University for the Integrated Pest Management Collaborative Research Support Program, Blacksburg

Harilal KN, Kanji N, Jeyaranjan J, Eapen M, Swaminathan P (2006) Power in global value chains: implications for employment and livelihoods in the cashew nut industry in India: summary report. International Institute for Environment and Development (IIED), London

Hawkes C, Ruel MT (2011) Value chains for nutrition. 2020 conference brief 4. International Food Policy Research Institute, Washington, DC

Humphrey J, Schmitz H (2001) Governance in global value chains. IDS Bulletin 32.3.2001. Institute of Development Studies, University of Sussex, Brighton

IANWGE (Inter-Agency Network on Women and Gender Equality) (2011) Gender equality and trade policy. Resource paper, Women Watch. http://www.un.org/womenwatch/feature/trade/gender_equality_and_trade_policy.pdf

ILO (International Labour Organization) (1999) Country studies on the social impact of globalization: final report. Working party on the social dimensions of the liberalization of international trade (GB.276/WP/SDL/1), Geneva. http://ilo-mirror.library.cornell.edu/public/english/standards/relm/gb/docs/gb276/sdl-1.htm

International Potato Center (n.d.) Sweetpotato for profit and health initiative. Lima. Retrieved March 22, 2012, from http://sweetpotatoknowledge.org/sweetpotato-introduction/overview/sweetpotato-for-profit-and-health-initiative

Kaplinsky R, Morris M (2000) A handbook for value chain research. International Development Research Centre, London

Katungi E, Edmeades S, Smale M (2006) Gender, social capital, and information exchange in rural Uganda. CAPRi working paper no. 59. Consultative Group on International Agricultural Research, Washington, DC

Kennedy E, Cogill B (1987) Income and nutritional effects of the commercialization of sugarcane production in Southwestern Kenya. Research report 63. International Food Policy Research Institute, Washington, DC

Koczberski G (2007) Loose fruit mamas: creating incentives for smallholder women in oil palm production in Papua New Guinea. World Dev 35(7):1172–1185

Lopez-Carlos A, Zahidi S (2005) Women's empowerment: measuring the global gender gap. World Economic Forum, Geneva

Maertens M, Swinnen J (2009) Trade, standards, and poverty: evidence from Senegal. World Dev 37(1):161–178

Maquila Solidarity Network (n.d.) Toronto, Ontario, Canada. Retrieved March 22, 2012, from http://en.maquilasolidarity.org/

Mayoux L, Mackie G (2007) Making the strongest links: a practical guide to mainstreaming gender analysis in value chain development. International Labour Organization, Geneva

McCulloch N, Ota M (2002) Export horticulture and poverty in Kenya. IDS working paper no. 147. IDS Publications, University of Sussex, Sussex

Milberg W, Winkler D (2010) Economic and social upgrading in global production networks: problems of theory and measurement. Capturing the gains working paper 4. Department for International Development, London. http://www.capturingthegains.org/publications/workingpapers/wp_201004.htm

Olson J, Rubin D, Wangui E (2010) Gender, agriculture, and climate change: a regional analysis for USAID/East Africa. DevTech Systems, Rosslyn

Oxfam NOVIB (2004) Women's empowerment mainstreaming and networking for gender justice in economic development. Women's Empowerment: Mainstreaming and Networking, The Hague

Pietrobelli C, Rabelloti R (2006) Clusters and value chains in Latin America: in search of an integrated approach. In: Pietrobelli C, Rabelloti R (eds) Upgrading to compete: global value chains, clusters, and SMEs in Latin America. Inter-American Development Bank, Washington, DC

Porter G, Philips-Howard K (1997) Comparing contracts: an evaluation of contract farming schemes in Africa. World Dev 25(2):227–238

Quisumbing AR (ed) (2003) Household decisions, gender, and development: a synthesis of recent research. International Food Policy Research Institute, Washington, DC

Randriamaro Z (2006) Gender and trade: an overview. Institute of Development Studies, University of Sussex, BRIDGE, Brighton

Raworth K (2004) Trading away our rights: women working in global supply chains. Oxfam International, Oxford

Reardon T, Berdegué JA (2002) The rapid rise of supermarkets in Latin America: challenges and opportunities for development. Dev Policy Rev 20(4):371–388

Riisgaard L, Escobar Fibla AM, Ponte S (2010) Gender and value chain development: evaluation study. Danish Institute for International Studies/Danish International Development Agency, Copenhagen

Rubin D (2010) Gender assessment USAID/Senegal. DevTech Systems, Inc, Arlington

Rubin D, Me-Nsope N (2010) Gender integration in USAID/Mali accelerated economic growth program: report on technical assistance activity. Report prepared for USAID/Mali under the Michigan State University Project to Mobilize Food Security Initiatives in Mali—Phase II (PROMISAM II). Unpublished, Michigan State University, East Lansing, MI

Rubin D, Manfre C, Nichols Barrett K (2009) Promoting gender equitable opportunities in agricultural value chains: a handbook. Prepared under the Greater Access to Trade Expansion project of the WID IQC. U.S. Agency for International Development, Washington, DC

RUTA (n.d.) Cadenas de Valor con Enfoque de Género: Caja de Herramientos. RUTA, CATIE, GTZ. http://www.ruta.org/toolbox/toolboxGenero.html

Schneider K, Gugerty MK (2010) Gender and contract farming in Sub-Saharan Africa: literature review. Paper commissioned by the Bill & Melinda Gates Foundation. University of Washington, Seattle. Unpublished

United Nations Economic Commission for Africa/African Union Commission/African Development Bank (2010) Gender and intra-African trade: The case of West Africa. Chapter 12 In: Assessing regional integration in Africa VI: enhancing intra-African trade. UN Economic Commission for Africa, Addis Ababa

UN Women (2010) Unleashing the potential of women informal cross border traders to transform intra-African trade. New York. http://www.unwomen.org/en/digital-library/publications/2010/3/unleashing-the-potential-of-women-informal-cross-border-traders-to-transform-intra-african-trade

USAID MicroLinks Wiki (n.d.) How to conduct a value chain analysis. Retrieved March 22, 2012, from http://www.microlinks.org/good-practice-center/value-chain-wiki/32-value-chain-analysis

Van Ingen T, Kawu C, Wells S (2002) Gender equity in coastal zone management: experiences from Tanga, Tanzania. World Conservation Union, Nairobi

Ver Beek K (2001) Maquiladoras: exploitation or emancipation? An overview of the situation of Maquiladora workers in Honduras. World Dev 29(9):1553–1567

von Braun J, Meinzen-Dick R (2009) 'Land grabbing' by foreign investors in developing countries: risks and opportunities. IFPRI policy research brief 13. International Food Policy Research Institute, Washington, DC. http://www.ifpri.org/publication/land-grabbing-foreign-investors-developing-countries

von Braun J, Johm K, Puetz D (1994) Nutritional effects of commercialisation of a woman's crop: irrigated rice in the Gambia. In: von Braun J, Kennedy E (eds) Agricultural commercialisation, economic development, and nutrition. Johns Hopkins University Press for the International Food Policy Research Institute, Baltimore, pp 343–364

Von Bulow D, Sorensen A (1993) Gender and contract farming: tea outgrower schemes in Kenya. Rev Afr Pol Econ 20(56):38–52. http://www.informaworld.com/index/769331381.pdf

WEMAN (Women's Empowerment Mainstreaming and Networking for Gender Justice in Economic Development) (n.d.) Retrieved March 22, 2012, from http://www.wemanglobal.org/1_WEMANVision.html

White S (1999) Women's employment in the agro and food processing sector: south Asia and East Africa. Aga Khan Foundation Canada, Ottawa. Women in Informal Employment: Globalizing and Organizing (WIEGO), April, http://previous.wiego.org/publications/99report.pdf

Wold BK (1997) Supply response in a gender perspective: the case of structural adjustment in Zambia. Reports-statistics Norway 97/23. Statistisk sentralbyrå, Oslo

World Bank (2001) Engendering development through gender equality in rights, resources, and voice. World Bank policy research report. Management 1: report no. 36546-MW. World Bank, Washington, DC

World Bank (2007) World development report 2008: agriculture for development. World Bank, Washington, DC

World Bank/IFAD/FAO (International Fund for Agricultural Development/Food and Agriculture Organization for the United Nations) (2009) Gender in agriculture sourcebook. World Bank, Washington, DC

Chapter 13
Mainstreaming Gender Sensitivity in Cash Crop Market Supply Chains

Ruth Vargas Hill and Marcella Vigneri

Abstract Gender-specific constraints on the production and marketing of cash crops have important implications for the ability of men and women to participate in market-oriented agricultural growth and development. This chapter analyzes how gender inequalities in resources result in different levels of participation, methods of production, and modes of marketing cash crops. Two empirical case studies of traditional perennial export cash crops—cocoa in Ghana and coffee in Uganda—provide empirical evidence on the effects of such constraints. Women cocoa farmers in Ghana face barriers in accessing input markets, particularly markets for labor and non-labor inputs, influencing their choice of production technology. In Uganda, the low quantities marketed, and lack of access to bicycles, limit female coffee farmers to marketing channels that have very low transaction costs, but which receive lower prices. To enable women to engage in cash crop production, the authors provide three context-specific recommendations: (1) improving women's access to land and encouraging better integration of food markets through improved roads and increased mobile networks; (2) strengthening female farmers groups or marketing groups to which female farmers can belong so that women may achieve scale in marketing; and (3) improving access to credit and extension services to relieve female farmers' constraints in purchasing quantity- or quality-enhancing inputs. Further work in assessing the patterns and underlying determinants of female engagement in a wide variety of cash crop markets will be needed to better identify the most appropriate interventions.

R.V. Hill (✉)
Senior Economist, World Bank, 1818 H St. NW, Washington, DC, USA
e-mail: Rhill@worldbank.org

M. Vigneri
Independent Development Economist (Specialising in Agricultural Development and research collaborator for IFPRI 's Ghana Strategy Support Programme [GSSP]), 31 Stone Meadow, Oxford OX2 6TD, UK
e-mail: Marcella.vigneri@gmail.com

A.R. Quisumbing et al. (eds.), *Gender in Agriculture: Closing the Knowledge Gap*, DOI 10.1007/978-94-017-8616-4_13,

Keywords Gender • Cash crops • Production • Marketing • Constraints and barriers

13.1 Introduction and Rationale

This chapter reviews the impact of gender-specific constraints on the production and marketing of cash crops, which we define as those crops that are grown primarily for marketing rather than for household consumption. A rich literature illustrates the existence of structural socioeconomic barriers for women's ability to access land, markets, education, and networks which often add more time pressure to the complex workload of women in rural areas. Moreover, several studies have explained how these constraints have an impact on women farmers' lower levels of input use and their lack of technology adoption (World Bank et al. 2009; Morrison et al. 2007; Doss 2001; Quisumbing 1994).

Our specific research interest is to consider the impact of such constraints on cash crop production. The chapter focuses on analyzing how gender inequalities in resources result in different levels of participation, methods of production and modes of marketing cash crops. This in turn bears consequences for women's potential outcome in the cultivation of these high value crops. The chapter also looks at constraints faced by women participating in cash crop markets through a combination of literature review and original data analysis. It examines the root causes of these constraints, whether they arise as a result of discrimination in input and output markets for cash crops themselves or as a result of constraints in assets and other resources.

Cash crop production differs from general agricultural production in that it entails engaging in output markets to make sales. This requires reliable access to these markets, and has implications on the scale and quality of production. Being able to produce at scale is important in cash crop production for two reasons: (1) when households engage in cash crop production, they are exposed to price volatility in the cash crop and the food crop they may wish to purchase; as a result, it is often only those farmers that can achieve food security first that choose to engage in cash crop production (Fafchamps 1992), and (2) engaging in markets involves some fixed costs, such as searching for a buyer, and a minimum quantity needs to be sold to make it worth incurring these costs (Key et al. 2000).

In a number of contexts "cash" crops also differ from "food" crops in that social norms dictate that they traditionally imply more male involvement in some of the decisionmaking, production, and sale processes. Evidence suggests that female participation in cash crop markets is often lower than male participation (World Bank et al. 2009). As an example, women only represent 20 % of cocoa farmers in Ghana (Vigneri and Holmes 2009), and female-headed households are significantly less likely to farm coffee than households headed by men in Uganda. Assessing the nature of female involvement in cash crop production is important, not just because it differs from the production of other crops, but because cash crop production holds significant potential as a means by which rural households can improve their welfare.

To shed light on the above issues, this chapter begins with a literature review that analyzes the major constraints women face in cash crop production. It then discusses two case studies that provide suggestive evidence on the effects of such constraints: women cocoa farmers in Ghana face barriers in accessing input markets, particularly markets for labor and non-labor inputs, and this influences their choice of production technology. Women farming coffee in Uganda adopt differential types of transactions for selling their crop relative to male farmers. The final section concludes with recommendations on how to increase women's access to cash crop markets.

13.2 Literature Review

The following subsections assess what constraints women face in producing and accessing cash crop markets, specifically those constraints that women face in increasing the scale of their sales, in achieving quality output, and in minimizing the costs of market transactions. Given the existence of a comprehensive literature analyzing gender biases among women farmers (see Quisumbing and Pandolfelli 2009; World Bank et al. 2009; Morrison et al. 2007; Doss 2001; Quisumbing 1994), this chapter will focus on reviewing the constraints that relate to scale, input, and output market access.

13.2.1 Male and Female Crops

One frequent distinction made in the literature is that cash crops and export crops are male crops, while subsistence crops are female crops (for example, Koopman 1993; Kumar 1987, and Randolph and Sanders 1988). Evidence suggests that men may take over production and marketing, even of traditional women's crops, when it becomes financially lucrative to do so (World Bank et al. 2009). A standard explanation for the division of crops by gender is that women are responsible for feeding the family and thus grow subsistence crops. On the other hand, men are responsible for providing cash income and to this end they grow cash and export crops (Doss 2001).

Doss (2002), using empirical data from Ghana, argues that it is not possible to divide crops into "men's" and "women's" crops because social norms dictate the types of crops they can grow. Social norms also constrain access to land, availability of labor, access to extension and credit, and access to output markets (Doss 2001). It is important to note that social norms relating to women's and men's crops change over time. There are a number of examples of crops or commodities that started in the women's domain but became controlled by men as they were commercialized (Kasente et al. 2001; Doss 2001; Lilja and Sanders 1998; von Braun and Webb 1989). However, this is not always the case. Saito et al. (1994) noted that traditional pattern of intrahousehold rights and obligations may change in response to evolving social and economic circumstances and migration of men in search of more remunerative activities elsewhere. Saito et al. (1994) found that the gender-specific nature of African farming was disappearing as women were growing high value crops, taking on tasks traditionally performed by men (such as land clearing), and making decisions on the daily management of the farm and household.

13.2.2 Access to Land and Labor

Access to land, a primary factor of production, is often more constrained in female-headed households and also for women in households headed by men. A rich literature reports that regardless of how access to land is gained, female-headed households tend to have smaller landholdings than households headed by men (Morrison et al. 2007; Doss 2001). In addition, women's landholdings may be less fertile and more distant from the homestead (Doss 2001). However, direct empirical evidence on the gender-disaggregated effects of land on the probability of producing cash crops does not exist. A number of studies, however, have found that households with smaller parcels of land are less likely to engage in cash crop production (Fafchamps 1992) and this can be partly understood by the need for scale.

Having smaller plots thus disadvantages women. Additionally, women who do access cash crop markets often cultivate smaller plots of land (Vigneri and Holmes 2009), which has a bearing on the type of fixed marketing costs it makes sense for them to incur. This in turn may have implications for the marketing channel chosen to sell their crops.

Limited landownership inhibits access to credit for inputs (see Ghana case study below). Tenure insecurity also impairs women's investment incentives (Morrison et al. 2007), as shown by Goldstein and Udry (2005) in Ghana. They find that individuals in positions of power in the local political hierarchy have more secure land rights and, as women are rarely in positions of power, they face more insecure property rights.

Labor, a second primary factor of production, is also often more constrained in female-headed households and also for women in households headed by men. Labor availability depends on the amount of household labor that can be mobilized for agriculture and on the labor that can be hired in local labor markets. Female-headed households may have less access to labor because they include fewer men and may have fewer resources for hiring non-family labor. Within male-headed households, women who manage agricultural activities may also have difficulty in mobilizing labor due to social constraints.

An important gender division of labor also exists among various agricultural tasks (Blackden and Wodon 2006). Women are primarily responsible for food processing, crop transportation, and weeding and hoeing, while men do most of the land clearing. This is inevitably a limiting factor in the amount and quality of time women can allocate to look after their farms. Moreover, women in poor households face particularly serious time constraints because of their various livelihood activities and childcare responsibilities (Quisumbing and Pandofelli 2009).

Paolisso et al. (2002) evaluate the effect of the Vegetable and Fruit Cash Crop Program in Nepal and find that men and women spend roughly the same average time in cereal and livestock production, although women spend more time caring for children under 5 years of age, while men spend more time in fruit and vegetable production. Von Braun and Webb (1989) also find that the adoption of new technologies in The Gambia led to increased work on communal plots for both men and women, with relatively larger increases for women than men.

Differential access to labor affects the scale and efficiency of production. Udry (1996) finds that within households, the lower productivity observed on female plots compared to male plots results from labor and fertilizer (manure) being more intensively applied on men's plots. Similarly, Holden et al. (2001) find that female-headed households in Ethiopia have lower land productivity due to insufficient access to male labor and oxen and low substitutability among factors of production.

It is worth noting that land and labor constraints may change over time and may also be affected by participation in cash crop markets. In Ghana's Western region, Quisumbing et al. (2004) find that women's active participation in cocoa production has challenged and changed the norms by which women usually acquire land. Land is being transferred from husband to wife when the wife helps to establish cocoa fields. Similarly, the adoption of labor-intensive cocoa farming increased the demand for women's labor in Ghana (Quisumbing et al. 2004).[1]

13.2.3 Input Use and Technology Adoption

A recent review of the literature that assesses the use of fertilizer, sprays, and new varieties of seeds (Peterman et al. 2010, also see Chap. 7) shows that while rates of adoption tend to be lower for women than for men, in more than half of the studies reviewed, differences in human capital, access to credit, extension, and networks explain gender differences in adoption. Once these factors are accounted for using multivariate regression analysis, gender differences disappear. The overall evidence suggests that many of the constraints to access inputs and adopt new technologies are not related to the characteristics of the input or technology per se but instead originate in other markets that are relevant for the adoption decision—land, labor, credit, and information (Morrison et al. 2007). Gladwin (1992) finds that lack of access to credit and cash, rather than the farmer's sex, is the critical factor that significantly limits fertilizer application. However, since female farmers have less access than males to credit and cash, they apply less fertilizer, resulting in lower yields. Doss and Morris (2001) find that women's and men's differentials in planting improved varieties of maize are explained by gender differences in accessing complementary inputs, especially land and extension services. Once those inputs were controlled for, the sex of the farmer is no longer statistically significant in

[1] The positive impact of this increased demand of labor needs to be discussed since it may cause an increase in the total working hours of women.

explaining adoption decisions. Quisumbing (1994) finds that farmers with larger areas cultivated and higher values of farm tools are more likely to adopt new technology. To the extent that women farmers may have less education, less access to land, and own fewer tools, they may be less likely to adopt new technologies.

The existing evidence also suggests that women farmers who are also household heads are more likely to be affected by constraints on accessing inputs and technology adoption. Croppenstedt et al. (2003) shows that female-headed and male-headed households with equal factor endowments do not differ in their adoption and intensity of fertilizer use; however, female-headed households are generally at the lower end of the endowment distribution and any differences are driven by this fact. Doss and Morris (2001) report that female farmers residing in male-headed households in Ghana are just as likely to adopt new technologies as male farmers; however, holding everything else equal, female farmers in female-headed households are less likely to adopt than male farmers. This may arise because female-headed households tend to be, on average, smaller and their incomes lower than male-headed households.[2]

13.2.3.1 Human Capital

Most of the available evidence suggests that education (usually defined as formal schooling or literacy) is an important determinant of the decision to adopt new technologies (as well as of the decision to adopt new technologies early), since it increases the ability of the individual to process relevant (new) information (Morrison et al. 2007). As suggested by Morrison et al. (2007), to the extent that women are less educated than men, they are more likely to delay adoption or to forgo it entirely. The adoption of new technologies is important for accessing cash crop markets since these markets often require better quality products.

Since cash crops may require a higher level of technology adoption, the low ability of women farmers to process relevant and new information might also constrain their access to cash crop markets. The importance of own-schooling for adoption is probably greater in the case of female-headed households, where the potential for positive education spillovers to other household members (male members) is reduced (Morrison et al. 2007). In their Ghana study, Doss and Morris (2001) find that female farmers in male-headed households tend to have less formal schooling than male farmers, and that female farmers in female-headed households have even less. Similarly, Croppenstedt et al. (2003) find that very few female-headed households are literate, and virtually none have four or more years of formal schooling.

[2] However, as stated by Doss and Morris (2001), it is difficult to disentangle the causal relationships among these factors. To the extent that household size and composition affect productivity, female-headed households will be less productive. Reverse causality may also apply: a household may be female-headed because the farm had low productivity and the male head left to find better opportunities.

13.2.3.2 Credit

The ability to obtain rural credit is often correlated with land tenure and agricultural productivity (see, for example, Hoff and Stiglitz 1990, and Bell 1990). When some land is titled, however, it may be difficult for a farmer without land title—a common circumstance for many smallholders—to obtain credit (Doss 2001). Credit may also be tied to the lender's perception of the farmers' ability to repay the loan. For example, farmers have to prove their ability to produce a marketable surplus in order to receive credit, and this in turn depends on the type and size of the land they work. Therefore, where women have less quality land and are perceived to produce more for home consumption and less for the market, they may find it harder to obtain credit when these criteria are employed. In addition, an institutional bias exists toward providing financial services to the head of the household who holds a title, and this represents a discriminating factor for women who neither are the head of the household nor hold title to property (Vigneri and Holmes 2009).

Women farmers may be better off adopting high-value crops that do not require large initial investments or asset ownership, as these will limit their ability to access credit (Quisumbing and Pandofelli 2009).

13.2.3.3 Extension Services

Agricultural extension services are an important instrument for the provision of information on new technologies and crops (Anderson and Feder 2003; Evenson 2001; Doss and Morris 2001; also, see Ragasa, Chap. 17). However, extension services often fail to reach female farmers, in particular female-headed farming households (Doss and Morris 2001; Quisumbing 1994; Saito et al. 1994), even though female farmers often indicate a strong demand for such services (Saito et al. 1994). However, Doss and Morris (2001) argue that the differential pattern of extension contact by gender may have less to do with gender per se and more with extension agents approaching more frequently farmers with relatively better access to land, labor, and capital (both human and financial) as well as farmers with a history of adopting technological innovations. Because women are underrepresented among these better-off farmers, extension agents are more likely to overlook them in their programs.

13.2.3.4 Networks and Information

Learning about a new technology and its use from other farmers in the community (via imitation or information exchange within social networks) has been shown to be an important determinant of the adoption decision (Morrison et al. 2007). Conley and Udry (2010) find that farmers in Ghana are more likely to have information links with other farmers of the same sex, clan, and age, and that these links were important for technology diffusion. Similarly, Weir and Knight (2000) found that

88 % of adopters indicated that their decision was influenced by somebody of the same sex. To the extent that female farmers have less-extensive or poorer-quality information networks, knowledge transfer through these networks are more likely to be impaired (Morrison et al. 2007).

Since cash crops often require more information to adopt improved varieties, new technologies or inputs, and to obtain price information, women's reduced access to information can constrain their access to cash crop markets. The importance of information constraints is probably greater in the case of female-headed households, as suggested by Saito et al. (1994), who noted that extension agents often prefer to talk to women in male-headed households rather than those in female-headed households. Thus, a bias might not simply be based on gender, but also on status and household structure.

13.2.4 Accessing Output Markets

Transaction costs influence the ease with which households and individuals access markets. Households are almost always able to physically access some output market for their crops; however, if the costs of doing so are prohibitively high, this will limit market access.

Transaction costs vary with the type of market channel and the crop being sold. They also vary considerably across individuals. Physical distance from markets is important, but so is access to transport assets, or sources of market information (radios, mobile phones, personal relationships with traders). Skills and human capital characteristics—such as a capacity for contract negotiation—will also have an impact on the cost of transacting.

Research in the United States (Fu et al. 1988; Edelman et al. 1990; Fletcher and Terza 1986) has shown that farmer characteristics influence farmers' choice of sale mechanism. They find that the profile of producers associated with newer forms of market organization largely coincides with the expected profile of early adopters of new methods and technology (i.e., relatively more educated, diligent information seeking, and willingness and ability to take risks). Fafchamps and Hill (2005) find that for coffee farmers in Uganda, wealthy farmers are less likely to sell at the market if they are selling a small amount, but are more likely to sell at the market the higher the quantity sold, reflecting their greater ability to pay for public transportation to the nearest market. Owning a bicycle is also found to be a significant determinant of transporting coffee to the nearby market.

A number of studies show that farmer characteristics determine whether or not they enter contracts. Warning et al. (2003) suggest that there is less access for smaller farmers to contracts, as do Balsevich et al. (2003, 1149), who found that in Costa Rica 80 % of the volume of vertical arrangements comes from medium- and large-scale producers and packers. However, Warning and Key (2002) find that rich and poor farmers have equal access to contract farming arrangements in Senegal.

Similarly, smallholders in Indonesia, unlike in Latin America, are found to be well integrated into the modern value chains (Chowdhury et al. 2005).

When female farmers have differential levels of wealth, access to or ownership of means of transportation, knowledge of trader networks and access to market information, significant differences exist in the extent and nature of their transactions in output markets. Moreover, the higher the fixed costs of transacting, the larger the scale of production is required. Female farmers also face many gender-specific constraints for accessing cash crop output markets (Morrison et al. 2007). These include (1) physical harassment by market or health officials when the high cost of permits leads women to market their wares outside market boundaries; (2) time burdens that constrain women from seeking the best prices for their output; (3) marital conflict if fluctuating prices lead a husband to believe that his wife is withholding money from him because she brought home more money on previous trips to the market. Women's farmer groups are also less successful than men's groups both at searching for and accessing new output markets for their existing products and at pursuing new products under contract arrangements. This is because men are more likely to be approached for their products by agricultural companies (or other chain actors) who wrongly assume that men are the primary producers in the household (Barham and Chitemi 2009).

13.3 Case Study: Producing Cocoa in Ghana

This section presents an important example of how gender barriers in cash crops production affect the productivity outcomes of women cultivating cocoa in Ghana. Using original data for a period of observed production expansion that occurred between 2002 and 2004, this section offers insight on how male and female cocoa farmers raised the land productivity on their managed farms, given their different levels of inputs use.

13.3.1 Background on Cocoa in Ghana

The cocoa sector of Ghana is reputedly considered an engine of growth for the country's economy. Exports generate revenues that are second only to gold (in 2005 alone, cocoa beans and cocoa products jointly accounted for about 28 % of total exports). Since 2001, a significant share of the country's agricultural productivity gains has been generated by the export crop (World Bank 2008), with official production figures more than doubling between 2001 and 2003 alone. Cocoa accounts for 10 % of total crop and livestock production values (World Bank 2007), contributing to 28 % of agricultural growth in 2006 (Breisinger et al. 2008), and providing livelihoods for over 700,000 smallholders.

Table 13.1 Average weekly hours spent on domestic chores, by sex

	Fetch wood and water	Cleaning	Cooking	Errands	Child-care	Elderly and sick care	Other	Total house-holds	Total work	Total
Women	1.27	1.32	8.04	1.01	5.12	1.46	0.13	26.06	36.72	62.78
Men	0.48	0.40	1.24	0.93	2.83	0.40	0.98	10.22	38.55	48.77

Source: Author calculation from GLSS-V (Ghana Living Standards Survey)

Cocoa, a perennial tree crop with a life-cycle of 25–30 years, is characterized by a production technology requiring the use of working capital mainly to hire labor for clearing and weeding the land, and to purchase chemicals for controlling the spread of pests and diseases.

The key productive assets are land and labor. Changes in the mode of land acquisition have taken place in Ghana since the mid-1980s, where an intestate law was introduced to allow individuals to leave parts of their cultivated land in inheritance to both their spouse and children. Quisumbing et al. (2001) report that in Ghana's Western region, land is now transferred from husbands to wives and children as gifts in return for the time spent to establish men's cocoa fields. Once this land is given, it cannot be taken away by other family members, and this has partly contributed to increasing women's bargaining power in the sector.

Labor employed on cocoa farms, the second pivotal input to production, is clearly gender differentiated by farming tasks. While male labor is essential for clearing and tree felling, female labor is used for less physically demanding tasks such as weeding and harvesting. Asymmetric divisions of labor in the household, however, also mean that women are required to allocate a substantial amount of time to domestic chores. Extensive responsibilities in the household, combined with demands for working on husband's land or farming activities, limit the time women spend on their own productive economic assets, or it means they work many more hours a day than men (Baden et al. 1994; Sarpong 2006).

Table 13.1 gives an illustration of such time imbalances by looking at intrahousehold activities and employment commitments among women cultivating cocoa.[3] Female farmers spend on an average week up to 1.5 more time on domestic work than their male counterparts, and up to 29 % more time than men working between household duties, and farm and nonfarm employment. This will inevitably be a limiting factor in the amount and quality of time they can allocate to look after their cocoa farms, a point to which we return below.

Among contracted labor types, annual labor is a comparatively cheaper way to maintain a farm, as payment can be deferred until harvest (MASDAR 1998). Yet, the precarious state of farmers' finances means that many have become reluctant to

[3]The information is drawn from the fifth round of the nationally representative Ghana Living Standards Survey and was matched for comparability to the data employed in the rest of the case study by looking at the same three regions (Ashanti, Brong Ahafo, and Western), and using the same definition of cocoa farmers (see discussion in data section below).

enter into such contracts, and daily wage contracts are the most frequently used. Hiring labor outside the household requires availability of cash, which farmers are, in general, very short of. As will be shown further below, women farmers are often more cash constrained than their male counterparts, implying that they are likely to face more stringent constraints on this key resource, unless they are able to source it from other household members.

There is, however, another type of nonhousehold labor known as *nnoboa* groups. These are labor exchange groups, which are typically used more frequently by poorer farmers who cannot afford to pay cash to obtain needed farm labor. There is an important difference in labor deployment strategies between male and female farmers: male farmers generally tend to use more *nnoboa* labor, while female farmers rely more on wage labor. This is because female farmers cannot obtain male labor through *nnoboa*, for which men and women form separate groups. Farmers need male labor for strength-demanding tasks such as tree felling; consequently, female farmers in the lower wealth ranks who have no other means of procuring male labor, have to rely on wage or annual labor.

13.3.2 Data

The dataset used in this case study is the Ghana Cocoa Farmers Survey (GCFS). This was first collected in 2002, and had a follow-up visit in 2004, which generated the 2-year panel described below.

The GCFS covered a diverse range of instruments on land use, inputs, production, and marketing choices (Teal et al. 2006). The original sampling frame for the 2002 baseline survey was the 1999 Ghana Living Standard Survey (GLSS), from which a representative cross section of cocoa farmers was identified and compared with the production records of the Cocobod (the state-run marketing board). The sampling methodology described above generated a sector-representative survey for the three most important areas of production—Ashanti, Brong Ahafo, and Western (Vigneri 2005).

An important feature of the GCFS is how cocoa farmers were identified. These were identified as the individuals managing all aspects relating to cocoa production (that is, the amount of land, labor, and non-labor inputs used, the share of land allocated to the cultivation of the tree crop, and the marketing channel choice), which did not necessarily coincide with the owner of the land.

13.3.3 Characteristics of Female and Male Managers

Cocoa has traditionally been considered a "men's" crop: because of the high returns it generates and the intensity of the labor use requirements, male farmers have always dominated the composition of the cocoa farming population. More recently, however, with the progressive individualization and commercialization of land rights (Quisumbing et al. 2001), cultivating cocoa trees has become more gender balanced; women are also able to acquire land rights, they manage their own farms, and retain control of the income from sales.

Cocoa production is particularly good for female farmers for two reasons: (1) it can provide women a more secure way to gain rights to land; and (2) it provides economic security as it is known to represent over 75 % of income to its smallholder producers (Teal et al. 2006; Vigneri 2005).

Table 13.2 below describes the profile of male and female farmers across the two rounds of the GCFS, pointing to a number of differences in how men and women engage in the production of the crop.

The first feature in the sample is the high proportion of landowners (on average 84 % in 2002 and 89 % in 2004). This is the case for both male and female farmers, and largely mirrors the ownership status of smallholders observed in larger representative samples of cocoa farmers' population. There are marked differences between male and female farmers. Female farmers are older and markedly less educated than their male counterparts. They produce less cocoa on systematically smaller farms. They are noticeably more cash constrained,[4] apply lower levels of fertilizer and insecticide, and use less agricultural equipment.

This evidence confirms the existence of important gender inequalities in the use of productive resources. However, the one indicator in which male and female cocoa farmers do not show any statistically significant difference is land productivity: the levels observed are comparable across men and women.

The second half of the table further highlights the differences in the composition of labor on male and female farms, the labor-to-land ratios, and labor productivity. In both years, women employ more household labor than men on each unit of farm land. What is interesting, though, is the use of hired labor on the intensive margin and its productivity (output per unit of hired labor input) on women-controlled farms, which are comparable to the figures observed on male-managed farms. In 2004, the productivity of hired labor on women-managed farms is almost double that observed on land controlled by male farmers.

This suggests the existence of a gender difference in the allocative efficiency of productive inputs, a point to which we return in greater detail in the empirical section below.

The third feature of the descriptive statistics is the use of fertilizer. Between 2002 and 2004 both female and male cocoa farmers increased the amount of fertilizer used by a factor of nine. Adoption rates, however, have not risen at par for women compared to men. The percentage of women using fertilizer has gone up by 25 percentage points, while that of men has increased by 42 percentage points. This is a remarkable difference, which indicates the persistence of substantial gender differences in the access to and use of productive inputs.

[4] By which we identify all farmers who do not have a bank account.

Table 13.2 A profile of cocoa farmers, by sex

	Survey year	Women	Men	Total	Statistically significant differences
Sample size		75	353	428	
Age farmer	2002	54.17	50.40	51.06	
	2004	56.17	52.40	53.06	
Education farmer (number of years)	2002	3.75	7.25	6.64	
	2004	3.24	7.08	6.41	
Share of smallholders who own cocoa farm	2002	93	82	0.84	
	2004	96	88	0.89	
Kilograms of cocoa produced	2002	860	1,364	1,276	***
	2004	1,040	1,855	1,712	***
Farm size (hectare)	2002	4.93	6.54	6.26	***
	2004	5.43	7.93	7.49	***
Yields (kilograms cocoa/ha)^	2002	154.44	185.33	180.18	
	2004	182.52	213.84	205.92	
Share of farmers cash constrained	2002	59	39		***
	2004	60	36		***
Kilograms fertilizer	2002	14.42	28.35	25.91	
	2004	145.96	286.54	261.91	***
Percentage of farmers using fertilizers	2002	0.12	0.08	0.09	
	2004	0.37	0.50	0.47	
Liters insecticide	2002	6.53	12.93	11.81	
	2004	4.82	10.04	9.13	
Percentage of farmers using insecticide	2002	45.83	49.60	48.94	
	2004	99.92	99.95	99.94	
Real value of agricultural equipment^	2002	65,000	97,000	92,000	
	2004	54,348	86,957	79,710	
Total person days on cocoa	2002	251.93	342.16	326.35	
	2004	618.47	736.35	715.69	
Total labor productivity (kg cocoa/total person days)	2002	6.66	9.33	8.86	
	2004	2.72	4.40	4.10	
Household person days/ha	2002	43.83	20.64	24.70	***
	2004	102.24	85.88	88.74	
Household labor productivity (kg cocoa/household days)	2002	23.25	33.05	31.44	
	2004	4.61	8.14	7.53	***
Hired person days/ha	2002	38.23	36.16	36.52	
	2004	70.20	61.16	62.74	
Hired labor productivity (kg cocoa/hired days)	2002	16.12	18.88	18.41	
	2004	23.53	12.42	14.23	***

Source: Authors' calculations from the Ghana Cocoa Farmers Survey, 2002 and 2004 rounds
Notes: *** Indicates that the differences between men and women in the starred indicators are statistically significant at the 1 % level. ^ These are median values reported in place of the mean values to counter the effect of outliers in the distribution of these variables

Table 13.3 Yield regressions—fixed effects model

Dependent variable is: kg cocoa/ha	(1) Full sample	(2) Men	(3) Women	(4) Men – disag. lab	(5) Women – disag. lab
Cocoa farm size	−0.45***	−0.44***	−0.60**	−0.41***	−0.68***
	(0.10)	(0.11)	(0.23)	(0.12)	(0.23)
Person days/ha	0.01	−0.00	0.17*		
	(0.04)	(0.04)	(0.09)		
Household person days/ha				0.06	0.04
				(0.04)	(0.07)
Hired person days/ha				−0.02	0.09*
				(0.03)	(0.05)
Kilos fertilizer/ha	0.04	0.06	−0.05	0.06	−0.04
	(0.09)	(0.11)	(0.13)	(0.11)	(0.14)
Liters insecticide/ha	0.16**	0.17**	0.17	0.15**	0.26
	(0.07)	(0.07)	(0.17)	(0.07)	(0.17)
Real value equipment/ha	0.05*	0.07**	−0.06	0.07**	−0.05
	(0.03)	(0.03)	(0.05)	(0.03)	(0.05)
Rainfall	0.72***	0.59**	1.49**	0.58**	1.44**
	(0.22)	(0.24)	(0.64)	(0.24)	(0.66)
Farm quality controls⁺	Yes	Yes	Yes	Yes	Yes
Constant	−3.10	−3.90	−5.86	−1.11	−8.14
	(3.31)	(3.92)	(6.79)	(3.90)	(8.19)
Observations	795	658	137	658	137
Within group sample size	428	353	75	353	75
R-squared	0.27	0.25	0.50	0.26	0.50

Notes: All variables are in logs. Dummy variables to control for farmers not using inputs (fertilizer, insecticide, agricultural equipment) where used in all regressions but are not reported. Robust standard errors are in parentheses. Statistical significance levels are marked as follows: * significant at 10 %; ** significant at 5 %; *** significant at 1 %. ⁺ These include dummies for farms that have been treated against pests, as well as the mean age of all cocoa farms managed by the same individual

13.3.4 Production in Female- and Male-Managed Plots

We now examine differences in the factor proportions used by male and female smallholders on the cocoa farms that they manage and test for the existence of gender differences in the efficiency of production. Table 13.3 presents the results from estimating an intensive production function using a fixed-effects (FE) model, first pooling the sample, then disaggregating the estimation between female and male managed farms. Columns 4 and 5 are comparable estimations where the labor variable is disaggregated into household and hired labor.

Three important observations emerge from the regression results. First, the size-productivity relationship (as expressed by the coefficient of land) is consistently negative and statistically significant across all regressions, suggesting that—once all other productive inputs are controlled for—yields are higher on smaller plots.[5]

Second, columns 2 and 4, and 3 and 5, explore the variation in yields between men- and women-managed farms for comparable specifications of the production function. The elasticity of yield with respect to land (the rate at which yield falls as land size increases) on women controlled farms is 16–27 % higher than that observed on men controlled farms. Given that constant returns to scale are not rejected in any of the models, this suggests the existence of gender differences in factor proportions, although this difference is not significant.

Third, labor is the only variable input with a positive and significant contribution to women's yields. In columns 2 and 3, the estimated coefficient on labor, which aggregates household and hired labor, shows that a 10 % raise in total labor contributes to a 1.7 % increase in yields on female-managed plots. The size of the elasticity of labor on yields on male managed cocoa farms is, on the other hand, marginal and statistically insignificant.

In columns 4 and 5, we further tease out this result by disaggregating labor input; the results show that hired labor is responsible for the positive and significant contribution to yields on female-managed plots (with a 0.09 elasticity coefficient). We further disaggregate the labor employed on male and female managed cocoa farms in Table 13.4.

While the limited number of observations in this case study does not allow us to conduct further regression analysis on the "labor" effect that we find, this table provides some indications. Both male and female cocoa farmers have increased the labor deployed on their farms over the period of increased expansion. Male farmers increased the input from male members in their households by almost a factor of four and increased other sources of family and hired labor more moderately (between 35 and 64 %), even decreasing their use of *nnoboa* labor. Female farmers, on the other hand, substantially increased the amount of female household labor as well as all three components of hired labor. In particular, increases in annual and *nnoboa* labor—the sources of hired labor less subject to cash constraints—were large (247–271 %).

In sum, the Ghana case study on cocoa farming provides three important lessons about gender differences in high value cash crops. First, while women farmers remain a minority group among smallholders engaged in the cultivation of these crops, there are some clear signals that the conditions underlying this imbalance are changing. The progressive, more individualized evolution of land rights—which is observed in the Ghanaian case presented above—provides a clear illustration of these benefits. Relaxing the barriers faced by women farmers in securing privileges over the farms they manage will generate higher yields, which in turn translates into higher cash incomes controlled by women. Second, and in line with what is known

[5] We also note that the same qualitative results were obtained after estimating the model for female farmers against a subsample of male farmers managing the same size of cocoa farms as the sampled women. This additional step was taken as observed differences could be driven by the different distribution of the land variable between men and women in the original sample.

Table 13.4 Labor employed on male- and female-managed cocoa farms

	Person days employed		
Type of labor	Year of survey	Female-managed cocoa farms	Male-managed cocoa farms
Household men	2002	56.12	38.63
	2004	75.17	186.05
	Difference (2004–2002)	0.34	3.82
Household women	2002	28.40	47.68
	2004	75.63	78.18
	Difference (2004–2002)	1.66	0.64
Household child	2002	9.68	8.49
	2004	11.15	11.48
	Difference (2004–2002)	0.15	0.35
Annual	2002	16.12	43.97
	2004	56.01	59.81
	Difference (2004–2002)	2.47	0.36
Daily wage	2002	130.85	168.10
	2004	226.08	274.55
	Difference (2004–2002)	0.73	0.63
Nnoboa	2002	10.76	35.04
	2004	39.87	26.18
	Difference (2004–2002)	2.71	−0.25

Source: Ghana Cocoa Farmers Survey, 2002 and 2004 rounds

in the literature, female-managed farms are as productive as male-managed ones. The data used in this study show that, if anything, female farmers became more productive than their male counterparts at a time of expansion in the sector as a whole. Third, and more important for the thrust of this chapter, we have shown that the "all else equal" statement recurrent in gender-focused productivity analysis conceals important differences in how women actually access productive inputs. In the case of cocoa, we have shown that when female farmers are able to hire in more labor, their productivity increases. However, they do so by increasing use of a particular type of non-household labor (*nnoboa*) that is not subject to cash constraints. We have also shown that women farmers also are not able to increase substantially their use of nonlabor inputs—possibly as a result of being cash constrained, which may explain their adoption of production technologies that are intensive in labor, but not in the use of modern inputs.

13.4 Case Study: Marketing Coffee in Uganda

This section presents data on coffee transactions for 300 coffee farmers in the central and western districts of Uganda. A quarter of the households in the sample are female-headed, and the following analysis presents data on differences in coffee production and harvesting, access to market information, and the nature of coffee market transactions between male- and female-headed households.

13.4.1 Background on Coffee in Uganda

Coffee is Uganda's largest export good, comprising 26 % of export earnings in 2000/2001, and providing direct and indirect, partial employment to an estimated five million people (Bank of Uganda 2001; Kempaka 2001). Robusta coffee accounts for nearly 90 % of Uganda's coffee production and is predominantly grown in lowland areas in central and southwestern Uganda. Like so much commodity crop production in low income countries, production of coffee is concentrated among smallholder farmers. In Uganda, coffee is usually intercropped with staple crops—often *matooke* (a banana-like staple), beans, sweet potatoes, and maize. The production technology is basic. Few farmers use purchased inputs such as fertilizer or pesticides, and few use modern farming methods such as irrigation.

Compared to other crops that Ugandan coffee farmers can grow, coffee is relatively profitable. The Uganda Coffee Development Authority (UCDA) provides some estimates of the costs of growing coffee, which can be used to estimate the return to coffee production for an average farmer.[6] Estimates suggest a return of $197 per ha, which compares favorably to a return of $150 for growing *matooke*. However, there is a large degree of variation in this price as Fig. 13.1 shows—much more so than for other crops. Prices during the year in which data were collected were much lower than average: the median price recorded for a kilo of unmilled Robusta (*kiboko*) was 16 cents, which would imply a per hectare return of $57.89.

In 1992 the export and domestic marketing of coffee was liberalized. The market for coffee is one of the most liberalized in the world, with few regulations or barriers to entry at any point in the marketing chain. Farmers usually make individual sales at the farmgate to small traders who tour the countryside on bicycles or motorcycles and act as aggregators either for bigger independent traders or for exporters and their agents. The majority of Ugandan producers sell their coffee in the form of dry cherries, locally known as *kiboko*, which are then milled (the cherry is separated from the husk) by the traders who buy the coffee. Milled coffee of average quality is referred to as fair average quality (FAQ) coffee.[7] Sometimes farmers sell their coffee at the nearest market and mill the coffee themselves before selling it. There is considerable competition reported at the primary marketing level, and so farmers should have little problem selling their coffee at competitive market prices.

[6] Details on how these estimates were calculated can be found in Hill (2009).

[7] Well looked-after, healthy trees produce a ratio of 0.6 kg of Fair Average Quality (FAQ) coffee cherries for 1 kg of *kiboko*, while old and diseased trees produce *kiboko* with a lower ratio that can reach as low as 1:0.4.

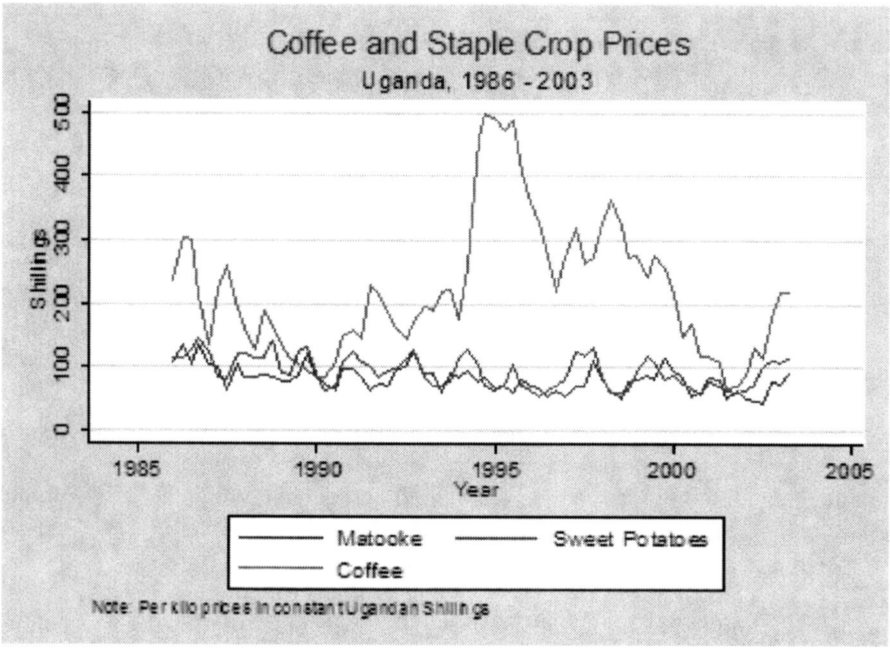

Fig. 13.1 Coffee and staple crop prices 1986–2003 (Source: Henstridge 1997 (UCDA and UBOS data))

13.4.2 Data

In early 2003, 300 coffee farming households were sampled from the 1999/2000 Uganda National Household Survey and revisited with a survey asking many of the same questions (thereby creating a small panel) plus additional questions on coffee sales made in the previous year. The households surveyed came from five districts—Bushenyi, Kayunga, Luwero, Masaka, and Mukono—that together comprise half of Uganda's Robusta coffee production. Almost one in four of the households surveyed were female-headed (23 %). These female-headed households largely comprise widowed women (68 %) but also include unmarried, separate, and divorced women. In nearly all cases, the respondent was the head of the household, as the survey enumerators were directed to speak to someone who was knowledgeable about the production and marketing of the crop, typically the household head.

While detailed data on the nature of coffee sales were collected, the survey did not collect information on who made the sale. We assume that in male-headed households, the male head made the sale, and in female-headed households, the female head. The analysis thus relies on a comparison of male- and female-headed households. While this is, in general, an imperfect proxy of intrahousehold decisionmaking on coffee sales, in this context it is a justifiable assumption, because coffee sales are in general handled by the head of the household.

Table 13.5 Basic characteristics of female- and male-headed households

	Female-headed households	Male-headed households	T-test
Working age household members	3.1	3.7	−2.41**
Total household days spent farming	466	573	−2.54**
Land owned (acres)	3.6	6.2	−2.72***
Coffee trees owned	189	650	−2.59***
Proportion of trees…			
in productive stage of life	0.84	0.77	2.50***
too young to produce	0.15	0.23	−2.43***
Log of asset wealth	7.05	7.68	−4.08***
Education	2.8	6.4	−6.02***
Age	57	50	3.29***
Share of labor spent on coffee	0.36	0.34	0.56
Proportion of productive trees harvested	0.83	0.81	0.50
Quantity harvested (kg, total)	47	151	−2.89***
Quantity harvested per tree (kg)	1.35	1.52	−0.51

Notes: *** Diff. significant at 99 %; ** Diff. significant at 95 %

13.4.3 Characteristics of Female- and Male-Headed Households

Table 13.5 compares basic characteristics of female- and male-headed households. A number of basic differences exist between these households. Female-headed households have less labor, land, and coffee trees than male-headed households. In particular, the difference in the mean size of land owned and thus the number of coffee trees owned is substantial. To determine whether this is driven by outliers, we also compare the median of these two variables. The median is still higher among men and a Pearson Chi-squared test shows that the null of equal medians can be rejected at 1 % degree of significance for both variables.

Female-headed households also tend to have lower levels of wealth and lower levels of education. Women household heads tend to be older, because they assume headship once their male partner has died. As a result of these basic differences in scale, liquidity, and human capital, we may expect crop choice, production methods, and access to markets to be quite different for male- and female-headed households.

Table 13.5 also presents some information on the nature of coffee production. First, only 15 % of the trees women own are too young to produce compared to 23 % of trees owned by male-headed households. This implies that women are less likely to plant trees; if they planted trees on a regular basis, they would have younger trees in similar proportions with men. The share of labor allocated to coffee production and the proportion of trees harvested are comparable between these two types of households, as is the yield per tree (counting only productive trees). However, because female-headed households farm on a much smaller scale than male-headed households, the quantities sold by women are much smaller than the quantities sold

Table 13.6 Access and use of markets, by female- and male-headed households

	Female-headed households	Male-headed households	T-test
Transportation to sales market			
Distance to coffee market (miles)	11.1	10.8	0.17
Ownership of bicycle (number owned)	0.19	0.68	−6.02***
Access to market information			
Received price information from other than buyer	0.42	0.47	−0.73
Knew name of buyer of coffee	0.42	0.54	−2.19**
Marketing channels used			
Proportion of sales of dry coffee	0.83	0.85	−0.49
Proportion of sales at the market	0.07	0.15	−2.07**
Proportion of sales of milled coffee	0.00	0.03	−1.96*

Notes: *** Diff. significant at 99 %; ** Diff. significant at 95 %; * Diff. significant at 90 %

by men. Transactions made by female-headed households are 47 kg, on average, compared to 151 kg for men. Again we also compare the median values of quantities sold. The median quantity (of FAQ equivalent) sold by women is 32.4, and 54 kg for men. A Pearson Chi-squared test shows that the null of equal medians can be rejected at 1 % degree of significance (not reported here).

13.4.4 Marketing Patterns of Female- and Male-Headed Households

We use the marketing focus of this survey to provide some information on differences in market access and marketing patterns of male- and female-headed households. Table 13.6 shows the results of access to and use of markets by female-headed and male-headed households. First, we note that while there is no difference in the location of the households (both are equidistant from sales markets), accessibility to these markets differs as a result of differences in bicycle ownership. Women are much less likely to own a bicycle than men and as a result, the time taken to travel to market will, on average, be much higher for women than for men.

Respondents were asked whether they received price information from anyone in addition to the trader who purchased their coffee. For just under half of the households interviewed, price information was received from someone in addition to the buyer of coffee, and little difference was observed between male- and female-headed households. However, it does appear as though female-headed households have less access to trader networks than male-headed households. Respondents were asked whether or not they knew the name of the person that had bought their coffee, and were asked to name the buyer when they did; 54 % of male households were able to name the trader who had purchased their coffee, while only 42 % of female-headed households were able to (see Table 13.6).

We are ultimately interested in whether these or other differences between female- and male-headed households result in different use of market channels. This information is reported in the bottom panel of Table 13.6. As discussed in Sect. 13.4.1, the usual mode of selling coffee consists in farmers drying their Robusta and making sales of unmilled Robusta (*kiboko*) at the farmgate. Some farmers elect to transport their coffee to market in return for receiving a higher price, and in addition, a few farmers will choose to mill their coffee once at the market before selling it (for an additional premium). No significant difference exists in the proportion of households that dry their coffee before making the sale. However, there are substantial differences in the proportion of women that travel to sell their coffee and in the proportion of women that mill their coffee before sale. In 15 % of the transactions made by male-headed households, the coffee is sold at the nearby coffee market. This proportion is less than half (7 %) for the transactions made by female-headed households. A very low proportion (3 % of transactions) was for milled coffee, and these transactions were all made by male-headed households.

These descriptive statistics suggest that female-headed households sell smaller quantities of coffee (as a result of owning fewer trees and producing less coffee) and engage in less value addition (transporting to market, milling) than male-headed households. We now proceed to use the analysis of choice of market channel presented in Fafchamps and Hill (2005) to explore the determinants of gender differences in choice of marketing channel.

Fafchamps and Hill (2005) show that the choice of market outlet (farmgate or market) is determined by the quantity of coffee being sold, the time it would take a household to transport that coffee to market (determined by the distance of a household from the market and whether or not a household owns a bicycle), and the wealth of the household. A household's wealth proxies both the opportunity cost of time for transporting the coffee, and the ease with which a household can access the liquidity that may be needed to transport large quantities of coffee or to transport coffee over large distances (both of which would require paying for bus transportation or hiring a truck). Thus the impact of wealth varies with the amount of coffee being sold and the distance of an individual from market.

Tables 13.5 and 13.6 highlight a number of these determinants of market choice that vary between female- and male-headed households. Female-headed households are poorer, sell smaller quantities of coffee, and are less likely to own bikes than male-headed households. Any or all of these differences could drive the gender differences we observe in choice of market channel. To determine the relative importance of these factors, we re-estimate the Fafchamps and Hill model augmented with a dummy that equals 1 if the household is headed by a female. Results are presented in Table 13.7. The first column shows the simple gender difference in the probability of selling at the market and the farmgate, which is significant at 5 %. The final column (column 6) shows the estimates of the full model with the gender dummy. Once distance, quantity, and wealth are included, there is no significant difference in the way female- and male-headed households market. Gender differences thus arise as a result of differences in these other characteristics.

Table 13.7 Determinants of choice of market channel

	(1) Simple gender difference	(2) Gender difference controlling for distance to market	(3) Controlling for bike ownership	(4) Controlling for quantity	(5) Controlling for wealth	(6) Full model
Female household head	−0.437	−0.420	−0.206	−0.104	−0.369	−0.024
	(0.196)**	(0.205)**	(0.226)	(0.224)	(0.219)*	(0.257)
Distance to market		−0.021	−0.019	−0.017	−0.169	−0.184
		(0.007)***	(0.007)***	(0.007)**	(0.078)**	(0.061)***
Bike dummy			0.484			0.350
			(0.181)***			(0.222)’
Log (quantity sold)				0.500		−1.596
				(0.133)***		(0.418)***
Residual for quantity				−0.290		1.086
				(0.146)**		(0.569)*
Log (wealth)					0.035	−1.494
					(0.104)	(0.347)***
Log (quantity) * log (wealth)						0.299
						(0.068)***
Residual for quantity * wealth						−0.189
						(0.089)**
Distance * log (wealth)					0.023	0.025
					(0.011)**	(0.009)***
Constant	−1.014	−0.562	−0.981	−2.658	−0.863	7.262
	(0.078)***	(0.146)***	(0.230)***	(0.598)***	(0.750)	(1.974)***
Number of observations	489	489	489	486	489	486
Wald Chi-squared	4.98**	4.54***	5.80***	7.32***	3.94***	6.35***
Pseudo R-squared						

Notes: District and season dummies included from column (2) on, but not shown. Standards errors are in parentheses. *** Variable significant at 99 %; ** Variable significant at 95 %; * Variable significant at 90 %; ’ Variable significant at 85 %

Table 13.8 Price received

	(1) Female-headed household	(2) Male-headed household	(3) T-test	(4) T-test controlling for marketing channel
Price received (US$ per kilo)	0.14	0.15	−1.67*	−0.09

Note: Both t-tests include district and month of sale dummies * Diff. significant at 90 %

Columns 2–5 separately include distance to market, ownership of bike, quantity, and wealth to identify which factors render the gender dummy insignificant. While physical distance to market does not explain the gender difference, owning a bike does. Once the number of bikes owned is included, the gender dummy is no longer significant. The quantity of coffee marketed is also an important determinant of the gender difference in marketing channel. Wealth reduces the significance and magnitude of the gender dummy, but a gender difference still remains even when controlling for differences in wealth. It thus appears that gender differences in marketing are largely explained by women marketing smaller quantities of coffee and not owning bicycles. Before concluding, we discuss the ultimate impact of gender on the price received. Female-headed households receive a lower price for coffee than men. On average, women received 14 cents per kilo of *kiboko* while men received 15 cents. However, this is entirely explained by the difference in marketing channels. Table 13.8 shows a significant difference between female- and male-headed households when only district and month dummies are included (column 3). However, column 4 shows there is no gender difference in the price once we control for how they sell. This is an important point, as it reflects the fact that the main constraint women face is in accessing marketing channels that allow value addition, rather than facing any discrimination in the marketing channel in which they are engaged.[8]

13.5 Conclusion

This chapter contributes to the understanding of the constraints women face in engaging in cash crop markets. First, it confirms a point largely documented in the existing literature: women are as productive as men and receive comparable prices to those received by men when they farm with *equal* resources and sell their crops in the *same way*. Second, rarely do women have similar access to assets and markets as men and this has non-trivial implications for how they produce and market cash crops.

[8] It is worth noting that Table 13.8 presents data on gross returns. Households that engage in value addition incur costs of time, fuel costs (if applicable), and milling fees that will reduce the overall price differential reported. Given that we do not have data on time spent marketing, we cannot estimate the net return.

In the Ghanaian case, limited access to liquidity for purchasing inputs induces women to adopt sub-optimal production technologies (i.e., technologies that are labor and land using; the two factors that are more costly/scarce to farmers).

In the Uganda case study, the low quantities marketed, and lack of access to bicycles, limit female farmers to marketing channels that have very low transaction costs, but which receive lower prices.

While our conclusions are based on two specific crops and contexts (both cocoa and coffee are traditional perennial export cash crops, quite different in nature from newer cash crop markets, such as horticulture or flower exports), we propose three context-specific recommendations.

1. In food crop markets are characterized by poor integration and relatively high price volatility, scale in production will be an important determinant of whether or not individuals produce cash crops, disproportionately affecting female farmers that produce at a smaller scale. Both improving women's access to land, and encouraging better integration of food markets through improved roads and increased mobile networks, will enable women to engage in cash crop production.
2. Interventions that strengthen female farmers' groups or marketing groups to which female farmers can belong, allow women to achieve scale in marketing. These interventions could include group leadership training, financial management training, training group leaders on how to find buyers, or introducing local buyers to female marketing groups. Directly reducing transaction costs that are specifically faced by women—in the Uganda case by encouraging female use of bicycles to be more socially acceptable—is also essential.
3. When purchasing quantity or quality enhancing inputs is difficult, female farmers may compensate by increasing resources that do not require upfront payment in cash such as labor. Improving access to credit (through contract farming targeted at female farmers, improved access to microfinance for women, or female savings schemes and credit associations) and extension services becomes even more important in ensuring cash crop production in these situations.

Further work in assessing the patterns and underlying determinants of female engagement in a wide variety of cash crop markets is needed to better identify the most appropriate interventions. This will require piloting some of the suggested interventions and evaluating their impact on improving access to inputs, on lowering transaction costs and in achieving scale in production and marketing. This chapter provides evidence that designing, piloting, and potentially scaling up such interventions has merit: cash crop production by women can be highly profitable provided that many of the constraints discussed in this chapter are lifted.

References

Anderson J, Feder G (2003) Rural extension services. Policy research working paper 2976. World Bank, Washington, DC

Baden S, Green C, Otoo-Oyortey N, Peasgood T (1994) Background paper on gender issues in Ghana. Report prepared for the West and North Africa Department, Department for Overseas Development, UK. Bridge, Brighton

Balsevich F, Berdegue JA, Flores L, Mainville D, Reardon T (2003) Supermarkets and produce quality and safety standards in Latin America. Am J Agric Econ 85(5):1147–1154

Bank of Uganda (2001) Annual report 2000–2001. Kampala

Barham J, Chitemi C (2009) Collective action initiatives to improve marketing performance: lessons from farmer groups in Tanzania. Food Policy 34(1):53–59

Bell C (1990) Interactions between institutional and informal credit agencies in rural India. World Bank Econ Rev 4(3):297–328

Blackden CM, Wodon Q (eds) (2006) Gender, time use, and poverty in Sub-Saharan Africa. World Bank working paper no. 73. World Bank, Washington, DC

Breisinger C, Diao X, Kolavalli S, Thurlow J (2008) The role of cocoa in Ghana's future development Ghana. IFPRI background paper 11. International Food Policy Research Institute, Washington, DC

Chowdhury S, Gulati A, Gumbira-Sa'id E (2005) High value products, supermarkets, and vertical arrangements in Indonesia. Markets, Trade, and Institutions Division discussion paper 83. International Food Policy Research Institute, Washington, DC

Conley T, Udry C (2010) Learning about a new technology: pineapple in Ghana. Am Econ Rev 100(1):35–69

Croppenstedt A, Demeke M, Meschi M (2003) Technology adoption in the presence of constraints: the case of fertilizer demand in Ethiopia. Rev Dev Econ 7(1):58–70

Doss CR (2001) Designing agricultural technology for African women farmers: lessons from 25 years of experience. World Dev 29(12):2075–2092

Doss C (2002) Men's crops? Women's crops? The gender patterns of cropping in Ghana. World Dev 30(11):1987–2000

Doss CR, Morris ML (2001) How does gender affect the adoption of agricultural innovations? The case of improved maize technology in Ghana. Agric Econ 25(1):27–39

Edelman M, Schmiesing B, Olsen D (1990) Use of selected marketing alternatives by Iowa farmers. Agribusiness 6(2):121–132

Evenson R (2001) Economic impacts of agricultural research and extension. In: Gardner B, Rausser G (eds) Handbook of agricultural economics, vol 1. Elsevier, North Holland, pp 573–628

Fafchamps M (1992) Cash crop production, food price volatility, and rural market integration in the Third World. Am J Agric Econ 74(1):90–99

Fafchamps M, Hill RV (2005) Selling at the farmgate or traveling to market. Am J Agric Econ 87(3):717–734

Fletcher S, Terza J (1986) Analysing farmer's selection of available marketing alternatives using the multivariate probit model. Can J Agric Econ 34(2):243–252

Fu T, Epperson J, Terza J, Fletcher S (1988) Producer attitudes towards peanut market alternatives: an application of multivariate probit joint estimation. Am J Agric Econ 70(4):910–918

Gladwin CH (1992) Gendered impacts of fertilizer subsidy removal programs in Malawi and Cameroon. Agric Econ 7(2):141–153

Goldstein M, Udry C (2005) The profits of power: land rights and agriculture investment in Ghana. Economic growth center discussion paper 929. Yale University, New Haven

Henstridge NM (1997) The reconstruction of a macroeconomic dataset for Uganda. Centre for the Study of African Economies working paper series no. 98–3. Institute of Economics and Statistics, University of Oxford, Oxford

Hill RV (2009) Using stated preferences and beliefs to identify the impact of risk on poor households. J Dev Stud 45(2):151–171

Hoff K, Stiglitz J (1990) Imperfect information and rural credit markets—puzzles and policy perspectives. World Bank Econ Rev 4(3):235–250

Holden S, Shiferaw B, Pender J (2001) Market imperfections and land productivity in the Ethiopian highlands. J Agric Econ 52(3):53–70

Kasente D, Lockwood M, Vivian J, Whitehead A (2001) Gender and the expansion of nontraditional agricultural exports in Uganda. In: Razavi S (ed) Shifting burdens: gender and agrarian change under neo-liberalism. Kumarian Press Inc, Bloomfield, pp 35–66

Kempaka G (2001) Coffee and its impact and relevance to PEAP (Poverty Eradication Action Plan). In: The coffee yearbook 2000–2001. Uganda Coffee Trade Federation, Kampala

Key N, Sadoulet E, de Janvry A (2000) Transactions costs and agricultural household supply response. Am J Agric Econ 82(2):245–259

Koopman J (1993) The hidden roots of the African food problem: looking within the rural household. In: Folbre N, Bergmann B, Agarwal B, Floro M (eds) Women's work in the world economy. New York University Press, New York, pp 82–103

Kumar SK (1987) Women's role and agricultural technology. In: Mellor JW, Delgado CL, Blackie MJ (eds) Accelerating food production in Sub-Saharan Africa. Johns Hopkins University Press, Baltimore, pp 135–147

Lilja N, Sanders J (1998) Welfare impacts of technological change on women in Southern Mali. Agric Econ 19(1):73–79

MASDAR Ltd (1998) A socio-economic study of the cocoa farming community. Hampshire.

Morrison A, Raju D, Sinha N (2007) Gender equality, poverty, and economic growth. Policy research working paper 4349. World Bank, Washington, DC

Paolisso M, Hallman K, Haddad L, Regmi S (2002) Does cash crop adoption detract from childcare provision? Evidence from rural Nepal. Econ Dev Cult Chang 50(2):313–337

Peterman A, Behrman J, Quisumbing A (2010) A review of empirical evidence on gender differences in non-land agricultural inputs, technology, and services in developing countries. Background paper prepared for The State of Food and Agriculture 2010–11, Food and Agriculture Organization of the United Nations, Rome

Quisumbing AR (1994) Improving women's agricultural productivity as farmers and workers. Education and Social Policy Department discussion paper series 37. World Bank, Washington, DC

Quisumbing A, Pandofelli L (2009) Promising approaches to address the needs of poor female farmers: resources, constraints, and interventions. IFPRI discussion paper 00882. International Food Policy Research Institute, Washington, DC

Quisumbing AR, Payongayong E, Aidoo JB, Otsuka K (2001) Women's land rights in the transition to individualized ownership: implications for the management of tree resources in Western Ghana. Econ Dev Cult Chang 50(1):157–182

Quisumbing A, Payongayong E, Otsuka K (2004) Are wealth transfers biased against girls? Gender differences in land inheritance and schooling investment in Ghana's Western Region. Food Consumption and Nutrition Division discussion paper 186. International Food Policy Research Institute, Washington, DC

Randolph S, Sanders R (1988) Constraints to agricultural production in Africa: a survey of female farmers in the Ruhengeri Prefecture of Rwanda. Stud Comp Int Dev 23(3):78–98

Saito K, Mekonnen H, Spurling D (1994) Raising the productivity of women farmers in Sub-Saharan Africa. Africa Technical Department series no. 230. World Bank, Washington, DC

Sarpong GA (2006) Improving tenure security for the rural poor: Ghana case study. LEP working paper number 2. Food and Agriculture Organization of the United Nations, Rome

Teal F, Zeitlin A, Maamah H (2006) Ghana cocoa farmers survey 2004: report to Ghana cocoa board. Centre for the Study of African Economies, Oxford University. http://www.gprg.org/pubs/reports/pdfs/2006-04-teal-zeitlin-maamah.pdf

Udry C (1996) Gender, agricultural production, and the theory of the household. J Polit Econ 104(5):1010–1046

Vigneri M (2005) Trade liberalisation and agricultural performance: micro and macro evidence on cash crop production in Sub-Saharan Africa. D. Phil thesis, Oxford University, UK

Vigneri M, Holmes R (2009) When being more productive still doesn't pay: gender inequality and socio-economic constraints in Ghana's cocoa sector. Paper presented at the FAO-IFAD-ILO workshop on gaps, trends, and current research in gender dimensions of agricultural and rural employment, Rome

von Braun J, Webb P (1989) The impact of new crop technology on the agricultural division of labor in a West African setting. Econ Dev Cult Chang 37(3):513–534

Warning M, Key N (2002) Social performance and distributional consequences of contract farming: an equilibrium analysis of the Arachide de Bouche program in Senegal. World Dev 30(2):255–263

Warning M, Key N, Soo Hoo W (2003) Small farmer participation in contract farming. Working paper, University of Puget Sound, Tacoma, WA, US

Weir S, Knight J (2000) Adoption and diffusion of agricultural innovations in Ethiopia: the role of education. Centre for the Study of African Economies working paper 2000–5. Institute of Economics and Statistics, University of Oxford, Oxford, UK

World Bank (2007) World development report 2008: agriculture for development. World Bank, Washington, DC

World Bank (2008) Country brief: Ghana. World Bank, Washington, DC

World Bank/IFAD/FAO (International Fund for Agricultural Development/Food and Agriculture Organization for the United Nations) (2009) Gender in agriculture sourcebook. World Bank, Washington, DC

Chapter 14
Gender Inequalities in Rural Labor Markets

Jennie Dey de Pryck and Paola Termine

Abstract This chapter explores gender differences in rural employment that hinder the achievement of food security and the reduction in poverty. This chapter begins with a review of employment statistics to uncover gender differences in rural employment with particular reference to traditional agriculture and modern agro-industries, identifying gender gaps in wages, working conditions, and occupational segregation as key challenges to overcome. The chapter argues that the barriers to gender equality in rural labor markets are socially constructed and primarily stem from systemic *institutional* gender inequalities. These institutions include both social norms and the structure of labor market organizations. While institutional change cannot take place overnight, particularly with regard to changing social norms, much can be done to improve gender equity in rural labor institutions through government policies, corporate social responsibility programs, and building the strength of women in labor organizations. This involves overcoming vested interests in the status quo that provides a supply of cheap labor, and therefore requires both political will and resources.

Keywords Gender • Employment • Labor markets • Labor force participation • Social norms

---

J. Dey de Pryck (✉)
Global Forum on Agricultural Research (GFAR), Gender in Agriculture Partnership (GAP), c/o Food and Agriculture Organization of the United Nations (FAO), Viale delle Terme di Caracalla, Rome 00153, Italy
e-mail: jenniedeydepryck@yahoo.com

P. Termine
Child Labour in Agriculture, Fundamental Principles and Rights at Work (FPRW), International Labour Organization, 4 route des Morillons, CH-1211, Genève 22, Switzerland
e-mail: termine@ilo.org

A.R. Quisumbing et al. (eds.), *Gender in Agriculture: Closing the Knowledge Gap*, DOI 10.1007/978-94-017-8616-4_14,

14.1 Introduction

Rural areas are home to three out of four of the developing world's poor, most of whom depend directly or indirectly on agriculture for their livelihoods. Many are small producers in crops, livestock, fisheries or forestry, while a significant number have no resources other than their own labor and work as wage laborers in the fields, at sea or on the shores, in artisanal agro-processing or in agriculture-based industry. Many agricultural households supplement their income with wage work in their own villages or (directly as migrants or indirectly through remittances) from work in urban or other rural areas in their own countries or abroad.

To tackle rural poverty, *employment* must play a key role. Because the majority of the poor are working but are in vulnerable employment, the overriding imperative is not simply to create new jobs but to create *quality* work with higher and more stable incomes and with safer and healthier working conditions. This challenge is captured by the International Labor Organization's (ILO) Decent Work Agenda, which has been recognized by the United Nations as crucial to the achievement of the Millennium Development Goals (MDGs). This promotes rights at work, decent and productive employment and income for women and men, social protection for all, and social dialogue, with gender equality and nondiscrimination as cross-cutting priorities (ILO 1999).[1]

The structure of rural employment has been evolving in response to technological innovation in agriculture, expansion of nonfarm entrepreneurial activities, increasing out-migration, higher educational attainments, changing international and domestic demand patterns for developing country agricultural products due to changing food habits, and the emergence and consolidation of global agricultural value chains that increasingly integrate production with processing and service provision. All these changes have major implications for rural employment and income, which vary by region, farming system, value chain, extent of market penetration, and political stability. They can bring significant gains or losses to different socio-economic groups in rural areas as well as agricultural companies and their shareholders, and do not necessarily advance the goal of decent work nor reverse the near-universal heritage of marked gender inequality in rural labor markets.

Within this context, this chapter addresses the following questions:

- What are the gender differences in rural employment with particular reference to traditional agriculture and modern agro-industries?
- What are the barriers to gender equitable employment in rural labor markets and their underlying causes?
- What "good practice" policies, laws and other instruments can be replicated or scaled up to tackle these barriers? How can women fight these barriers themselves?

[1] MDG1 now includes a target to "achieve full and productive employment and decent work for all, including women and young people" (ILO 2010, xi).

14.2 The Gender Structure of Rural Employment

14.2.1 Participation and Employment Rates

Gender differences in labor-force participation rates are striking, with female participation much lower than male participation worldwide across all sectors. ILO data show considerable increases in female participation rates worldwide over the last 60 years, rising from 35.9 % in the 1950s to 47.9 % at the end of the 1990s, while male rates remained fairly stable at 93.7 % and 90.1 %, respectively (Tzannatos 1999). More recent ILO studies using different measurement techniques estimated the female participation rate at 50.3 % compared with 76.5 % for men in 2012 (ILO 2014, Table A8). While male rates have declined significantly as more young men pursue higher education and older men retire earlier on pensions, female rates have continued to rise, partly reflecting changes in definitions of work and better data collection methods, although other factors are also at play (Tzannatos 1999; Doss, Chaps. 3 and 4).

Comparing participation rates (that include persons actively looking for work) with employment rates (i.e., persons who are actually working), the gender gap narrows slightly for employment rates, although there are significant variations among regions and countries (Table 14.1). Male employment rates ranged between 67.4 % in the Middle East to 78.5 % in South Asia in 2010. Female rates varied considerably more, from around 60 % in East Asia and Sub-Saharan Africa, to 19.2 % in North Africa and 14.3 % in the Middle East, although these data most likely fail to capture much of women's employment (see Doss, Chaps. 3 and 4, for possible reasons).

Agriculture accounted for the dominant share of employment for both men and women in 2012 in South Asia and Sub-Saharan Africa, while the agricultural share of employment was only slightly less than for services for both sexes in South-East Asia and the Pacific, and for women in East Asia (ILO 2014, Table A10). Agricultural share of employment was particularly high in Sub-Saharan Africa where both men's and women's shares were 60.2 % and 62.2 % respectively, and in South Asia where

Table 14.1 Employment-to-population ratio, by sex, world and developing regions (percent)

	2000		2010		2013[a]	
World and regions	Males	Females	Males	Females	Males	Females
World	73.9	48.5	72.1	47.1	72.2	47.1
East Asia	78.3	66.8	73.6	60.7	73.9	60.9
South-East Asia and the Pacific	78.4	55.6	78.3	56.0	78.7	56.6
South Asia	79.5	33.1	78.5	30.1	77.7	29.1
Latin America and the Caribbean	74.8	42.7	74.9	48.2	75.3	49.3
Middle East	67.1	12.8	67.4	14.3	68.5	14.9
North Africa	65.4	16.7	68.1	19.2	67.6	18.9
Sub-Saharan Africa	71.2	57.1	70.1	59.3	71.3	59.6

Source: ILO (2014, Table A5)
[a]2013 are preliminary estimates

women's share was 66.4 % compared with 45.5 % for men, despite the steady decline in the share of employment in agriculture between 1999 and 2012 in all regions and for both men and women.

Globally, women were engaged to a slightly greater degree in agriculture than men, at 37.6 % compared to 33.3 % for men, despite considerable regional differences. For example, women's share of employment in agriculture was higher than men's in East Asia, South Asia, North Africa, and strikingly so in the Middle East, and similar in Sub-Saharan Africa. This contrasts sharply with the Latin America and Caribbean region, where women's share of employment in agriculture was small. Men's share of employment in agriculture and industry was much higher than women's, while their share in services was some 25 % less (FAO 2011).

Despite the declining share of employment in agriculture, there were an estimated 1.000 billion workers worldwide in the agricultural sector in 2013 and the *actual* number of workers grew between 1999–2009, only to fall slightly in 2012 and 2013. Among the developing economies, the number of workers in agriculture declined considerably in East Asia and very slightly in South Asia and Latin America and the Caribbean, but rose in the other developing regions. Between 1999 and 2009, for example, growth in agricultural employment accounted for half of all employment growth in Sub-Saharan Africa and 33 % in South Asia (ILO 2014).

14.2.2 Disparities in Employment Status

In the last decade, most developing country agricultural employment was self-employment and relatively few men, and even fewer women, worked for a wage (Table 14.2). Notable exceptions were the relatively large share of agricultural wage earners in South Asia (21.8 % male and 11.4 % female) and male wage earners in Latin America (20.9 % male but only 2.3 % female). With regard to nonagricultural rural work, a considerably higher percentage of men than women were wage earners or self-employed (except for East Asia and the Pacific (excluding China), where the figures were similar). Women played a very small role in nonagricultural rural self-employment and wage work in Sub-Saharan Africa, South Asia, and the Middle East and North Africa.

These regional averages mask differences between and within countries. Fontana and Paciello (2010) cited data for 2005 showing that both men and women worked much more than the average in Tanzania (78 % and 81 % worked in agriculture, respectively), whereas in India, male rates were consistent with regional averages but female casual labor (32.6 % of the rural female labor force) greatly exceeded the averages.

14.2.3 Vulnerable Employment and Working Poverty

ILO data indicate that women are overrepresented in vulnerable employment, defined as "the sum of own-account workers and unpaid family workers"

Table 14.2 Rural employment, by sex and employment status, developing regions (percent)

Employment status	Sub-Saharan Africa	South Asia	East Asia and the Pacific (excluding China)	Middle East and North Africa	Europe and Central Asia	Latin America and the Caribbean
Agriculture						
Self-employed						
Males	56.6	33.1	46.8	24.6	8.5	38.4
Females	53.5	12.7	38.4	38.6	6.9	22.8
Wage earner						
Males	4.0	21.8	9.4	9.4	10.1	20.9
Females	1.4	11.4	5.7	1.0	5.4	2.3
Nonagriculture						
Self-employed						
Males	6.9	11.8	11.5	8.8	7.4	9.2
Females	6.8	2.9	11.3	2.8	1.6	11.7
Wage earner						
Males	8.6	15.4	17.4	30.9	31.3	17.2
Females	2.8	2.7	8.4	3.9	18.1	11.5
Non-active or not reported						
Males	21.7	14.6	14.4	26.0	27.5	13.4
Females	32.7	64.3	35.5	53.3	46.9	51.2

Source: Adapted from World Bank (2007, Table 9.2)

Notes: Regional averages based on household and labor force surveys for 66 countries, which account for 55 % of the population in Sub-Saharan Africa, 97 % in South Asia, 66 % in East Asia and the Pacific (excluding China), 47 % in the Middle East and North Africa, 74 % in Europe and Central Asia, 85 % in Latin America and the Caribbean. Activity refers to the individual's reported principal activity. Data are for 2000 or nearest year (ages 15–64)

(ILO 2011, 21). Such workers are typically in informal work arrangements where they lack adequate social protection and social dialogue mechanisms and receive low pay (ILO 2010, 2011). Although at a global level there was a slight fall in vulnerable employment between 2000 and 2012 (with women's share declining from 55.3 to 49.3 % and men's from 50.5 to 47.1 %), these figures are nonetheless alarming as they show that nearly 48 % of the world's workforce—some 1.49 billion workers—were in vulnerable employment in 2012 (ILO 2014, Tables A12 and A13). The global averages mask variations between countries: for example, in some nine countries with latest year data for at least 2000, the share of women in vulnerable employment was above 75 % (ILO 2010). ILO regional data for 2012 (ILO 2014, Tables A10 and A12) and for Sub-Saharan and North Africa for 1998–2008 (ILO 2009a) suggest that agricultural and vulnerable employment change in parallel.

14.2.4 Gender Differences in Employment in Nontraditional Agro-industries

Most wage employment in agriculture—whether seasonal or permanent—is in traditional private farms or plantations. However, employment is increasing rapidly in nontraditional agricultural export (NTAE) industries, such as flowers, horticulture, livestock (chickens) and fish/seafood, where the work is often seasonal. The actual employment figures are not known at country or value-chain levels. Country study reviews by Dolan and Sorby (2003) for Asia, Latin America, and Sub-Saharan Africa, and Deere (2005, 2009) for Latin America, showed that women occupied at least half and often as much as 60–90 % of the jobs in these industries.

While acknowledging that sector employment estimates vary considerably, Dolan (2005) estimated that approximately 50,000 wage workers were engaged in the Kenyan fresh produce industry, of whom 70–80 % were women. In Chile, by the early 1990s, fruit production/processing absorbed about 25 % of agricultural labor, approximately 30 % of which was female (Jarvis and Vera-Toscano 2004). Maertens (2010) reported that female workers represented 90 % of the approximately 12,000 employees in French beans in the Niayes area of Senegal in 2005 and 60 % of the 3,000 employees in the tomato agro-industry in the Senegal River Delta in 2006. Female employment in these industries had increased rapidly since 1999 to involve about 30 % of local households. Approximately 90 % of these women workers had never worked outside the household farm before. In Sri Lanka, over 90 % of tuna plant workers were women (De Silva and Yamao 2006, in Okali and Holvoet 2007). While female employment in the Latin American fisheries sector was only 38 % for whitefish in Argentina (Patagonia), it was higher for seafood, ranging from 52 % in Uruguay, to 57 % in Brazil (Rio Grande do Sul), and 72 % in Argentina (Mar del Plata) (Josupeit 2004).

14.3 Manifestations of Gender Inequalities in Rural Labor Markets

Gender inequalities in rural employment are pervasive across all regions and cultures, sharing commonalities or reflecting differences, which we explore below.

14.3.1 Occupational Segregation

Despite narrowing gender gaps in educational attainments in many developing countries, women continue to be disadvantaged by horizontal occupational segregation with women clustered in fewer sectors and occupations than men in both agricultural and nonagricultural rural employment and by vertical segregation where women are mainly confined to lower skilled, manual work within occupational hierarchies (United Nations 2009; Fontana and Paciello 2010). Traditional patterns of segregation are perpetuated in NTAE industries where women are typically concentrated in certain activities (e.g., processing/filleting seafood, packing, labelling, and bar-coding produce), supported by gender stereotypes of "feminine" traits such as conscientiousness and dexterity or different skills than men who largely monopolize managerial and skilled technical posts or undertake the physically heavier work (Dolan and Sorby 2003; Josupeit 2004; FAO et al. 2010a). However, a study of Nicaraguan cotton, coffee, and tobacco plantations found that occupational segregation by gender broke down under conditions of (male) labor shortage (Deere 2005).

Segregation in low-technology occupations limits women's opportunities to develop new skills, hindering future professional growth and reinforcing the perception of these female jobs as low-pay and low-status occupations. Occupational segregation also represents a barrier for women to move to better jobs in other sectors, but does not prevent them from losing existing work if it is mechanized or becomes more remunerative and is taken over by men (Fontana and Paciello 2010).

14.3.2 Informal Contractual Arrangements

Rural wage employment in traditional agriculture is characterized by a high prevalence of seasonal, temporary (short-term), and casual (daily) jobs for both men and women, especially women who often work part-time (FAO 2011). Usually performed under informal contractual arrangements, such work offers neither protection against occupational hazards and risks, nor social benefits, and involves very little bargaining power. Workers are thus usually forced to accept low wages. Violence and sexual harassment are more common under these conditions (World Bank et al. 2009).

Plantations and NTAE industries are more likely to have a small cadre of permanent staff handling the managerial, supervisory, administrative, or skilled technical work. These permanent employees generally have employment contracts that provide job security and access to pensions, health and injury insurance, maternity, holiday, or sick leave, and other benefits. Men invariably predominate in these jobs, while women are commonly employed on short-term, informal agreements. For instance, Barrientos (2007) reported that 75 % of women workers in the South African fruit industry were temporary or casual, some 65 % of women employees were temporary in the Kenyan flower industry, and in the Zambian vegetable industry, 60–65 % of women were temporary workers (see, also, Deere 2005, 2009; Dolan and Sorby 2003; Elson 1999; Jarvis and Vera-Toscano 2004; Standing 1999).

Informal workers in plantations and NTAE industries rarely receive benefits or compensation for work-related injuries, even if such benefits are stipulated by national labor law, due to widespread nonenforcement of these laws, especially in rural areas. Ethical codes of practice under the "Corporate Social Responsibility" (CSR) movement typically exclude workers on informal contracts (Dolan 2005), although some companies provide nonwage benefits such as healthcare, transport or transport allowances, childcare facilities, and education for informal workers and support for their children's education (Dolan and Sorby 2003). Informal contractual arrangements bring high levels of insecurity and wages can vary daily or weekly, depending on seasonal demand (Dolan 2005; Barrientos 2007). Informal contracts usually exclude opportunities for training, which might lead to more skilled and better paid work (Dolan 2005).

14.3.3 Increasing Labor Flexibility in NTAE Industries

Although NTAE industries generally provide better employment conditions than traditional agriculture, many are increasingly shifting from permanent employment to informal, flexible work arrangements and systems of remuneration and/or externalizing production through contract smallholders and employment relations through unregulated contract labor (Dolan and Sorby 2003). This is, in part, a response to the cost-cutting competitiveness spurred by increasing globalization, resulting in a search for ways of lowering labor costs, with firms putting a premium on workers prepared or forced to accept low-wage jobs (Standing 1999). Such flexible labor also enables international and national agro-industries and supermarket chains to better withstand production risks (due to climatic variations, pests, and diseases) and commercial risks (stringent standards, changing demand and "just-in-time" production methods), allowing them to vary their employment levels rapidly while keeping labor costs down (Dolan and Sorby 2003; Dolan 2005; Barrientos 2007). However, the costs of this flexibility are passed on to the precarious workforce.

Because women need flexibility to balance their productive and reproductive roles, they are often pushed into informal, part-time work (ILO 2009b). However, as Standing (1999) commented, intermittent, casual, and partial work patterns are not *intrinsically* bad, *if* the surrounding conditions are appropriate. Thus, as Barrientos (2007) and Okali and Holvoet (2007) pointed out, despite the problems, many women prefer this type of work, since the wages bring them more independence and influence within their households, and they can still handle their domestic responsibilities. Similarly, Maertens (2010) attributed her findings that women temporary employees in the Senegal French bean and tomato industries worked, on average, 1 or 2 months a year less than men to women's preference for some flexibility in employment to accommodate their household responsibilities rather than discrimination.

Unfortunately, Standing (1999) also observed that men's employment tended toward more flexible and informal work, a trend also noted in NTAE industries (Appendini 2002; Dolan and Sorby 2003). While signalling a greater convergence between male and female labor patterns, this may also indicate a weaker position of labor in general.

14.3.4 Wage Gaps

Although gender wage gaps have narrowed in much of the developing world since the 1970s (Tzannatos 1999), throughout most regions and many occupations, women's earnings typically average about two-thirds of men's, and they are usually paid less than men for the same or comparable work (Standing 1999; Tzannatos 1999; ILO 2009a).

Noting the scarcity of comparable sex-disaggregated data on rural earnings, Fontana and Paciello (2010, Table I-9) reviewed a variety of case studies showing that women were generally paid less in both agricultural and nonagricultural work. The sharpest gaps were in Afghanistan and Pakistan, where women's agricultural wage rate was 50 % of men's, and even lower in Afghanistan for nonagricultural rural work. Whitehead (1996) found that female agricultural wage laborers were paid between one-third and one-half the male rate in North-Eastern Ghana. Several Latin American studies (Deere 2005) found that women were typically paid less than men for agricultural and nonagricultural wage work. However, studies indicate that gender gaps are smaller in some NTAE industries (Fontana and Paciello 2010; FAO 2011).

Wage gaps can depend on whether workers are paid hourly, daily, or piece rates (Fontana and Paciello 2010). Dolan and Sorby (2003) found women doing piece-work in South African fruit often earned more than men. Jarvis and Vera-Toscano (2004) found women working on piece rates in grape packing sheds in Chile had higher average daily earnings during the peak season than workers in comparable wage jobs that tended to be male. However, women earned considerably less than men in wage employment, with a gender wage differential of about 25 %. Piece rates benefit employers because these incentivize intense work over a long day and reduce supervision costs, but are more stressful for the workers. The women would have preferred regular, year-round jobs, even with monthly wages lower than their peak season monthly piece-rate earnings, as it would be easier to manage their money over the whole year and the work would be less strenuous.

The salient issue is whether these wage differentials are due to gender discrimination or can be explained by factors such as women's disadvantages in terms of education and skills, lack of organized representation and bargaining power, more limited labor market mobility because of their reproductive roles and/or cultural norms, or because they tend to predominate in part-time and casual jobs (United Nations 2009). Women's wages may also be lower because of job and training discrimination, occupational segregation, direct wage discrimination, or because women are willing to work for less, having lower "aspiration wages" (Standing 1999).

The evidence points to marked gender wage discrimination. Citing empirical studies that analyzed the relative wages of women and men in 19 industrialized and 42 developing countries, the World Bank (2001) noted that differences in observed worker and job characteristics explained only about 20 % of the gender wage gap. The rest of the gap resulted from factors that were difficult to measure directly, such as differences in workers' abilities or differences in labor market treatment (discrimination).

Hertz et al. (2010) undertook a comparative analysis of urban and rural wage data in FAO's Rural Income Generating Activities (RIGA) dataset, covering 14 developing and transitional economies. Average male wages were higher than average female wages everywhere except for rural Panama and the "non-explained" gender rural wage gap averaged about 25 %. The gap tended to fall in conditions of economic growth and was larger in countries with substantial occupational segregation. They also found a relationship between the unexplained gender wage gap and the overall level of unexplained wage inequality, suggesting that where labor market institutions work to reduce overall wage inequality, for example, via minimum wages or the effects of unionization, the unexplained gap between male and female wages is also reduced. However, a more detailed, context-specific understanding of institutional, economic and social factors affecting the gender wage gap is needed to address these differences. For example, if wage differentials arise from education differentials, the remedy would be better access to education for women and girls, whereas if the prime cause is segregation even of educated women into low-paying occupations, then changes in social norms and/or the enforcement of nondiscrimination in hiring would be required. Another option would be legislation on equal pay for men and women for comparable jobs, although enforcement is likely to be difficult in rural areas, and rural women are less likely to be aware of their rights or be organized to claim them.

Tzannatos (1999, 559–560) argued that discrimination against women in employment leads to productive inefficiency and thus constitutes a loss to the economy as well as having adverse effects on welfare. His simulations of the potential gains in output from eliminating gender occupational and wage differences using data from 11 Latin American countries suggested that eliminating wage discrimination would lead to a one-time gain equivalent to approximately 6 % of GDP. This would require a redistribution of GDP of approximately 30 %, as about one-fifth of the labor force would have to be reallocated to reduce segregation. Welcoming Tzannatos' recognition of the efficiency losses entailed by gender discrimination in labor markets and noting that economists often assume that market forces will eliminate gender discrimination, Elson (1999) commented that male employers are not just "economic men" interested in maximizing profits. Altering perceptions of women's productive potential might threaten male employers' advantages in the wider system of male social and political power.

14.4 What Are the Reasons for Gender Inequalities in Rural Labor Markets?

14.4.1 The Role of Institutions

The barriers to gender equality in rural labor markets are socially constructed and primarily stem from systemic *institutional* gender inequalities. Institutions comprise two components (Baas et al. 2008): (1) social norms, values, beliefs, customs, traditions, and practices that determine how people are expected to act/behave, and (2) organizations and their capacities to operate according to these "rules." Organizations, many of which have direct or indirect effects on rural labor markets, include formal institutions (e.g., ministries of agriculture or labor), formal membership organizations (e.g., workers' and employers' organizations), informal institutions (e.g., exchange or contract labor groups), and formal/informal institutions (e.g., markets, apprenticeship systems).

Institutions embody incentives that can encourage or discourage prejudice and discrimination (North 1990; World Bank 2001). Neither component of "institutions" is static and both evolve in response to new economic and technological opportunities, exposure to new ideas and ways of behavior through migration or modern media and communication technologies, and external shocks. However, institutions often adapt more slowly to transformation than would be economically efficient, because the deeply embedded social norms they perpetuate are advantageous to some socioeconomic groups and for men in general.

Gender inequalities operate at three interrelated levels—the household/community, labor market organizations, and the public sphere. At the household/community level, social norms and practices constrain women's access to livelihood capitals and support gender stereotypes about "suitable" work for men and women. Labor-market organizations often build on these stereotypes to develop and/or perpetuate rules and procedures that keep women out of membership and leadership roles, stifle their "voice," and consign them to inferior jobs. The failure of public institutions to provide and/or enforce policies, legislation, and other measures to ensure gender equitable rights in the world of work is largely due to male dominance in leadership positions and gender stereotypes that (falsely) recognize men as the main breadwinners, with women playing only secondary, supporting roles.

14.4.1.1 Gender Asymmetries in Productive and Reproductive Work

Elson (1999, 611–612) stressed that "labor markets are gendered institutions operating at the intersection of the productive and reproductive economies." They operate in ways that fail to acknowledge the contribution of the reproductive economy and disadvantage those who do most of this reproductive work. Poor rural women are particularly disadvantaged in this regard. Their almost total responsibility for reproductive work undermines their ability to engage in productive work, including wage labor. Their reproductive work is time-consuming and physically tiring. Owing to their lack of access to basic facilities such as electricity, running water and domestic appliances, they generally work longer hours than men if their productive and reproductive hours are summed (Fontana and Paciello 2010; FAO 2011). Second, their caring roles oblige them to seek flexible work near their homes, especially if there are no childcare facilities. Such work is more likely to be part-time or casual with lower pay and no social benefits. Finally, these time constraints, compounded by social restrictions on women's travel and interaction with men, often limit women's ability to organize, network, and participate in training opportunities, thus reducing their chances to develop their human and social capital.

14.4.1.2 The Social Norm, Human Capital and Poverty Nexus, and Gender Labor Market Inequalities

Social norms and practices often act as barriers to women entering labor markets. Das (2006) found female participation rates very low in North India and Pakistan, where female seclusion is common, and much higher in Southern India. These norms can break down in conditions of stress or poverty. In rural Vietnam, men predominated in wage labor and the few women undertaking such work were mostly from female maintained, landless households (Kabeer and Thi Van Anh 2002). In Uganda, where only a few men and women did agricultural wage work for neighboring households, the motive of the men, who were mainly young, was to establish themselves financially, while the women were usually divorcees or widows without land or adult male support. These women's wage work was thus distress work, and since this weakened their position, they often received lower wages (Kasente et al. 2002). In contrast, women seasonal workers in Mexican agro-industries had varying motives: widows supporting young families resorted to wage labor out of desperation, whereas others worked to accumulate savings or contribute to household finances, earning greater financial independence or voice within the family (Appendini 2002).[2]

Rural girls' and women's disadvantages in terms of educational enrollments and attainments in many developing countries[3] often serve as a significant entry barrier to the labor market, reinforced by occupational segregation and lower wages for women's work, which provide an additional disincentive to invest in girls' education. Gender disparities in education reduce the pool of talent from which employers can draw, lowering the average ability of the workforce, while confining women to less skilled, lower paid work (Klasen and Lamanna 2003).

Relatively high female educational attainments can be discounted in the labor market by cultural norms, exemplified by India and Pakistan (Das 2006). After controlling for husband's income, Indian women's education takes a U form, with high labor participation by uneducated women, falling to the lowest participation for women who have completed primary education, and rising again with post-primary education. Since the supply of well-paying, secure jobs for educated women is particularly low in rural areas, educated rural women generally prefer to opt out of the labor market rather than accept low status (manual) jobs. On the demand side, cultural values of status, seclusion, and family honor prevent higher status women from working outside the home. Nonetheless, more educated women may be forced to work during times of crisis, suggesting some flexibility in cultural values.

---

[2] Evidence across countries is mixed as to whether earning a wage empowers women, since social norms often require them to hand over their earnings to their husbands or other (male) relatives (Elson 1999, 614–616).

[3] This generalization is not always valid (see Hausmann et al. 2006).

14.4.1.3 Gender Asymmetries in Social Capital, Bargaining Power, and Rural Labor Markets

Institutions also shape power relations within the family, society, and the economy (World Bank 2001). However, these are not static as Kabeer (2010, 106–108) emphasized in her gendered theory of change in power relations, which is based on the "intertwined notions of *structure*, the institutionalized constraints [such as norms, rules, and practices]. . . on human action that give rise to durable forms of inequality, and *agency*, the role of human actors and their efforts to reproduce, modify, or transform structural inequality." While rural women are likely to share certain inequalities by virtue of their gender, women are not homogeneous and different groups of women also share interests (and advantages and disadvantages) with men from the same class, caste, or racial group, although any disadvantages/inequalities are likely to be intensified for women. Women's agency to effect changes depends on their capacity for "voice" and their capacity to "exit" (i.e., withdraw or withhold cooperation). These capacities depend on the resources they can mobilize, which can be individual (land, wages, equipment, human capital) or collective, such as their social capital gained through membership in social groups and networks, and their *strategic potential* to bring change. The attitudes of men and their willingness to support women are likely to be crucial to success. Social groups and networks overlay and reinforce business or employment-related networks. As Meinzen-Dick et al. (Chap. 10) note, rural men and women tend to belong to different types of groups, with men prioritizing production- and market-oriented groups and women preferring, self-help, civic, and religious groups. Das (2006) also noted that women's social capital and networks in South Asia were grounded in their communities and not in the market. Women's limited social capital in productive, work-oriented groups represents a significant entry barrier to paid work and reduces their bargaining power to negotiate fair wages, decent labor conditions, and safety nets for periods of unemployment. On the other hand, women's increasing labor-market participation can (although not necessarily) lead to expanding female membership in work-related networks (Elson 1999). It can also have a positive impact on women's freedom and agency (Sen 1999), although women can also suffer negative impacts (Koggel 2005).

The lower strategic potential of women's social capital is particularly visible in formal enterprises such as plantations, where women's representation in trade unions is weak. The reasons reflect cultural norms that ascribe leadership roles to men and women's predominance among temporary workers who are excluded from union membership. Interestingly, there has been an erosion of trade unionism in the newer NTAE industries, often due to the companies' deliberate sabotage of unions and/or the predominance of casual, temporary, and seasonal labor, which is a result of both the nature of the products and management decisions (World Bank et al. 2009). Women are more likely to be affected, due to their predominance in these casual, temporary jobs. However, the lack of unionization is not necessarily detrimental to labor rights in NTAE companies or traditional plantations, as the emerging fair trade and CSR movements as well as women's own organizations, such as the Self-Employed Women's Association (SEWA) in India, are increasingly influencing labor conditions (World Bank et al. 2009).

14.4.2 Policy Failures

Public action to reverse gender disparities in rural labor markets is crucial, since social institutions that perpetuate these disparities are difficult for individuals to change. Yet policy failures stemming from ignorance of the nature and costs of gender labor-market inequalities to the well-being of individual workers and their families and to the economy are common. Some policymakers are reluctant to address issues associated with social norms and religious and cultural traditions, believing these should be addressed by advocacy rather than policy (World Bank 2001).

Experiences with targeted reforms (such as legislation on equal wages for equal work or for work of comparable value, maternity benefits, pensions, hazardous work) have had mixed gender outcomes. A major handicap is the limited implementation capacity in many developing countries, particularly in rural areas and the informal economy, further complicated by the existence of multiple and often inconsistent legal systems—statutory, customary, and religious (World Bank 2001; FAO 2006). State action needs broad social support to make lasting and deep changes—and its effectiveness is greater when civil society groups, especially women's organizations, participate actively in open dialogue. Such a process can also stimulate markets to operate more openly, facilitating information exchange and encouraging choices on the basis of efficiency rather than gender, ethnicity, or age (World Bank 2001).

14.5 How Can These Barriers Be Removed?

Despite the existence of social barriers to women's engagement in economic activities, rapid changes are occurring in gendered work patterns in response to alterations in incentives, economic needs, and opportunities. Complementary *public action* is needed, not only to accelerate rural women's insertion into these evolving labor markets, but also to ensure that this process occurs with the closing, rather than the widening, of current gender disparities. Viewing public action as "not merely what is done *for* the public by the State, but also what is done *by* the public for itself" (Drèze and Sen 1991, 28–29), means recognizing and supporting civil society, workers' and employers' organizations not only in identifying gender discrimination in rural labor markets and pressuring government to take action, but also in taking independent action.

The major challenge is to integrate the actions by different stakeholders, so they perform complementary, positive roles. The state's role needs to be balanced to avoid furthering the interests of powerful stakeholders (especially the business sector) at the expense of the weakest (i.e., rural male and female workers). The state will need to ensure an enabling environment to foster private-sector commitment to decent, gender-equitable rural employment through, for example, fiscal and other incentives or quotas, while providing a labor policy and regulatory system with appropriate enforcement mechanisms in both formal/informal and rural/urban labor markets.

In this age of globalization, the United Nations, and especially the ILO, have important roles to play in shaping employment policies that ensure gender equity, by providing relevant conventions and recommendations, collecting, analyzing, and disseminating gender-disaggregated labor market data, and providing policy and technical support. The ILO's Decent Work Agenda, with its focus on promoting opportunities for productive work, workers' rights, social protection, and social dialogue, provides a valuable framework that we use below to review "good practices" in "public action" in tackling these gender barriers.

14.5.1 Increase Rural Women's Employment Opportunities

Because the share of agricultural employment is declining globally relative to other sectors, opportunities for expanding employment in traditional agriculture are contracting. More promising strategies emphasize complementary labor-market policies to improve rural workers' education, skills, and opportunities for *decent* work in agriculture and rural nonfarm or agro-industrial enterprises or to facilitate their out-migration (World Bank et al. 2009). The gender impacts are not necessarily equitable.

14.5.1.1 Freeing Up Women's Time

Providing better services, labor-saving technologies and infrastructure, and/or encouraging a more gender equitable division of domestic and caring work may help free up rural women's time to take advantage of new employment opportunities (United Nations 2009). Progress in providing childcare facilities and labor-saving infrastructure such as clean running water and electricity is spotty. India's Plantation Labour Act requires employers to provide crèches when there are over 50 women workers (ILO 2003) and its National Rural Employment Guarantee Act (NREGA) programs are required to provide childcare facilities, but do not always do so (Khera and Nayak 2009). Some NTAE companies provide childcare facilities, for example, the Cargill Sun Valley poultry company in Thailand (Lawler and Atmiyanandana 2000) and some horticulture companies in Africa, especially in South Africa (Smith et al. 2004). Despite these encouraging examples, similar efforts are needed in most developing countries.

Improvements in redistributing reproductive work depend on such factors as changing social norms and institutional arrangements for parental leave. Positive signs can be found in Chile, where some husbands took on household tasks when their wives were working long days in the grape packing sheds during the peak season (Jarvis and Vera-Toscano 2004). Paternity leave is included in collective bargaining agreements (CBAs) negotiated by the National Union of Plantation and Agricultural Workers of Uganda (NUPAWU) and the Tanzanian Plantation and Agricultural Workers' Union (TPAWU) for the Tanzanian sugar industry

(ILO 2003). Other examples are civil society-driven, such as the Programa H in Brazil and Mexico to encourage young men's engagement in fatherhood, caregiving, and HIV/AIDS. The Sonke Gender Justice Network in South Africa strengthens men's ability and commitment to caring for rural children affected by HIV/AIDS (United Nations 2009).

14.5.1.2 Improving Female Education and Training and Breaking Down Occupational Stereotypes

Improving female education and reducing gender educational gaps in rural areas can be a significant pathway to increasing women's access to decent employment, particularly in nonagricultural work (Fontana and Paciello 2010). However, rural girls and boys are often treated differently and channelled into different subject areas, reinforcing gender labor-market segregation. Such biases need to be challenged through innovative teaching methods, teacher training and other dedicated initiatives.

Opportunities for training and promotion in NTAE industries are more common among technicians, machinery operators, managers, and administrative staff, who tend to be men (Dolan and Sorby 2003). Smith et al. (2004) found that although some horticulture companies with Codes of Practice provided women training, promotions to supervisor positions were rare. South African (male) employers thought that even with training, women would have difficulty managing male workers because of cultural biases favoring men as leaders. In Costa Rica, managers and women workers lacked confidence in women's ability to handle positions of authority. However, there are some positive practices, such as the Cargill poultry company in Thailand, which offered an educational program to assist (mainly female) employees to advance within the firm and a number of women have become supervisors (Lawler and Atmiyanandana 2000). Both male and female workers are given skills training in the São Francisco valley grape farms in North East Brazil, and many women are "monitors," although the higher level management positions are mainly male (Selwyn 2010). These examples suggest that while it may be easier for women to supervise female workers, targeted programs are needed to give women training and guidance to handle more responsibility and to sensitize male managers and workers to support women in these roles.

14.5.2 Advance Gender-Equitable Workers' Rights

14.5.2.1 Gender-Equitable Labor Legislation

Many features of existing labor laws, regulations, and institutions are obsolete in today's global economy, where work is increasingly informal and flexible (Chen 2009). There is a need for *appropriate regulations* that reflect current reality and address the biases that favor large over small enterprises, formal over informal workers, and men over women. In tackling these challenges, the state needs to play a pivotal role, starting with the ratification of the ILO Core Labor Conventions, updating or enacting appropriate national legislation to cover formal and informal workers, and setting up regulation and enforcement mechanisms.

Temporary and casual workers, who are predominantly female, are commonly excluded from existing labor laws, although some countries are taking measures to bring informal workers under labor legislation. Examples include Chile and South Africa (World Bank et al. 2009) and Ghana (Chen 2009), where the 2003 New Labor Act entitles temporary and casual workers to benefit from collective agreements on equal pay for work of equal value, access the medical provisions available to permanent workers, and receive the minimum wage for all days worked and for public holidays. Temporary workers employed by the same employer for a continuous period of 6 months or more must be treated as permanent workers. Labor Relations Regulations (HIV/AIDS) were adopted in Zimbabwe in 1998 (Chartier 2005), and laws protecting women plantation workers were introduced in Brazil in the mid-1990s (World Bank et al. 2009). Such legal protection for informal workers is crucial since, paradoxically, stringent labor laws benefiting formally employed workers, as in South Africa, can result in employers offsetting the higher labor costs through increased use of informal labor, especially contract workers, who are not covered by employment benefits (Barrientos and Kritzinger 2004).

By 2011, 117 countries had equal pay laws (UN Women 2011), but the pervasive problem is lack of enforcement of labor laws and standards, particularly in rural areas where much of the work is carried out in private farms or commercial enterprises or, especially in the case of women's paid postharvest processing or domestic work, within private homes (Fontana and Paciello 2010). Rural women face even greater barriers than men in accessing justice systems due to their lower autonomy, educational levels and awareness of their rights, and fewer networks beyond the family and local community (United Nations 2009). Although agribusinesses can circumvent the law using third party labor contractors, labor laws and standards are more likely to be enforced in larger plantations and NTAE companies where specific mechanisms encourage compliance: government mechanisms, collective bargaining agreements, and corporate social responsibility, which are discussed below.

14.5.2.2 Government Mechanisms

Appendini (2002) found that Mexican exporting companies running packinghouses had to be registered and comply with government fiscal and labor regulations. In the larger avocado packinghouses, male and female workers received the minimum wage and were enrolled in the social security program, although payments for overtime and Sunday work were sometimes less than legally stipulated. However, the indemnity for working over 3 months was not paid, workers had no written contracts, and were dismissed verbally from day-to-day once the peak season was over.

Government work is more likely to respect labor regulations. Government-run cashew factories in India had better labor standards than private factories (Harilal et al. 2006). By absorbing surplus labor, public employment programs that respect labor standards and minimum wage rates can raise local wages and improve working conditions among other employers.

14.5.2.3 Collective Bargaining Agreements

Collective bargaining agreements (CBAs) between employers' and workers' organizations on terms and conditions of work are often negotiated to implement provisions in national laws that have not been enforced. They can also fill an important gap when labor legislation excludes informal work. CBAs can be gender-blind, although they have the potential to promote gender equality, eliminate discriminatory practices, and address women's specific needs (e.g., maternity leave). Barrientos and Smith (2006) found CBAs improved working conditions for male and female workers in South African fruit packinghouses and that the Ethical Trading Initiative (ETI) member company involved with Costa Rica banana producers had signed an International Framework Agreement with the IUF[4] and the regional trade union umbrella organizations, which led to some improvements. Such agreements are not necessarily won easily. It took a 22-day strike in June–July 2011 by Peru's La Ibérica chocolate factory's workforce of 100 (95 of whom were women), including a weeklong hunger strike, to arrive at a better, more equitable collective agreement (IUF 2011).

14.5.2.4 Corporate Social Responsibility and Codes of Conduct

The Corporate Social Responsibility (CSR) movement to which well-known multinational corporations belong is gaining considerable prominence, largely promoted by pressure from (Northern) consumers, NGOs, and governments in countries of origin or registration. The movement is being institutionalized through the adoption of various voluntary standards and codes of conduct that are often specific to particular companies, especially for NTAE products. CSR and its related standards and codes bring both advantages and problems (IFAD 2010). Among the "blind spots" in many CSR standards are a frequent neglect of employment creation and labor rights, especially their gender aspects. Dolan (2005) also pointed to the inadequate monitoring of code implementation and the failure to pick up sensitive issues such as gender discrimination and sexual harassment.

Since the codes are voluntary and not well monitored, adherence is likely to depend on incentives, such as increased productivity as in the Cargill poultry factory in Thailand (Lawler and Atmiyanandana 2000) or pressure from importing supermarket and agribusiness chains, which are influenced by civil society and

[4] International Union of Food, Agricultural, Hotel, Restaurant, Catering, Tobacco and Allied Workers' Associations (IUF).

consumer organizations. Few data exist on the gender impacts of CSR agreements and codes of conduct, although there is evidence of positive benefits for women workers. Maertens (2010) attributed the gender wage neutrality of the Senegal French bean and tomato agro-industries to the high standards in these export supply chains. The tomato industry, controlled by a single multinational certified by the ETI, paid wages that were 20–40 % higher than in the bean industry, where no explicit ethical standard was applied. Blowfield and Gallet (n.d.) found that the Ghanaian Volta River Estates Ltd. (VREL), which produces bananas, met the social and labor standards of the Fair Trade Organization and Max Havelaar. Men and women were treated equally; wages were above Ghana's minimum wage and permanent employees received paid annual and sick leave, and maternity leave. Although only 16 % of the 900 full-time workers were women, VREL was attempting to increase the female workforce.

In their study of the impacts of the ETI Code of Labor Practice in several countries, which included South Africa (fruit) and Costa Rica (bananas), Barrientos and Smith (2006) found that in general, permanent and regular workers benefited most, while migrant and contract workers experienced little change or poorer conditions. Although the codes led to some improvements for women (maternity benefits), they rarely addressed basic inequalities such as access to employment, promotion, and training. In Costa Rica, the codes brought significant positive changes in health and safety, compliance with minimum wages and legal entitlements, and working hours limited to 60 h per week (which reduced take-home pay but women appreciated the extra time with their families). Attitudes to women improved and efforts were made to promote them (for example, office work), but these opportunities were limited. South African labor legislation had higher standards than the ETI Base Code on most issues; sensitivity to gender issues was improving and one farm had developed an equal opportunities policy and another a policy on sexual harassment.

Some of the most effective codes were introduced by trade associations. For example, the Code of Practice developed by the Uganda Flower Exporters Association (UFEA) improved employment conditions (including wage rates, hours worked per day, and occupational health and safety) in an industry where 80 % of the workforce is female. Some companies also provided nonwage benefits such as accommodation (mainly for night-shift workers), free tea and lunch, medical attention, day shifts that finished at 5 p.m., leave, prompt payment of salaries, and salary advances in case of need (Asea and Kaija 2000). South Africa's Wine Industry Ethical Trade Association (WIETA)[5] adopted and monitors its own code of labor practice based on ILO conventions and national legislation. Its civil society members have played an important role in ensuring that the conditions of casual women workers are addressed in social audits (Barrientos 2007).

In other cases, NGOs have played a catalytic role. In the Colombia flower industry, NGO pressure was a key factor in eliminating gender wage disparities with all workers earning the minimum wage, improving working conditions and enhancing women's recruitment as supervisors (Deere 2009). Mozambique

[5] Renamed Agricultural Ethical Trade Initiative of South Africa in 2008 to facilitate work in the broad agricultural sector (WIETA website).

exemplifies a partnership approach between the government, NGOs, communities, and the private sector. A private entrepreneur received a loan guaranteed by the government cashew institute (INCAJU) to set up a cashew factory, with technical help from a USAID-financed NGO, TechnoServe, and assistance in marketing from the Dutch NGO, SNV. Labor rights were respected, and workers received a free meal at work and had written contracts providing health benefits, paid annual holidays, and severance pay in case of work-related illness or accidents. A trade union was set up and a crèche constructed, although the mothers had to arrange their own child carers (Kanji 2004).

However, many large companies have no commitment to CSR. This may reflect the type of product—where there is less consumer pressure or where market competition is so intense that globally retailers and importers/exporters are squeezing production costs, unconcerned about workers' labor rights or occupational safety and health. For example, the concentration of the India-U.K. cashew value chain among a small number of U.K. retail giants and the intense competition among them has driven down the terms on which Indian suppliers engage, particularly informalizing and casualizing the predominantly female labor force (Harilal et al. 2006, 4).

Despite many promising practices, these negative examples show that much more needs to be done: disseminating information on labor standards, employment rights and the content of codes of conduct; addressing gender better in codes of conduct including policies against sexual harassment; covering temporary workers in labor legislation and codes of conduct and involving labor contractors in formalizing these codes; and improving the monitoring and enforcement of labor legislation and ethical codes, involving CSOs/NGOs as well as trade unions and ethical organizations such as WIETA (World Bank et al. 2009).

14.5.3 Introduce Social Protection

Few rural workers have work-related social benefits such as health and disability insurance, pensions, or unemployment benefits. Some countries are trying to fill this gap by implementing social protection schemes for vulnerable workers, informal workers, the unemployed, and the poor in general, or by running employment guarantee schemes such as public works or social work programs.

14.5.3.1 Social Protection

Some of the existing schemes—for example, extensive social provisions under the 1951 Plantation Labor Act in India—are poorly enforced (ILO 2003). South Africa's system of state-run healthcare, pensions, and work-related injury benefits seems to be reasonably well implemented. In 1993 it extended basic conditions of employment (e.g., a maximum working week, maternity and sickness leave and benefits, and holidays) to agricultural workers and in 1998 further legislation extended some of these benefits, such as sickness benefits, to temporary and seasonal workers (ILO 2003). The Indian Government recently adopted the Unorganized Workers Social Security Act to provide informal workers with health and maternity benefits, life and disability coverage, and old-age protection. The Act also provided for setting up a National Social Security Board that would include representatives of informal economy men and women workers, but it is too early to assess its effectiveness (ILO 2009c).

State healthcare systems or health insurance schemes are being developed by governments, NGOs, trade unions, cooperatives, and agro-industries. Costa Rica and Tunisia provide almost universal health coverage (ILO 2003). China is expanding rural healthcare insurance through its New Rural Cooperative Medical System programs, with participation increasing since 2002 from 10 % to 80 % of the rural population (Liu et al. 2009). The Self-Employed Women's Association (SEWA), a trade union of women informal workers (including agricultural workers) in India, has a primary healthcare program for its members with life, asset, and health insurance (Chatterjee and Ranson 2003, in United Nations 2009). In Argentina, the *Unión Argentina Trabajadores Rurales y Estibadores* extended its health insurance and unemployment fund to large numbers of unregistered and unprotected agricultural workers (ILO 2003). Codes of conduct implemented in agro-industries frequently have provisions for health benefits and for ensuring a healthy work environment including occupational health and safety measures and HIV/AIDS awareness-raising (Barrientos and Smith 2006).

Other countries such as Botswana, Namibia, South Africa, Cuba, and Nicaragua have expanded their noncontributory pension schemes. Chile guarantees a solidarity pension for all citizens over 65 years of age who receive a pension of less than 150 % of the minimum wage, while in Brazil, the *Previdencia Social Rural* provides a monthly benefit equivalent to the minimum wage to male workers over 60 and female workers over 55 in subsistence agriculture, fishing or mining, and to their dependants on the death or disability of the main beneficiary (United Nations 2009).

Other schemes give cash benefits to mothers conditional on school attendance and clinic visits, regardless of the mother's employment status. Well-known examples are Mexico's *Progresa* (later renamed *Oportunidades*), Brazil's *Bolsa Familia* (United Nations 2009), and Argentina's *Plan Familias por la Inclusión Social* (Tabbush 2009).

While these examples are encouraging, similar schemes are urgently needed in other countries to protect poor and vulnerable rural men and women and reduce the burden that poor rural workers bear in caring for the elderly and sick.

14.5.3.2 Public Works Programs

A number of governments support public works programs to provide a "safety net" in times of crisis or address widespread unemployment and poverty. Such programs offer the opportunity to apply decent work standards and ensure gender equality. While the record in promoting gender equality is mixed, efforts are being made to recruit women (often through quotas), counteract gender stereotypes by encouraging women to take on "male" jobs, ensure gender equity in skills training, and equal wages for work of equal value. The Peruvian Rural Roads Maintenance Program (2003–2006) increased female participation from 3.5 % to 24 % by setting a female quota of 10 % combined with gender training for program stakeholders. Women undertook all maintenance activities, performing better than men in many of them, and also took on some managerial and supervisory roles (FAO et al. 2010b).

India's National Rural Employment Guarantee Act (NREGA) recognizes a "right to work" and provides targeted interventions aimed at asset creation and providing corrective measures for widespread rural unemployment and malfunctioning labor markets, including discrimination against women and scheduled tribes and castes. Guaranteeing 100 days of work at the statutory minimum wage to all rural households whose members are willing to perform unskilled manual labor, NREGA is self-targeting. At least one-third of the jobs should be reserved for women, and childcare facilities provided at the worksite if there are at least five children under the age of 6 at the site. Rural women appear eager to undertake NREGA work since, as government work, it is socially acceptable, locally available, with regular working hours limited to 7–8 h a day, and is paid at the statutory minimum wage. Women and men are paid the same wages, which has contributed to raising women's status and wage rate in the local market. The benefits have not been so positive in states where the female quota was not enforced and childcare facilities were not always available. Because of incidents of corruption, bank payments have sometimes been substituted for cash payments, excluding some (especially women) who do not have bank accounts (Khera and Nayak 2009).

In highly gender-segregated societies, separate programs are sometimes run for men and women, as in Bangladesh, where only women can work in the cash-for-work Rural Maintenance Program. Although both men and women participate in the Food for Asset Creation component of the Integrated Food Security Program, at least 70 % of the participants must be women. Women's intrahousehold decision-making power increased more, the higher the proportion of cash relative to food in the payment package. Although these programs have raised women's status within the household by challenging traditional norms of female seclusion, they sometimes worsen women's status within the community (Ahmed et al. 2009).

Public works programs commonly focus on physical infrastructure, which could be a disincentive to women's participation since such work is often seen as "male." Expanding cash-for-work programs to include social infrastructure and care services, for example, HIV/AIDS or caring for children, the sick, disabled or elderly, could enhance women's participation (United Nations 2009).

14.5.4 Promote Gender-Equitable Representation and Voice in Rural Workers' and Employers' Organizations

Increasing women's membership, managerial, and leadership roles in labor organizations can help women become more aware of, and able to fight for, their rights, while increasing male counterparts' awareness of the benefits of gender equality.

Progress is slow despite some noteworthy successes. For example, the National Union of Plantation and Agricultural Workers of Uganda (NUPAWU), which has some 15,000 women members representing about 32 % of the membership, established a Women Workers' Department in 1996 and amended its constitution in 2001 to reserve seats for women in the National Executive Council and each branch executive committee (ILO 2003). Its policy also requires that women comprise at least 30 % of any trade union training program.

From the early 1990s, the principal trade union operating in the São Francisco valley grape farms in North-East Brazil gave serious attention to improving women workers' conditions. Several women were elected to the leadership, including a women's officer, and by the early 2000s, over 30 % of trade union delegates were women. The CBAs negotiated with employers included the provision of crèche facilities; a paid day per month for women workers to visit doctors, and the right for women with babies in the crèche to breastfeed an hour per day, over and above the lunch hour; a 2-month period of paid maternity leave, and return to employment rights following such leave. However, since these gains contain significant non-wage costs to the employers, the latter are trying to reduce these costs by transferring more of the "female" work to men and reducing the proportion of women in permanent positions.[6] While this illustrates the complexity of the gains and losses for different categories of workers in global chains, the process is ongoing and active unions will doubtless campaign against this discrimination (Selwyn 2010).

When women face social and other obstacles to participating in mixed (male-dominated) groups, they may need to mobilize separately. SEWA, for example, has negotiated effectively with government and employers/contractors to obtain wage increases, annual bonuses, health benefits and/or pension contributions for informal workers including agricultural day-laborers (Chen 2009). In July 2011, the National Conference of Women Dairy Workers in India adopted an action plan to tackle discrimination between men and women workers and between permanent and contract workers; fight for equal opportunity and promotion rights for women dairy workers; get women workers' issues into the "top 3" priorities of their unions; and promote more active involvement of women in the unions, with a goal of 30 % of officers in union leadership at national level being women (IUF 2011).

Agricultural unions have often been undermined by global value chains either due to blatant management pressure or because intense competition driving down production costs gave little scope for meeting worker demands for better conditions. Harilal et al. (2006) found that although almost all the Indian women cashew workers they surveyed were trade union members, the unions had been weakened by the growing buyer power of the multinational importers and supermarkets and,

[6] In their study of Codes of Practice in the South African wine industry, Nelson et al. (2007) found that the proportion of women in permanent and casual employment in both adopting and non-adopting farms fell considerably between 2002 and 2004. Although they did not discuss the reasons, elsewhere in the paper they noted that the codes increased production costs which was a factor in the growing use of casual and contract labor that was not covered by the codes. Employers most likely reduced female permanent labor to avoid paying the maternity benefits and child care facilities required by the codes.

as in other globalized value chains, were relatively powerless to fight for higher wages and better working conditions. In contrast, Selwyn (2007) found that in the North-East Brazil grape export industry, workers were able to take advantage of their "associative power" (membership in relatively well-organized trade unions) and their "structural power" due to their position in the value chain and their power to disrupt it. The latter power included both "marketplace bargaining power" derived from the industry's need for skilled workers to meet the high quality standards of the northern importers and relatively tight labor markets of such workers, and "workplace bargaining power" in tightly integrated production processes where small disruptions can have disproportionate impacts on quality and meeting vital deadlines. Thus the unions were able to negotiate CBAs that brought gains for both male and female workers.

Initiatives to increase women workers' awareness of their legal rights are vital so that they can organize for effective action. In Kenya, Tanzania, Uganda, and Zambia, Women Working Worldwide, a U.K.-based network organization, together with local trade unions, promoted rights awareness among 6,000 permanent and casual female workers, which increased the women's confidence and ability to negotiate with employers. In Tanzania, training farm managers on women workers' rights resulted in general improvement in worker-management relations and greater space for gender concerns in CBAs (Women Working Worldwide 2007, 2008, in Fontana and Paciello 2010).

14.6 Conclusion

Although much of the attention on women in agriculture is devoted to agricultural producers, addressing gender disparities in agricultural wage employment is vital to reduce rural poverty. This is especially important for the millions of rural poor who lack assets other than their own labor. The gender gap in wages, working conditions, and occupational segregation are key challenges to overcome if agricultural labor is to play its role in improving the welfare of the rural poor. Decreasing the time burden of women's unpaid labor through improved rural energy, water supply, and health systems and redistributing this work within households is also needed to free women's time for more remunerative jobs.

The underlying causes of gender inequality in rural labor markets are institutional, including both social norms and the structure of labor market organizations. Institutional change cannot take place overnight. Changes in social norms are particularly complex and depend on many factors outside the scope of this chapter. However, much can be done to improve gender equity in rural labor institutions through government policies, corporate social responsibility programs, and building the strength of women in labor organizations. This involves overcoming vested interests in the *status quo* that provides a supply of cheap labor. It therefore requires both political will and resources.

Reducing rural poverty requires not just removing the barriers to women's participation in decent employment, but aiming for what Elson (1999, 622) calls "a transformatory employment policy; that is, a policy which helps to change peoples' perceptions of what is possible, beneficial, and fair; fosters cooperative action; and strengthens women's bargaining power in the workplace, the home, and the marketplace."

References

Ahmed A, Quisumbing A, Nasreen M, Hoddinott J, Bryan E (2009) Comparing food and cash transfers to the ultra poor in Bangladesh. Monograph 163. International Food Policy Research Institute, Washington, DC

Appendini K (2002) 'From where have all the flowers come?' Women workers in Mexico's non-traditional markets. In: Razavi S (ed) Shifting burdens: gender and agrarian change under neo-liberalism. Kumarian Press, Bloomfield, pp 93–108

Asea P, Kaija D (2000) Impact of the flower industry in Uganda. Working paper 148. International Labour Organization, Geneva

Baas S, Ramasamy S, Dey de Pryck J, Battista F (2008) Disaster risk management systems analysis: a guide book. Environment and natural resources management series no. 13. Food and Agriculture Organization of the United Nations, Rome

Barrientos S (2007) Female employment in agriculture: global challenges and global responses. Commonwealth Secretariat, London

Barrientos S, Kritzinger A (2004) Squaring the circle: global production and the informalization of work in South African fruit exports. J Int Dev 16(1):81–92

Barrientos S, Smith S (2006) The ETI code of labor practice: do workers really benefit? Report on the ETI impact assessment 2006. Part 1: main findings. Institute of Development Studies, Brighton

Blowfield M, Gallet S (n.d.) (but approximately 1998) Ethical trade and sustainable rural livelihoods: Volta River Estates fairtrade bananas case study. Draft. National Resources Institute, London

Chartier MC (2005) Legal initiatives to address HIV/AIDS in the world of work. International Labour Organization, Geneva

Chatterjee M, Ranson M (2003) Livelihood security through community-based insurance in India. In: Chen L, Leaning J, Narasimhan V (eds) Global health challenges for human security. Harvard University Global Enquiry Initiative, Asia Center, Cambridge, MA

Chen MA (2009) Informalization of labor markets: is formalization the answer? In: Razavi S (ed) The gendered impacts of liberalization: towards "embedded liberalism"? Routledge, New York/London, pp 191–218

Das MB (2006) Do traditional axes of exclusion affect labor market outcomes in India? Part I: Female labor force participation and the effects of gender. Part II: Caste, ethnicity and the Indian labor market. Social development papers, South Asia series, paper no. 97. World Bank, Washington, DC

De Silva DM, Yamao M (2006) The involvement of female labor in seafood processing in Sri Lanka: the impact of organizational fairness and supervisor evaluation on employee commitment. In: Choo PS, Hall SJ, Williams MJ (eds) Global symposium on gender and fisheries: seventh Asian fisheries forum, 2004. World Fish Center, Penang, pp 103–114

Deere CD (2005) The feminization of agriculture? Economic restructuring in rural Latin America. Occasional paper 1. UN Research Institute for Social Development, Geneva

Deere CD (2009) The feminization of agriculture: the impact of economic restructuring in rural Latin America? In: Razavi S (ed) The gendered impacts of liberalization: towards "embedded liberalism"? Routledge, New York/London, pp 99–130

Dolan CS (2005) Stewards of African labor: the evolution of trusteeship in Kenya's horticulture industry. Working paper in African studies no. 252. African Studies Center, Boston University, Boston

Dolan C, Sorby K (2003) Gender and employment in high-value agriculture industries. Agriculture and rural development working paper 7. World Bank, Washington, DC

Drèze J, Sen A (1991) Public action for social security: foundations and strategy. In: Ahmad E, Drèze J, Hills J, Sen A (eds) Social security in developing countries. Clarendon, Oxford

Elson D (1999) Labor markets as gendered institutions: equality, efficiency, and empowerment issues. World Dev 27(3):611–627

FAO (Food and Agriculture Organization of the United Nations) (2006) Gender and law: women's rights in agriculture. FAO legislative study 76, rev.1. FAO Legal Office, Rome

FAO (Food and Agriculture Organization of the United Nations) (2011) The state of food and agriculture 2011—women in agriculture: closing the gender gap for development. FAO, Rome

FAO/IFAD/ILO (Food and Agriculture Organization of the United Nations/International Fund for Agricultural Development/International Labour Organization) (2010a) Agricultural value chain development: threat or opportunity for women's employment? Gender and rural employment policy brief no. 4. Food and Agriculture Organization of the United Nations, Rome

FAO/IFAD/ILO (Food and Agriculture Organization of the United Nations/International Fund for Agricultural Development/International Labour Organization) (2010b) Women in infrastructure works: boosting gender equality and rural development! Gender and rural employment policy brief no. 5. Food and Agriculture Organization of the United Nations, Rome

Fontana M, Paciello C (2010) Gender dimensions of rural and agricultural and public works programmes. Experience from South Africa. Differentiated pathways out of poverty: a global perspective. In: Gender dimensions of rural and agricultural employment: differentiated pathways out of poverty, status, trends, and gaps. Food and Agricultural Organization of the United Nations, the International Fund for Agricultural Development, and the International Labour Office, Rome, pp 171–184

Harilal KN, Kanji N, Jeyaranjan J, Eapen M, Swaminathan P (2006) Power in global value chains: implications for employment and livelihoods in the cashew nut industry in India. International Institute for Environment and Development, London

Hausmann R, Tyson LD, Zahidi S (2006) The global gender gap report 2006. World Economic Forum, Geneva

Hertz T, de la O-Campos AP, Zezza A, Winters P, Quiñones EJ, Azzari C, Davis B (2010) Gender wage gaps in rural versus urban areas. In: FAO/IFAD/ILO (ed) Gender dimensions of agricultural and rural employment—differentiated pathways out of poverty: status, trends, and gaps. Food and Agriculture Organization of the United Nations, Rome, pp 140–154

IFAD (International Fund for Agricultural Development) (2010) New realities, new challenges: new opportunities for tomorrow's generation. Rural poverty report 2011. IFAD, Rome

ILO (International Labour Organization) (1999) Decent work. ILO, Geneva

ILO (International Labour Organization) (2003) Decent work in agriculture. Background paper for international workers' symposium on decent work in agriculture, Geneva, 15–18 September

ILO (International Labour Organization) (2009a) Global employment trends for women. ILO, Geneva

ILO (International Labour Organization) (2009b) Global employment trends 2009—update. ILO, Geneva

ILO (International Labour Organization) (2009c) Gender equality at the heart of decent work. Report of the Committee on Gender Equality, International Labour conference, 98th Session, Geneva, June

ILO (International Labour Organization) (2010) Women in labor markets: measuring progress and identifying challenges. ILO, Geneva

ILO (International Labour Organization) (2011) Global employment trends 2011: the challenge of a jobs recovery. ILO, Geneva

ILO (International Labour Organization) (2014) Global employment trends 2014: risk of a jobless recovery. ILO, Geneva

IUF (International Union of Food, Agricultural, Hotel, Restaurant, Catering, Tobacco and Allied Workers' Associations) (2011) http://www.iuf.org

Jarvis L, Vera-Toscano E (2004) The impact of Chilean fruit sector development on female employment and household income. World bank policy research working paper 3263. World Bank, Washington, DC

Josupeit H (2004) Women in the fisheries sector of Argentina, Uruguay, and southern Brazil. FAO fisheries circular 992. Food and Agriculture Organization of the United Nations, Rome

Kabeer N (2010) Women's empowerment, development interventions, and the management of information flows. In Pinto Y, Haddad L, Bonbright D, Lindstrom J (eds.), People-centred M&E: aligning incentives so agriculture does more to reduce hunger. IDS Bull 41(6):105–113

Kabeer N, Thi Van Anh T (2002) 'Leaving the rice fields, but not the countryside': gender, livelihoods diversification, and pro-poor growth in rural Vietnam. In: Razavi S (ed) Shifting burdens. Gender and agrarian change under neoliberalism. Kumarian Press, Bloomfield, pp 109–150

Kanji N (2004) Corporate responsibility and women's employment: the cashew nut case. Perspectives on corporate responsibility for environment and development no. 2. International Institute for Environment and Development (IIED), London

Kasente D, Lockwood M, Vivian J, Whitehead A (2002) Gender and the expansion of nontraditional agricultural exports in Uganda. In: Razavi S (ed) Shifting burdens: gender and agrarian change under neo-liberalism. Kumarian Press, Bloomfield, pp 35–66

Khera R, Nayak N (2009) Women workers and perceptions of the National Rural Employment Guarantee Act in India. Paper presented at the FAO/IFAD/ILO Workshop on gaps, trends, and current research in gender dimensions of agricultural and rural employment: differentiated pathways out of poverty, Rome, March 31–April 2

Klasen S, Lamanna F (2003) The impact of gender inequality in education and employment on economic growth in the middle east and north Africa. Mimeo. University of Munich, Munich

Koggel C (2005) Globalization and women's paid work: expanding freedom? In: Agarwal B, Humphries J, Robeyns I (eds) Amartya Sen's work and ideas: a gender perspective. Routledge, New York/London, pp 165–186

Lawler J, Atmiyanandana V (2000) Gender and agribusiness project (GAP) case study. Cargill Sun Valley, Thailand. International Programs and Studies, University of Illinois at Urbana-Champaign, Champaign

Liu H, Zhang L, Summerfield G, Shi Y (2009) A gendered view of reforming health care access for farmers in China. China Agric Econ Rev 1(2):194–211

Maertens M (2010) Are African high-value horticulture supply chains bearers of gender inequality? In: FAO/IFAD/ILO (ed) Gender dimensions of agricultural and rural employment—differentiated pathways out of poverty: status, trends, and gaps. Food and Agriculture Organization of the United Nations, Rome, pp 155–170

Nelson V, Martin A, Ewert J (2007) The impacts of codes of practice on worker livelihoods: empirical evidence from the South African wine and Kenyan cut flower industries. J Corp Citizsh 28(1):61–72

North D (1990) Institutions, institutional change, and economic performance. Cambridge University Press, Cambridge

Okali C, Holvoet K (2007) Negotiating gender changes within fisheries development. Sustainable Fisheries Livelihood Programme. Food and Agriculture Organization of the United Nations, Rome

Selwyn B (2007) Labor process and workers' bargaining power in export grape production, North East Brazil. J Agrar Chang 7(4):526–553

Selwyn B (2010) Gender wage work and development in North East Brazil. Bull Lat Am Res 29(1):51–70

Sen A (1999) Development as freedom. Oxford University Press, Oxford

Smith S, Auret D, Barrientos S, Dolan C, Kleinbooi K, Njobvu C, Opondo M, Tallontire A (2004) Ethical trade in African horticulture: gender, rights, and participation. IDS working paper 223. International Development Studies, Brighton

Standing G (1999) Global feminization through flexible labor: a theme revisited. World Dev 27(3):583–602

Tabbush C (2009) Gender, citizenship, and new approaches to poverty relief: conditional cash transfer programmes in Argentina. In: Razavi S (ed) The gendered impacts of liberalization: toward "embedded liberalism"? Routledge, New York/London, pp 290–326

Tzannatos Z (1999) Women and labor market changes in the global economy: growth helps, inequalities hurt, and public policy matters. World Dev 27(3):551–569

United Nations Division for the Advancement of Women (2009) World survey on the role of women in development: women's control over economic resources and access to financial resources, including microfinance. Department of Economic and Social Affairs, New York

UNW (United Nations Women) (2011) Progress of the world's women: in pursuit of justice. UNW, New York

Whitehead A (1996) Poverty in North-East Ghana. Report to ESCOR, Overseas Development Administration (ODA), London

WWW (Women Working Worldwide) (2007) Promoting women workers' rights in African horticulture. Overview of research into conditions on horticulture farms in Kenya, Zambia, Tanzania, and Uganda. Manchester

WWW (Women Working Worldwide) (2008) Promoting women workers' rights in African horticulture. End of Project Bulletin. Manchester

World Bank (2001) Engendering development through gender equality in rights, resources, and voice. World Bank policy research report. Management 1: report no. 36546-MW. World Bank, Washington, DC

World Bank (2007) World development report 2008: agriculture for development. World Bank, Washington, DC

World Bank/FAO/IFAD (Food and Agriculture Organization for the United Nations/International Fund for Agricultural Development) (2009) Gender in agriculture: sourcebook. World Bank/ FAO/IFAD, Washington, DC

Part V
Toward a Gender-Sensitive Agricultural Research, Development, and Extension System

Chapter 15
A System That Delivers: Integrating Gender into Agricultural Research, Development, and Extension

Ruth Meinzen-Dick, Agnes R. Quisumbing, and Julia A. Behrman

Abstract A paradigm shift is required in agricultural research, development, and extension (R, D, & E) systems in developing countries, from a focus on production toward a broader view of agriculture and food systems in which women's distinct role in ensuring the food security of their households is better recognized. The authors develop a conceptual framework linking various actors in the agricultural R, D, & E cycle that involves including women in agricultural priority-setting, conduct of research, development and extension, adoption and evaluation of new technologies, and impact assessment. It also entails recognizing women's roles throughout the value chain for both food and nonfood crops and for both marketed and nonmarketed commodities. Throughout the chapter the authors review each stage of the R, D, & E cycle, arguing that a number of key questions must be asked, including who are the actors? Who are the users of the technology? Whose needs are addressed at each stage, from priority setting, through implementation, to evaluation and impact assessment?

R. Meinzen-Dick (✉)
Environment and Production Technology Division, International Food Policy Research Institute (IFPRI), 2033 K Street, N.W., Washington, DC 20006, USA
e-mail: r.meinzen-dick@cgiar.org

A.R. Quisumbing
Poverty, Health, and Nutrition Division, International Food Policy Research Institute (IFPRI), 2033 K Street, N.W., Washington, DC 20006, USA
e-mail: a.quisumbing@cgiar.org

J.A. Behrman
Department of Sociology, New York University, 295 Lafayette Street, 4th floor, New York, NY 10012, USA
e-mail: jab965@nyu.edu

A.R. Quisumbing et al. (eds.), *Gender in Agriculture: Closing the Knowledge Gap*, DOI 10.1007/978-94-017-8616-4_15,

Keywords Gender • Agricultural research, development, and extension (R, D & E) • Priority setting • Conceptual framework • Adoption and evaluation of agricultural technologies

15.1 Introduction

Gender differences matter in agricultural production in various farming systems all over the world, where the ownership and management of farms and natural resources by men and women are defined by culturally specific gender roles. Gender differences are also important in the staffing and conduct of agricultural research institutions and extension systems. Whereas the fields of health, nutrition, and education have long acknowledged that explicitly addressing gender issues is one of the most effective, efficient, and empowering ways to boost development and address poverty, the field of agricultural research has lagged. In the world of national and international agricultural research, women continue to be underrepresented and underserved, and their contributions are not fully tapped. While progress has been made in developing extension systems that are more gender-sensitive, unless the source of new crop, fish, and livestock varieties and agricultural technologies takes women's different needs into account, the product that is being disseminated by extension systems may not meet different needs and preferences. A gender-responsive agricultural research, development, and extension system therefore needs to address women as well as men as both the clients and actors in agricultural research.

In this chapter we present a conceptual framework linking various actors in the agricultural research, development, and extension (R, D, & E) cycle, including a discussion of the implications of integrating gender into agricultural priority setting, conduct of research and development, extension, and the adoption and evaluation of new agricultural technologies. Ultimately, this calls for moving toward a broader view of agriculture and food systems that recognizes women's role throughout the value chain for both food and nonfood crops and for both marketed and nonmarketed commodities. Issues of gender staffing in research are then examined in more detail in Chap. 16 by Beintema, while Ragasa's Chap. 17 provides a review of gender issues in extension.

15.2 Envisioning Gender in Agricultural Research, Development, and Extension

A more gender-responsive agricultural research, development, and extension (R, D, & E) system calls for a comprehensive look at the system: who are the actors, who are the users of the technology, and whose needs are addressed at each stage, from priority setting, through implementation, to evaluation and impact assessment? In this section we provide a framework for considering these issues.

Many conventional analyses of agricultural research and development have used a pipeline analogy with *upstream* (basic) research feeding into *downstream* (adaptive) research to develop technologies, which are then passed on to extension systems to be disseminated and adopted by farmers, possibly followed by impact assessments to assess the payoffs to the research. The implicit image associated with this process is an improved staple crop variety being adopted by a male subsistence farmer. Although in recent years some attention has been given to involving women in participatory adaptive research and to examining gender differences in impacts of technologies, relatively little attention has been given to gender in the upstream priority setting and decisionmaking. In order to fully meet the needs of women and men as agricultural producers and consumers, it is imperative to go beyond mechanistic approaches and recognize that innovation systems are composed of multiple actors and linkages; these actors and linkages need to be considered at each stage of the R, D, & E process. Instead of a unidirectional flow between basic research, adaptive research, and end users, a more farmer-, consumer-, and gender-responsive agricultural research system would integrate gender into each stage of the process, starting with priority setting. Research and development would include women as well as men in the conduct of the research. Extension services, which are key to disseminating innovations, would serve women as well as men, employing female extension agents or new techniques of extension to do so. This, in turn, would ensure greater gender equity in adoption of innovations. Evaluation of technologies and programs would consider their differential impact on women and men. Finally, a responsive system would allow feedback from end users of the technology—men and women, farmers, and consumers—thereby creating an effective feedback loop, as illustrated in Fig. 15.1.

Furthermore, each stage in this cycle involves a combination of international and national agricultural research as well as the male and female farmers themselves. This differs from conventional views of priority setting and "upstream" research being done by international and national "scientists," with farmers only becoming involved in adaptive research and adoption. Instead, this acknowledges that farmers are also developing their own innovations, and should be involved throughout this cycle, including priority setting and evaluation. Indeed, it is how the work of these different actors (along with the private sector and NGOs) fits together that is crucial for the effectiveness of the overall agricultural sector.

The following are key elements of this process and examples of critical questions that are needed to evaluate the extent to which gender issues are being integrated.

15.2.1 Priority Setting

Effective integration of gender into the research, development, and extension cycle needs to start at the priority setting phase, where decisions are made regarding the kinds of agricultural R, D, & E that will receive investment. Engendering the priority setting process includes (1) consideration of the representation of women in

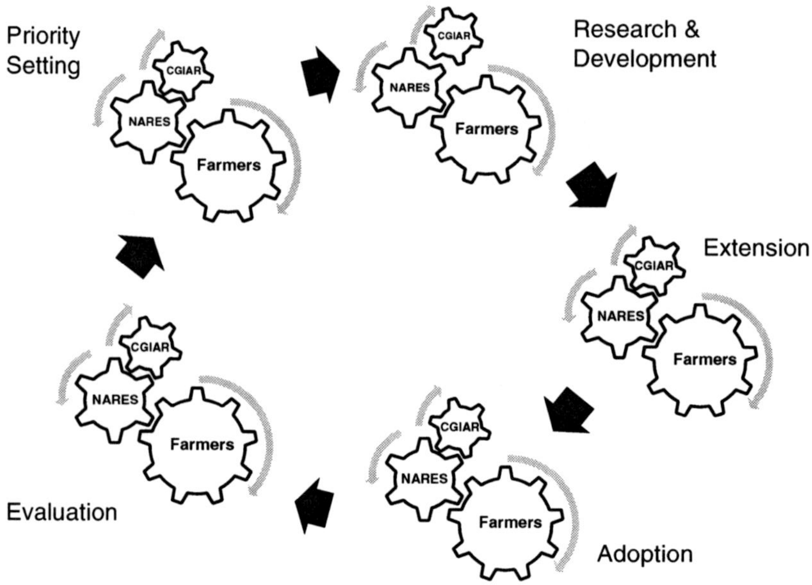

Fig. 15.1 Conceptual components of a gender-responsive agricultural research system (Source: Meinzen-Dick et al. 2010)

management research systems; (2) development of mechanisms to take the needs of women and men as producers and consumers into account; and (3) expansion of the definition of agricultural research beyond just production of field crops (often a male activity) to include homestead gardens, postharvest processing, supply chains, and consumption and nutrition outcomes, which are often of greater salience to women.

A gender-blind priority setting process is not likely to yield a gender-balanced agricultural R, D, & E portfolio. Therefore, the first question to ask is where and how are the differential needs, interests, and priorities of women and men reflected? This requires systematic gender analysis of needs in the field as well as of the balance of women's and men's voices in consultations at all levels of decisionmaking. For example, are women farmers' associations consulted at any point? Do female farmers have a voice in male-dominated farmer associations? If women themselves are not expressing their needs and priorities, how are these being taken into account?

The FAO Focus paper, "Women and Food Security," sums up the consequences of decades of ignoring women researchers' and women policymakers' critical role:

> While rural women are knowledgeable about and use a large amount of traditional technology, they have very little access to modern technology that could benefit them in their farm and household activities. This is due to women's lack of participation in setting research priorities or in generating and disseminating conventional technologies (FAO 2010).

15.2.2 Priorities of Women and Men

A wide-ranging body of empirical and theoretical literature shows that different household members have different preferences, incomes, resources, and needs, which often vary along gender lines (Haddad et al. 1997; Quisumbing and Maluccio 2003). In agriculture, these differences are relevant to the priorities of men and women as both producers and consumers. Like all gender differences, these differences are contextual and will vary among and within regions, countries, and communities and will change over time. However, we can identify some key questions to ask regarding the roles, resources, preferences, and needs as producers and consumers.

In terms of roles, women often have greater responsibility for family food production and processing, whereas men have greater involvement in market-oriented production. Even where women are engaged in markets, their responsibility for cooking food and serving it to their family is an important factor affecting preferences for certain crops (e.g., vegetable production for relishes) or varieties (e.g., with certain cooking traits). Men and women also play different roles in natural resource management, local organizations, and links to outsiders, which need to be considered in developing resource management strategies or group- and market-based programs. Moreover, women's responsibilities for childcare and domestic work create labor constraints, affecting the resources at their disposal for farming. However, in many regions, women are increasingly involved in agricultural production and the labor force as a result of male migration and occupational diversification, as well as with the growth of new agricultural value chains. In the dry forests of northern Mali, women have adopted new income-generating activities, such as charcoal production, in order to cope with their growing vulnerability to climate change (Brockhaus and Djoudi 2008).

Labor constraints and other differences in resources will affect men's and women's abilities to benefit from different types of agricultural technologies and innovations. Peterman et al. (2011) found that lower productivity is persistent on female-owned plots and in female-headed households in Nigeria and Uganda when accounting for a range of socioeconomic variables, agricultural inputs, and crop choice. Men and women also hold different types of assets, which play different roles within the household. Dillon and Quiñones (2010) found that women's assets grow more slowly than those of men over a long time period in northern Nigeria. Men's assets, primarily livestock, increased greatly in value over time; whereas women's assets, primarily durable goods and jewelry, increased at a much slower rate. In rural Bangladesh, husbands' and wives' asset stocks are drawn down for different kinds of shocks, with husbands' assets being liquidated to finance dowry and wedding expenses, and wives' assets being negatively affected by illness shocks (Quisumbing 2011).

Gender-based differences in task allocation within wage labor systems may result in differential health impacts on men and women. This is especially problematic when exposure to pesticides and other agrochemicals causes risks of reproductive difficulties, especially miscarriage and birth defects. Evidence from plantation systems indicates that women workers often receive less training and instruction than male workers for working with agrochemicals (Loewenson 2000). In Malaysian plantations, Oxfam (2007) finds a "gendering" of tasks whereby women, who are perceived to be more nimble and less capable of performing arduous tasks, are recruited as sprayers of chemical pesticides and herbicides and lack proper training and safety equipment. In a case study of biofuels plantations in Indonesia, Julia and White (2010) observe a similar gendering of tasks; women are assigned the tasks of spraying and fertilizer application, while men are assigned to harvesting. These "female tasks" require coming into contact with dangerous chemicals and productive gear is only purchased at the workers expense. A number of studies demonstrate a high correlation between the intense use of chemicals in the cut-flower industry and the pervasive negative environmental and health effects that affect female workers (and that of their children) in comparison with their male counterparts in the same working environment (Larrea and Maldonado 2005; Paz-y-Mino et al. 2002). Thus, research to reduce exposure to harmful agrochemicals is likely to be especially important for women.

Gender differences also play a key role in needs of men and women as consumers. Adolescent girls and women have a higher biological need for micronutrients, but culturally they are often prescribed to eat last or to curtail their consumption to ensure that others in the household have enough to eat. This can have long-lasting effects: not only is women's health affected, but effects are also transmitted to the next generation through low birth weight and malnutrition. For agricultural research to contribute to long-term poverty reduction of both the current and future generations, these differential needs of women need to be considered in the selection of varieties (e.g., through biofortification), crops (e.g., nutrient-rich vegetables), and processing (to preserve nutrients).

Considerable research on trait preferences by gender has been conducted, particularly from participatory research programs that have involved farmers in varietal selection. Although this downstream research is important, it begs the larger question of which crops, which agricultural systems, and which domains of action are addressed through agricultural research systems. The following are examples of how accounting for gender differences could shape priorities for different types of research:

- *Trait preferences*: The differential needs of male and female farmers are reflected in their different preferences for maturation periods, yields, tastes, and colors, and this affects adoption rates. Some progress in understanding and responding to gendered trait preferences in relation to crops has been made, but such progress is virtually absent in livestock and aquaculture research.
- *Crop, livestock, and aquaculture practices*: Beyond choosing particular traits, addressing gender issues in priority setting also requires examining *which* crops and animals are selected for research and improvement. The particular importance that male and female farmers place on different crops or species is culturally

specific, depending on the relative roles and resources of each gender; hence, it is crucial to have gender expertise in agricultural research centers to identify these roles.

- *Natural resource management*: Men and women both play crucial but different roles in natural resource management. For too long the agricultural sector has tended to focus on the activities in which men were more heavily involved, neglecting aspects, such as non-timber forest products, that are especially important to women, even if they have lower apparent market value.
- *Other domains of action*: The definitions of "agriculture" and "farmers" typically focus on activities conducted between the planting period and the harvest period, thus overlooking other activities, such as postharvest processing, where women are key actors. Women often provide the bridge between "productive" and "reproductive" domains; however, agriculture is often artificially defined as solely "productive." A renewed focus on nutrition and recognition that women are providers of family food even if they are not always direct agricultural producers points to another need to go beyond these linear definitions of "agriculture" and "farmers." A shift from thinking about "agriculture" (especially field crops) to thinking about "food" (including the processing and cooking) is very important in this regard, as well as getting beyond the "food" vs. "cash" crop divide.
- *Value chains and food systems*: Agricultural research is expanding from food production to income generation, and, in areas of limited female mobility, value chains and cash-oriented production are often dominated by men. Gender-based constraints affect the structure and relationships of value chains. Numerous studies of commercialization have shown that increases in cash income do not necessarily translate into gains for all household members. Where intrahousehold distribution is fairly equitable, substantial increases in household incomes need not have detrimental effects on either subsistence production or nutrition (von Braun et al. 1989). However, there are many cases of men taking over women's enterprises when the value of that produce increases. Participating in contract farming or warehouse programs may require using a bank account, which is often held in the man's name. Informal processing and food vending is an especially important source of income for women. Where there is more pooling (as opposed to separate "purses"), it will matter less who markets the product and receives the cash. Other measures regarding the way payments are made can help to ensure that women do not lose control over products and incomes when they are marketed. These include making payments into a woman's account or increasing transparency on prices paid at the market each day.
- *Agricultural institutions and policy research*: Research to identify an enabling environment for agriculture, including investment policy, property rights, infrastructure, and support services, needs to consider how governance structures affect women's access to and control over productive resources and incomes, and how policies will differentially affect men and women. Adaptive collaborative research can identify ways to strengthen the voice of women within local institutions and in agricultural policies.

All of this relates also to the way *agricultural research* is defined. Conventional definitions have been gender biased, focusing on the production of field crops, which are more likely to be male activities, and relatively neglecting homestead gardens, postharvest processing, supply chains, and consumption and nutrition outcomes, which are often of greater salience to women.

Instead of focusing on *agriculture*, thinking in terms of *food* leads to a more gender-balanced picture. Although crop research is important, and women are also involved in producing nonfood crops, the food sector—which is more relevant to this chapter—is broader, also including fish, livestock, garden production, water, trees, soil, and natural resources. In fish and livestock farming, it is important to ask whether species and varieties valued by women for their nutritional content, taste, or other uses receive the same attention as research on high-value, genetically improved varieties that bring high economic returns to male producers. Postharvest processing needs to be considered, not only for reaching high-value markets, but also for food safety and reduction of drudgery, which evidence indicates is borne particularly by women in the household. Various studies (e.g., McGuire and Popkin 1990; Levine et al. 2001) document the higher work burden of women in both domestic and productive activities in developing countries. A review of time allocation studies in World Bank (2001, 175–177) shows that women and girls are more involved in time-intensive activities such as fetching fuel and water. As subsistence farmers begin producing surpluses to be sold in the market, the distinction between food and cash crops breaks down; therefore, it is important to pay attention to the differential roles of women throughout the value chain, but particularly for nonmarketed crops, which are more often the domain of women subsistence producers. In addition to subsistence farming, it is important to look at the roles women increasingly play as traders, business owners, and laborers throughout the agricultural value chain in rural and urban areas. For properly balancing the agricultural research portfolio so that it reduces poverty and increases food security, it is important to value the nutritional and health benefits of such production and processing, and not only financial returns to marketed production (Meinzen-Dick et al. 2011a).

15.3 Research and Development

As with priority setting, it is important to consider who is conducting the research and how attuned researchers are to gender issues. A key aspect of this is the gendered staffing patterns of the CGIAR and national agricultural research systems (see Beintema, Chap. 16). However, it is important to look beyond these public-sector institutions as a source of innovation and to also consider private-sector R, D, & E, as well as the research conducted by farmers themselves, and the extent to which each of these address the needs of women.

15.3.1 Extension

In examining gendered patterns of extension, it is important to consider who delivers extension services (because female extension agents may be more likely to reach female farmers, especially in highly sex-segregated societies), who receives the extension services and information (only males or heads of households, or whether women are recognized as farmers and clients of the extension services), and how extension services are delivered (including individual- or group-based approaches, conventional extension, or farmer field schools). Of utmost importance is the issue of whether or not women are recognized as farmers and clients of the extension services. As with research and development, it is important to consider not only formal public extension services, but also private-sector and farmer-to-farmer dissemination, and how effective each of these is in recognizing and reaching women as producers and consumers (see Ragasa, Chap. 17).

15.3.2 Adoption of Innovations

All of the foregoing are likely to shape who can and will adopt agricultural innovations and benefit from them. Additional factors also can constrain adoption, such as lack of necessary cash, labor, skills, and property rights, and each of these may differ for men and women. Even after adoption, if particular innovations do not meet the needs or deliver results for women or men, the innovations may be dropped.

15.3.3 Evaluation and Impact Assessment

For gender concerns to affect future priority setting and conduct of agricultural R, D, & E, gender needs to be more systematically integrated into impact assessment and evaluation systems. An increased focus on gender equity requires a new approach to both ex-post and ex-ante impact assessment. This requires combining strong evaluation designs that generate good data, research methods that integrate economic and social analysis, and sufficient capacity to undertake the assessments (Adato and Meinzen-Dick 2007, 4).[1] Institutional willingness to undertake gender-sensitive impact assessment, and apply the results of such assessments to one's own research institution, is also important. Although the outcomes are likely to be context-specific, the following factors are likely to affect the extent to which agricultural R & D benefit women and men.

[1] Adato and Meinzen-Dick (2007) state this in the introduction of a volume on the social impact of agricultural research, in arguing for more poverty-focused impact assessment, but this argument can be applied equally well to increasing a focus on gender equity in the agricultural R, D, & E system.

Relative importance of production for the market vs. home for men and women. Because male and female farmers producing food crops have to decide whether or not to sell or consume the crop produced, the impact of agricultural technologies will depend on the relative importance of production for sale versus home consumption. The relative responsibility of men or women for food vs. cash crop production depends on the farming system. Within the African context, the standard explanation is that men are responsible for producing the cash crops, and women, the food crops (Koopman 1993). However, Doss (2002) has critiqued such characterizations by providing evidence that both men and women are involved in cash and food crop production. Moreover, food crops can be sold for cash if marketable surpluses exist. Nonetheless, it is clear that differential gender preferences do exist, and need to be considered when introducing new technologies.

Differences in trait preferences. Given that male and female farmers have different roles and responsibilities in providing for the household's food security, it is not surprising that research indicates they also have different preferences when evaluating new technologies or practices for potential adoption. Preferences are conditioned by the end use of the crop, whether it will be sold right away (yield and profitability) or used for home consumption (storage, taste, and processing). Bellón et al. (2007), in examining men's and women's differential preferences for grain characteristics in Oaxaca and Chiapas, Mexico, found that traits related to vulnerability (tolerance to drought, resistance to rot, and resistance to pests) are significantly more important to poor female farmers than to their male counterparts. In general, consumption characteristics were more relevant for women than for men, a reflection of the women's role as subsistence farmers and household food providers. Smale's (1995) work on farmer preferences in Malawi found that while hybrid maize improved yields for sale, traditional maize stores better and ultimately provides better yields for household consumption. Given that households produce for both sale and personal consumption, there are obvious trade-offs.

Labor and employment impacts. While gender-related determinants of the marketed surplus are a relatively new area of analysis, the gender-differentiated impact of agricultural technology adoption on labor and employment is even less studied. While the relationship between poverty and landlessness varies across regions (Ahmed et al. 2007), employment is an important source of income for the working poor (de Villard et al. 2010).[2] Worldwide, the working poor (those earning less than

[2] In a study based on the analysis of household data and review of empirical research in 20 countries, Ahmed et al. (2007) find that, globally, there does not seem to be a uniform pattern of higher landlessness among the poor, although the relationship varies among Sub-Saharan Africa, Latin America, and Asia. In all parts of Asia, those who are landless are the poorest. In Sub-Saharan Africa, however, little difference was found between the incidence of landlessness among the poorest and less poor households, and, in some cases, the reverse pattern was found. This corresponds to the findings of other studies that in Sub-Saharan Africa, the poorest often own some land (usually very small plots), but they lack access to markets and other key resources, such as credit and agricultural inputs. In Latin America, although the incidence of landlessness is high, it was actually found to be higher among those who live on more than $1 a day than among those living on less than $1 a day. This suggests that in Latin America, the poorest are more likely to be self-employed

$2 per day) account for 40.6 % of total employment, with substantial regional variation. The working poor account for 82 % of total employment in Sub-Saharan Africa and 80.9 % in South Asia, but only 33 % in East Asia. In the world generally and in the poorest regions in particular, female workers have a significantly higher probability than male workers of being involved in vulnerable forms of employment (de Villard et al. 2010).[3]

New technologies may affect men and women in different ways, even within the same socioeconomic class, due to initial differences in their involvement in agricultural fieldwork and non-fieldwork, especially domestic work and childcare; the extent of their control over and the patterns of distribution of household earnings and expenditures; and the extent of their direct access to productive resources, especially land. Land availability and the structure of land rights in agricultural-based countries influence the form of employment that rural women have access to, with a prevalence of wage labor and unpaid family contributions in South Asia (a land-scarce region) and mostly smallholder self-employment in Sub-Saharan Africa (a land-abundant region). Latin America, which is the most urbanized of all developing regions (and also has the most equal educational levels by sex), is the only region where the ratio of rural women's nonagricultural employment to agricultural employment is higher than the corresponding rural men's ratio (Fontana and Paciello 2010).

Since a majority of the poor—and women—in Africa and Asia derive incomes from labor on their own and others' farms, the employment effects of new technologies are important factors determining changes in incomes and welfare. This is illustrated by the studies of the adoption of irrigated rice and high-yielding or modern varieties (HYV or MVs) in Asia and Africa in the 1970s and 1980s, as well as the introduction of other crops in Africa (see, for example, Unnevehr and Stanford 1985).

For new technology to increase employment opportunities for women, there must be a concurrent increase in the demand for women's labor. In contexts where there is a growing supply of landless women's labor, women will benefit only if productivity increases are accompanied by increased labor demand or if productivity increases free up women's time for leisure, self-care, or other, more remunerative tasks. Furthermore, women's ability to benefit from technical change depends largely on their control over valuable resources. When women have at least some control over income derived from land, they stand to benefit from technical change that will increase productivity of household labor and land. On the other hand, for women who lack control of proceeds from land, labor becomes their primary resource. In this case, neutral or labor-using technical changes will increase demand for their labor; however, labor-saving technical change will reduce employment opportunities (Unnevehr and Stanford 1985). Finally, the effect of technology adoption may not be discerned in the period after immediate adoption, since the diffusion of agricultural innovations is a long-run process.

cultivators rather than the non-poor, perhaps lacking employment opportunities in nonagricultural sectors.

[3]Vulnerable employment is defined as persons who are less likely to have formal work arrangements or access to benefits or social protection programs (ILO 2009).

Some of these long-run adjustment effects may involve the movement of labor from agriculture to nonagriculture.

Impacts of technologies on bargaining power, control of resources, and time burdens. Technologies that increase returns to women's labor may strengthen their bargaining power. Doss (2001) points out that in certain instances, increases in women's labor and time availability come with a corresponding increase in responsibility and control over output. For example, in western Ghana, Quisumbing and colleagues (2001) found that a new land transfer practice has resulted where husbands transfer land to their wives in exchange for labor on cocoa fields. This change has come about as a result of increased incentives to adopt cocoa, which uses women's labor intensively, owing to increased profitability.

However, it is difficult to predict, ex ante, the impact of new technologies on bargaining power and control of resources, because gender roles are dynamic. As women's activities become more lucrative as a result of adoption of new technologies, traditionally female tasks may be taken over by men—or women may move into spheres formerly controlled by men. Unfortunately, examples of the former are more common, as illustrated by an example from The Gambia, where Schroeder (1993) found that women lost control of communal vegetable garden plots after an environmental stabilization intervention. Following the intervention, men asserted control over those plots, a traditionally female domain, because of the lucrative new fruit trees, fenced enclosures, and improved soil. As a result, women lost an important source of income and bargaining power.

Several studies indicate African women's time burdens actually increased with the adoption of new technology (Suda 1996; Berio 1984). Authors suggest that with the onset of new technology, women must take on additional and highly time-consuming tasks or process increased levels of output. For example, in Malawi and Zambia, women, who are in charge of processing, reported hybrid maize was more difficult to pound and this became a more time-consuming, arduous task (Jha et al. 1991; Hirschmann and Vaughan 1984).

Environmental impacts. In addition, the use of new technologies, such as pesticides, may have serious potential health effects to which women may be more vulnerable. Evidence from tomato processing plants in Mexico indicates that protective equipment is not adequate and illness due to the ingestion of pesticides and other agrochemicals is common (Barron and Rello 2000). Likewise, in Kenya's fresh vegetable industry, chemicals used for storage, mixing, and spraying have led to skin allergies, headaches, and fainting (Dolan and Sutherland 2002). These health effects may differentially affect men and women as there is evidence that women workers in plantations often receive less training and instruction than male counterparts, do repetitive work that can result in health difficulties, and face reproductive difficulties as a result of exposure to agrochemicals (Loewenson 2000). For example, Oxfam (2007) finds that in Malaysia, women plantation workers are often recruited as sprayers of chemical pesticides and herbicides and are not given proper training and safety equipment.

Ultimately, it is difficult to predict the impacts of agricultural technologies and development interventions without a thorough knowledge of the culture and context. That is why it is essential to have evaluation systems that will identify the positive and negative impacts of agricultural R, D, & E on gender-specific outcomes, including gender relations, and can feed back that information to help adapt priorities for future agricultural programs. Moreover, capturing the full impact of agricultural technologies on lives and welfare requires going beyond narrow indicators of productivity to broader indicators of well-being, particularly in capturing the differential impact on men and women.

15.3.4 Toward Indicators for Gender-Equitable Agricultural Research

The many studies that have found gender-differentiated determinants of technology adoption as well as differential impacts of new technologies by gender provide compelling justification for the adoption of gender-sensitive indicators for prioritizing technologies for development and dissemination. While the specific criteria will vary by culture, context, and agroclimatic zone, among others, the most important overarching principles for evaluation are as follows:

- The extent to which women are involved in the crop or sector in terms of production, marketing, or processing has not decreased (or has increased) as a result of the program
- Reduction of gender disparities in access to productive resources and control of incomes as a result of the program
- Improvements in diets or nutritional status of individuals, particularly in areas where there are marked gender disparities in nutritional status/nutrient adequacy.

It is important to consider both what is evaluated and how it is done. Evaluation, adoption, and impact assessment studies have often focused on household-level indicators and collected the data from male heads of households, often using standard and predetermined indicators. In a project in Malawi, Njuki et al. (2008) report using community indicators to evaluate research for a development program. Men and women had different indicators for similar objectives and different perceptions of the extent to which the project had achieved these objectives. This underscores the need in evaluation and impact assessment studies to interview both men and women and to have gender-specific indicators.

15.4 A Revitalized Agricultural Research and Extension System

In 1976, Margaret Mead, as head of the American Association for Advancement of Sciences, noted:

> The Euro-American tendency to attribute the concern with agricultural production (with food before it leaves the harvest field) to men and to attribute the concern with food after it

> leaves the harvest to women led to the dual assumption that scientific agriculture was a male field and food knowledge (food preservation, nutrition, child rearing, and home management) was a female field.... What is needed are departments or schools in which all the skills related to food—including plant genetics, animal husbandry, veterinary skills, nutrition, child development, food management, etc.—are taught without discrimination to both men and women. Only in this way can there be any hope of including women at every level of the decisionmaking process and of restoring the concept that the primary function of food is to feed people, and to feed them well.... (Mead 1976, 11).

Almost forty years later, her concerns still apply. Reorienting the agricultural research system to be more gender responsive requires being more aware of the different needs and preferences of male and female farmers; the different roles that men and women play in the production and marketing process; differential access to and control of productive resources; differential constraints that female farmers may face in adopting new technologies, including time constraints owing to domestic responsibilities and nonmarket production; the representation of male and female scientists and extension agents in the agricultural research and extension systems, among others. In most cases, the distribution of private and public resources has ignored or disadvantaged female farmers. But there are ways to improve the gender responsiveness of the agricultural research system, which, in turn, will not only improve the productivity of women farmers, but, because of the particular role of women in household food security, also improve the welfare of their whole families. Key areas for attention include

- Identifying the strategic priorities for gender-equitable agricultural research and extension. In many cases these strategic priorities may lead to new emphasis, for example, more on foods contributing to diverse and nutritious diets, or require addressing underlying gender inequalities in access to resources in order to unleash the full productivity of millions of women agricultural producers.
- Fully integrating gender into the agricultural R, D, & E system, including priority setting, conduct of research and development, extension, adoption, and evaluation of outcomes.
- Transforming the enabling conditions, including institutional structures and policies for gender-equitable agricultural research.

Research on gender mainstreaming across a range of development organizations has found that to be successful, four enabling factors are necessary: political will, technical capacity, accountability, and organizational culture (James-Sebro 2005). *Political will* refers to the ways in which an organization's leadership conveys the importance of, and expresses its support for, the integration of gender, including the inclusion of gender in policy documents and the allocations of funds. *Technical capacity* refers to the professional qualifications and skills staff have to integrate gender into their work. Yet even if these skills are present, *accountability* mechanisms need to be in place to ensure that staff operationalize the institute's commitment to gender integration. Such mechanisms include monitoring and evaluation of gender results and staff incentives. Finally, *organizational culture* refers to creating an environment supportive of gender integration, one in which staff are encouraged to share lessons learned on gender and to ask questions about its relevance to their work.[4]

Creating a gender-responsive agricultural research system means going beyond the traditional boundaries of crop-oriented research and revising the way people think about gender roles throughout the agricultural sector. A gender-responsive agricultural system not only addresses the gender differences in needs and priorities in all aspects of conventional agricultural research and development, but it is also able to stimulate thinking beyond production agriculture to consider the following issues.

Gender roles in natural resource management. A narrow focus on production technology often neglects the natural resource base—trees, soils, water, agrobiodiversity, and other natural resources—that men and women manage. But here there is a need to look beyond the narrowly defined *agricultural* uses of these resources to also consider domestic uses of water, the energy needs of women for cooking fuel, and how these impact forest use, carbon emissions, and the like. Although outsiders may segment these into different departments, for rural people, and women in particular, the lines between *productive* and *domestic* uses of resources are not distinct.

An expanded concept of the food sector. The food sector is broader than crop production, also including fish, livestock, garden production, and water. Most agricultural research is devoted toward increasing yields of staple crops, often neglecting vegetables grown in home gardens, despite the important contributions these make to household consumption, food security, and nutritional status.

Postharvest processing. Postharvest processing needs to be considered, not only for reaching high-value markets, but also for reducing food losses, preserving nutrient content of food, ensuring food safety, reducing drudgery, and releasing women's time for other activities.

Value chains. Even though most of the leading donor institutions have adopted value-chain approaches as a strategy for enhancing economic growth and reducing poverty, until recently, very few have considered how gender issues affect value-chain development (see Rubin and Manfre, Chap. 12). It is now increasingly recognized that the introduction of new technologies can affect the on-farm division of labor and that the adoption of high-valued crops can alter men's and women's control of resources within the household; however, the gender dimensions of the link between household and market is relatively less understood. Even if the agricultural research system is not involved in all stages of the value chain, understanding gender issues in value chains can help identify leverage points at which interventions can avoid transferring income or control from women to men, and even generate positive gender outcomes, while meeting the goals of improved efficiency and poverty reduction.

Linkages to health and nutrition. A gender-responsive agricultural research system recognizes the strong linkages between agriculture, research, and nutrition. Agriculture can play a critical role in improving the nutritional quality and diets of

[4] Drawn from the *Gender Audit Questionnaire Handbook*. 2003. Washington, DC: InterAction.

the poor by recognizing that men, women, and children have different biological needs for macro- and micronutrients. Agricultural research can improve access to—and utilization of—inexpensive, nutritious, and diverse foods to improve nutrition outcomes, while also improving food security and health outcomes. Agricultural research can also pay closer attention to agriculture–health linkages, particularly to help fight infectious diseases. Most of the world's emerging diseases are zoonotic, transmitted between animals and people. Animal diseases that decrease meat and milk production also strongly have an impact on human health. Recognizing the important roles of men and women in livestock production would help mobilize them to prevent the spread of zoonotic diseases, or arrive at more gender-equitable risk-mitigation mechanisms. Similarly, better water management can reduce water-borne diseases or those, such as malaria and schistosomiasis, with water-related vectors. Reducing the burden of ill health also alleviates women's time burdens.

Supporting policies and institutions. A supportive institutional and policy environment is also important for successful agricultural development as well as agricultural research. Strengthening women's property rights or rights under family and civil law can give women greater incentive and ability to invest in the land, have bank accounts, or obtain credit. Collective action institutions can play a major role, either through women's organizations or through ensuring that women are fully included in farmers' associations, water user groups, forest committees, or local decisionmaking bodies that manage natural or financial resources and services.

All of this is an ambitious agenda. Serious work for poverty reduction must be ambitious and multifaceted. Addressing gender in agricultural research and development must be a shared endeavor. No single type of organization can be solely responsible, but neither should any be exempt from responsibility for considering how their work will affect women as well as men.

The first step is to increase awareness that gender issues are not peripheral to agriculture but are fundamental to increasing productivity, incomes, nutrition, food security, sustainability, and ultimately the contribution of agriculture to poverty reduction. Both research and firsthand experience play an important role in generating this awareness. Statistical and impact assessment agencies need to be involved to ensure that the data and methods are developed to capture gender differences in needs, contributions, and outcomes.

The second step is to ensure that those who set priorities, those who implement and disseminate research, and those who evaluate the impacts of agricultural R, D, & E can identify the relevant gender dimensions of their work. As a result, paying attention to gender will no longer be seen as the responsibility of a small group or something that people do in their spare time as an addition to their "real" work, but will be seen as an integral part of excellence in agricultural R, D, & E. This, in turn, requires strengthening the capacity of all involved, linking contextual knowledge about gender relations to broader patterns and even global lessons.

Political will and supportive structures are needed that create accountability; make financial, human, and time resources available for this; and recognize and reward excellence in these endeavors. There are costs to addressing gender and expanding the clientele of the agricultural research and development community to include women farmers, consumers, traders, and business owners on a par with men. However, the returns are also significant, not only in terms of productivity, but also food security, nutrition, environmental sustainability, and long-term poverty reduction. Mechanisms are needed to share lessons from countries and programs that have made significant strides toward gender equity: what motivated these changes; what key changes were made; and what outcomes have they seen for women, their families, and society as a whole? The agricultural sector is not alone in this; much can be learned from experiences with gender integration in other sectors and development agencies (e.g., Moser and Moser 2005; Rao and Kelleher 2005) that share with agricultural research and development the objectives of fighting poverty and hunger while conserving the environment.

Acknowledgments This chapter draws from Meinzen-Dick et al. (2011b), which was first presented at the Global Forum for Agricultural Research 2010. We acknowledge funding from the Global Forum for Agricultural Research as well as fruitful discussions that shaped the direction of this research.

References

Adato M, Meinzen-Dick R (eds) (2007) Agricultural research, livelihoods, and poverty: studies of economic and social impacts in six countries. Johns Hopkins University Press for the International Food Policy Research Institute, Baltimore

Ahmed AU, Hill RV, Smith LC, Weismann D, Frankenberger T (2007) The world's most deprived: characteristics and causes of extreme poverty and hunger. International Food Policy Research Institute, Washington, DC

Barron MA, Rello F (2000) The impact of the tomato agroindustry on the rural poor in Mexico. Agric Econ 23(3):289–297

Bellón M, Adato M, Becerril J, Mindek D (2007) Improved maize germplasm, creolization, and poverty: the case of *Tuxpeño*-derived material in Mexico. In: Adato M, Meinzen-Dick R (eds) Agricultural research, livelihoods, and poverty: studies of economic and social impacts in six countries. International Food Policy Research Institute, Washington, DC, pp 238–284

Berio A-J (1984) The analysis of time allocation and activity patterns in nutrition and rural development planning. Food Nutr Bull 6(1):53–68

Brockhaus M, Djoudi H (2008) Adaptation at the interface of forest ecosystem goods and services and livestock production systems in northern Mali. Center for International Forestry Research (CIFOR) Information Brief. CIFOR, Bogor, Indonesia. http://www.cifor.cgiar.org/publications/pdf_files/Infobrief/019-infobrief.pdf. Accessed 19 Apr 2010

de Villard S, Dey de Pryck J, Suttie D (2010) Consequences of gender inequalities and policy options for gender equitable rural employment: workshop contributions. In: FAO/IFAD/ILO (ed) Gender dimensions of agricultural and rural employment: differentiated pathways out of poverty: status, trends, and gaps. Food and Agriculture Organization of the United Nations, Rome

Dillon A, Quiñones E (2010) Asset dynamics in Northern Nigeria, Discussion paper 1049. International Food Policy Research Institute, Washington, DC

Dolan C, Sutherland K (2002) Gender and employment in the Kenya Horticulture value chain. Globalisation and poverty working paper. Overseas Development Group, University of East Anglia, Norwich, UK

Doss CR (2001) Designing agricultural technology for African women farmers: lessons from 25 years of experience. World Dev 29(12):2075–2092

Doss C (2002) Men's crops? Women's crops? The gender patterns of cropping in Ghana. World Dev 30(11):1987–2000

FAO (Food and Agriculture Organization of the United Nations) (2010) Women and food security. FAO focus. Food and Agriculture Organization of the United Nations, Rome. http://www.fao.org/docrep/013/i2050e/i2050e.pdf. Accessed 31 Jan 2010

Fontana M, Paciello C (2010) Gender dimensions of rural and agricultural and public works programmes. Experience from South Africa. Differentiated pathways out of poverty: a global perspective. In: FAO/IFAD/ILO (ed) Gender dimensions of rural and agricultural employment: differentiated pathways out of poverty, status, trends, and gaps. Food and Agricultural Organization of the United Nations, the International Fund for Agricultural Development, and the International Labour Office, Rome, pp 171–184

Haddad L, Hoddinott J, Alderman H (eds) (1997) Intrahousehold resource allocation in developing countries: methods, models, and policy. Johns Hopkins University Press for the International Food Policy Research Institute, Baltimore

Hirschmann D, Vaughan M (1984) Women farmers of Malawi: food production in the Zomba District. University of California, Berkeley

ILO (International Labour Office) (2009) Guide to the new millennium development goals employment indicators: including the full set of decent work indicators. International Labour Office, Geneva

James-Sebro M (2005) Revealing the power of gender mainstreaming: enhancing development effectiveness of nongovernmental organizations in Africa. InterAction's Commission on the Advancement of Women, Washington, DC

Jha D, Hojjati B, Vosti S (1991) The use of improved agricultural technology in Eastern Province. In: Celis R, Milimo JT, Wanmali S (eds) Adopting improved farm technology: a study of smallholder farmers in Eastern Province Zambia. International Food Policy Research Institute, Washington, DC, pp 173–201

Julia, White B (2010) Agro-fuels, enclosure and incorporation: gendered politics of oil palm expansion in a Dayak Hibun community in West Kalimantan. Working paper. International Institute of Social Studies, Rotterdam, the Netherlands

Koopman J (1993) The hidden roots of the African food problem: looking within the rural household. In: Folbre N, Bergmann B, Agarwal B, Floro M (eds) Women's work in the world economy. New York University Press, New York, pp 82–103

Larrea ML, Maldonado P (2005) Circuito Espacial de Producción de la Floricultura de Exportación, Caso Ecuatoriano. Programa EcoSalud CEAS/CIID, Quito

Levine JA, Weisell R, Chevassus S, Martinez CD, Burlingame B, Coward WA (2001) The work burden of women. Science 294:812

Loewenson R (2000) Occupational safety and health for women workers in agriculture. International Labour Organization. Labour Educ 1–2(118–119):35–45

McGuire J, Popkin B (1990) Helping women improve nutrition in the developing world: beating the zero sum game. World Bank technical paper no. 114. World Bank, Washington, DC

Mead M (1976) A comment on the role of women in agriculture. In: Tinker I, Bo Bramsen M (eds) Women and world development. Overseas Development Council, Washington, DC, pp 9–11

Meinzen-Dick R, Quisumbing A, Behrman J, Biermayr-Jenzano P, Wilde V, Noordeloos M, Ragasa C, Beintema N (2010) Engendering agricultural research. Discussion paper 00973. International Food Policy Research Institute, Washington, DC

Meinzen-Dick R, Behrman J, Menon P, Quisumbing A (2011a) Gender: a key dimension linking agricultural programs to improved nutrition and health. 2020 conference brief 9. International Food Policy Research Institute, Washington, DC

Meinzen-Dick R, Quisumbing A, Behrman J, Biermayr-Jenzano P, Wilde V, Noordeloos M, Ragasa C, Beintema N (2011b) Engendering agricultural research, development, and extension. Monograph 176. International Food Policy Research Institute, Washington, DC

Moser C, Moser A (2005) Gender mainstreaming since Beijing: a review of success and limitations in international institutions. Gend Dev 13(2):11–22

Njuki J, Mapila M, Kaaria S, Magombo T (2008) Using community indicators for evaluating research and development programmes: experiences from Malawi. Dev Pract 18(4):633–642

Oxfam (2007) Biofueling poverty: why the EU renewable-fuel target may be disastrous for poor people. Oxfam Briefing Note, Oxford

Paz-y-Mino C, Bustamante G, Sanchez ME, Leone PE (2002) Cytogenetic monitoring in a population occupationally exposed to pesticides in Ecuador. Environ Health Perspect 110(11):1077–1080

Peterman A, Quisumbing AR, Behrman J, Nkonya E (2011) Understanding the complexities surrounding gender differences in agricultural productivity in Nigeria and Uganda. J Dev Stud 47(10):1482–1509

Quisumbing AR (2011) Do men and women accumulate assets in different ways? Evidence from rural Bangladesh. Discussion paper 1096. International Food Policy Research Institute, Washington, DC

Quisumbing AR, Maluccio J (2003) Resources at marriage and intrahousehold allocation: evidence from Bangladesh, Ethiopia, Indonesia, and South Africa. Oxf Bull Econ Stat 65(3):283–328

Quisumbing AR, Otsuka K, Suyanto S, Aidoo JB, Payongayong E (2001) Land, trees, and women: evolution of land tenure institutions in Western Ghana and Sumatra. Research report 121. International Food Policy Research Institute, Washington, DC

Rao A, Kelleher D (2005) Is there life after gender mainstreaming? Gend Dev 13(2):57–69

Schroeder RA (1993) Shady practice: gender and the political ecology of resource stabilization in Gambian garden/orchards. Econ Geogr 69(4):349–365

Smale M (1995) Maize is life: Malawi's delayed green revolution. World Dev 23(5):819–831

Suda CA (1996) Household labor use and changes in gender roles on small farms in Ndhiwa Division, Western Kenya: the challenge of comparing the contributions of different workers. Institute of African Studies, Nairobi

Unnevehr L, Stanford ML (1985) Technology and the demand for women's labor in Asian rice farming. In: International Rice Research Institute (ed) Women in rice farming: proceedings of a conference on women in rice farming systems. Gower Publishers, Brookfield, pp 1–20

von Braun J, Hotchkiss D, Immink M (1989) Nontraditional export crops in Guatemala: effects on production, income and nutrition. Research report 73. International Food Policy Research Institute, Washington, DC

World Bank (2001) Engendering development through gender equality in rights, resources, and voice. World Bank policy research report. Management 1: report no. 36546-MW. Washington, DC: World Bank.

Chapter 16
Enhancing Female Participation in Agricultural Research and Development: Rationale and Evidence

Nienke Beintema

Abstract Enhancing women's participation in agricultural research in developing countries can be an effective strategy for making agricultural research and development (R&D) systems more gender-aware. This chapter reviews the evidence on the trends in women's participation in agricultural research with more detailed analysis of Sub-Saharan Africa, for which more detailed information is available. The author makes use of the Agricultural Science and Technology Indicators (ASTI) initiative, which is one of the few sex-disaggregated data sources on agricultural researchers in developing countries. In developing countries, less than one out of four researchers is a woman, although large differences exist across countries. The share of women employed in agricultural research and development has been increasing in most countries, but their share disproportionately declines on the higher rungs of the career ladder. The chapter summarizes the various general human resource challenges in agricultural R&D that developing countries face, specifically in Sub-Saharan Africa, especially the challenges women face prior to and during their science careers. In addition to reviewing important data, the author makes an argument for why increased participation of women in science is important in the developing world.

Keywords Gender • Agricultural research and development (R & D) • Science and technology (S & T) • Agricultural research staff • Female participation

---

N. Beintema (✉)
Agricultural Science and Technology Indicators (ASTI) Initiative, Environment and Production Technology Division, International Food Policy Research Institute, 2033 K Street, Washington, DC 20006, USA
e-mail: n.beintema@cgiar.org

A.R. Quisumbing et al. (eds.), *Gender in Agriculture: Closing the Knowledge Gap*, DOI 10.1007/978-94-017-8616-4_16,

16.1 Introduction

Female farmers play an important role in agricultural production in many developing countries, but agricultural research and development (R&D)—as well as extension and higher education—are disproportionately led by men. Greater representation of women in agricultural research will offer a wider diversity of insights and perspectives to help research institutes to more fully address the unique and pressing challenges of both female and male farmers (IAC 2004a; EC 2008). The number of female scientists working in science and technology (S&T) research in industrialized and developing countries has increased substantially in recent decades, but the participation of women remains low in most countries. Moreover, most of the increases have occurred at the lower levels of R&D systems. Women are less represented in high-level research and management positions and as a result have less influence in policy- and decisionmaking processes (Huyer and Westholm 2007; IAC 2004a; Meinzen-Dick et al. 2011).

Sex-disaggregated data on participation in R&D, both over time and across countries, are extremely important for national and international decisionmakers, including research and human resources managers. Such data remain scarce, however. Even where available, information is often unusable due to inconsistent approaches, methodologies, samples, and time frames (Huyer and Westholm 2007; Otchet 2007).[1] In many countries, decisionmakers tend to give a low priority to gender policies and, therefore, sex-disaggregated data are not collected, which, in turn, results in lack of awareness of the importance of addressing the gender gap in S&T (Andres 2011). A number of international organizations have emphasized the need for sex-disaggregated data on capacity and have increased their own collection efforts.

One of the few sex-disaggregated data sources on agricultural researchers in developing countries is the Agricultural Science and Technology Indicators (ASTI) initiative,[2] which, as part of its overall data collection activities on agricultural R&D investments and capacities, has aimed to fill the data gap, at least for the agricultural sector. This chapter provides an assessment of gender staffing in agricultural R&D, based on the ASTI datasets. It will particularly focus on Sub-Saharan Africa for which more detailed information is available. The ASTI data will be linked to other available sources of information on female participation. The chapter addresses the challenges women face in conducting agricultural R&D and summarizes the general human resource challenges in agricultural R&D that developing countries face, specifically in Sub-Saharan Africa, as well as the challenges women face prior to and during their science careers.

---

[1] To facilitate cohesion in the collection of such statistics, the United Nations Educational, Scientific, and Cultural Organization (UNESCO) developed a tool kit on gender indicators in science, engineering, and technology, which outlines the issues and provides theoretical and methodological frameworks and guidelines, including a number of case studies (Huyer and Westholm 2007).

[2] ASTI comprises a large network of national, regional, and international partners, managed by the International Food Policy Research Institute (IFPRI). For more information see www.asti.cgiar.org.

16.2 Why Increased Participation of Women in Science Is Important

The emergence of knowledge-based, technologically advanced societies has led to greater recognition of the importance of S&T and a greater need for human, institutional, and infrastructural capacity in these disciplines. This is especially true in developing countries, which are shifting from simply being consumers of developed-country technologies to actually adapting those technologies and developing their own. Ultimately, countries that lack appropriate S&T capacity will fall behind (IAC 2004a; Huyer and Westholm 2007), and, unfortunately, capacity in many developing countries is limited due to lack of training and other opportunities. In addition, student enrollments in the fields of science and engineering have declined in a number of countries since the early 1990s. This situation has been exacerbated by high levels of human resource mobility from developing to developed countries, or from science to non-science and technical sectors. Attracting more women into science and engineering would be a highly beneficial way of reducing these shortages (Huyer and Westholm 2007). This lack of capacity has been a particular problem in agricultural research and higher education in Sub-Saharan Africa, where appropriate capacity is a necessary foundation for economic growth and to ensure food security (IAC 2004b; World Bank 2007).

In addition, women play an important role in S&T, not only as participants and implementers, but also as beneficiaries (Huyer and Westholm 2007). The InterAcademy Council (IAC) report on Women in Science (2004a) points out that greater participation in S&T by women will provide more diverse skills, experiences, perspectives, and working styles within the workforce. Higher rates of female participation in S&T systems can enhance the quality and competitiveness of research and innovation (EC 2008). The participation of women in S&T has, in fact, increased since the mid-1990s (including the numbers of girls being educated in S&T fields), but such increases have mostly occurred at the lower levels of S&T systems (Huyer and Westholm 2007; IAC 2004a).

Strengthening Africa's agricultural research capacity requires more than just increasing the number of women participating in absolute terms; it requires more women in senior, decisionmaking roles. Female farmers play an important role in African agriculture and addressing the needs of these farmers requires increased participation by female scientists, professors, and senior managers. Women also have different insights, which will support research institutes in more fully addressing the unique and pressing challenges of both female and male farmers in Africa (Goh et al. 2008).

16.3 Female Participation in Agricultural Research Globally

Since the early 1990s, ASTI has collected sex-disaggregated data on professional agricultural scientists by highest degree and institute type for more than 60 developing countries. These data are linked with the sex-disaggregated human resource

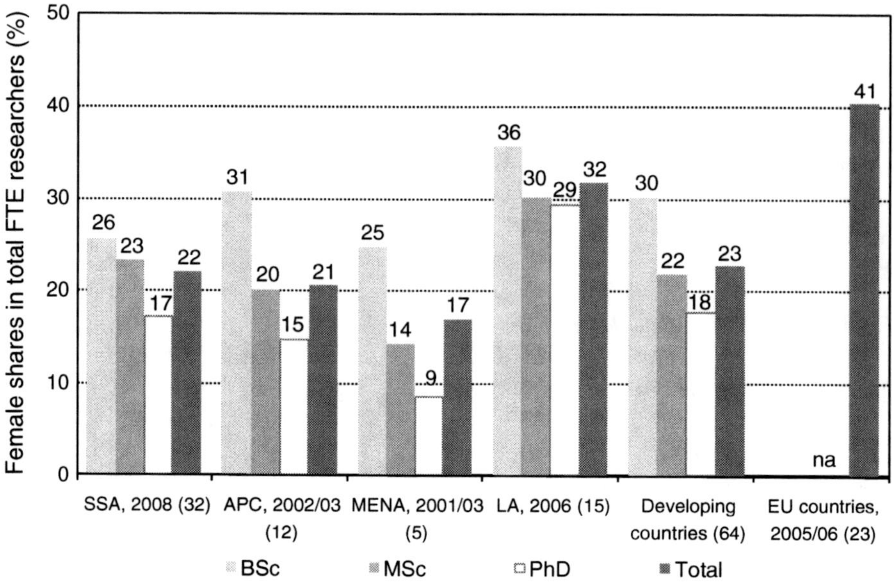

Fig. 16.1 Average female scientist shares in professional staff, by degree in 64 developing countries, 2001–08 (Sources: Calculated by author from ASTI datasets and EC (2009b). Notes: The number of countries included in the regional totals is shown in parentheses. SSA indicates Sub-Saharan Africa; APC, the Asia-Pacific region (here excluding China); MENA, the Middle East and North Africa; LA, Latin America; and EU, European Union member countries and associated non-EU countries. For countries included in the regional samples, see Beintema and Stads (2008 and 2011), EC (2009b), Stads and Beintema (2009). Data are presented in full-time equivalent (FTE) researchers)

indicators by field of science collected through the European Commission/Eurostat (Fig. 16.1). The human resource data presented in this chapter include only researchers in public agricultural R&D agencies. Public agricultural research comprises government agencies, higher education agencies, and nonprofit institutions. The data here, therefore, exclude the private for-profit sector. Furthermore, all data are based on full-time equivalent staffing, or FTEs, which take into account the proportion of time researchers spend on R&D activities (see Box 16.1 on ASTI's data collection procedures).

Currently, less than one out of four agricultural researchers in a sample of 64 developing countries were female. Across regions average shares of female scientists ranged from 17 % for the Middle East and North Africa, to 21 % and 22 % for Asia-Pacific countries and Sub-Saharan Africa, respectively, to 32 % in Latin America and the Caribbean.[3] Unsurprisingly, the share of females was higher in

---

[3] By way of comparison, the share of women in the total number of scientists employed at the 15 centers of the Consultative Group on International Agricultural Research (CGIAR) was 26 percent in 2008; an increase from the 20 percent in 2003 (Meinzen-Dick et al. 2011). The most recent UNESCO report shows that, in the 89 countries for which data were available, women account for an average of 27 percent of the total number of scientists, measured in head counts (UIS 2006).

Box 16.1 ASTI's Data Collection Procedures and Definitions

ASTI datasets are collected and processed using internationally accepted definitions and statistical procedures for compiling R&D statistics developed by the Organisation for Economic Co-operation and Development (OECD) and United Nations Educational, Scientific, and Cultural Organization (UNESCO). This is important to ensure the comparability of ASTI's datasets with others. For each country in which ASTI is active, the research team typically works with the national agricultural research institute or, in a few cases, with consultants. These national partners coordinate the implementation of the survey round, and coauthor and co-publish the resulting country briefs/notes and related fact sheets. To this end, three different survey forms were developed: one for government agencies and nonprofit institutions, one for higher education agencies, and one for the private sector. All forms have different sets of questions, with those for government agencies and nonprofit institutions requesting the most detail. Each form consists of four sections: (1) institutional details; (2) human resources, (3) financial resources; and (4) research focus. Over the years, the survey forms have been amended and improved, based on experiences and consultations with partners during the various national survey rounds. Agricultural research, as defined for ASTI purposes, includes research on crops, livestock, forestry, fisheries, natural resources, the use of agricultural inputs, and the socioeconomic aspects of primary agricultural production.

The human resources information collected through ASTI comprises the number of researchers and technicians by degree level and by sex and support staff by various categories. When reporting the survey results, ASTI bases its calculations of human resources on full-time equivalent staffing, or FTEs, which take into account the proportion of time researchers spend on R&D activities. University staff members, for example, spend the bulk of their time on nonresearch-related activities, such as teaching, administration, and student supervision, which need to be excluded from research-related resource calculations. As a result, four faculty members estimated to spend 25 % of their time on research would individually represent 0.25 FTEs and collectively be counted as 1 FTE.

the lower-degree qualification levels. An average of 30 % of the researchers with BS degrees was female, compared with 22 % and 18 % with MS and PhD degrees, respectively. By way of comparison, women accounted for 41 % of total agricultural research staff employed in the government sector in the 23 countries of the European Union in 2006 (EC 2009a), almost twice the 64-country average for the developing world.[4]

[4]The Women and Science Group of the European Union (EU), in collaboration with Eurostat and the national representatives of the *Helsinki Group*, developed a system of internationally comparable,

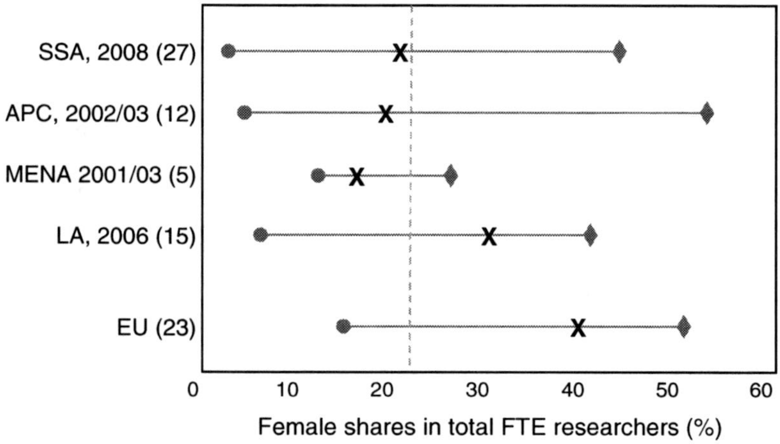

Fig. 16.2 Variation in female shares of agricultural research staff across countries within regions, 2003–08 (Sources: Calculated by author from ASTI datasets and EC (2009b). Notes: See Fig. 16.1; X denotes the regional average)

Unsurprisingly, large variations exist across countries within regions (Fig. 16.2). In the 32 Sub-Saharan African sample countries, female researchers represented between 3 % and 43 % of total agricultural research staff in 2008 (see next sections of this chapter for country-level details). The spread was even broader in the Asia-Pacific region. In 2002, female scientists in Pakistan and Nepal constituted only 6 % and 9 % of total research staff, respectively, while in Myanmar more than half of the agricultural researchers employed in 2002 were women. In contrast, differences across countries in the Middle East and North Africa are less pronounced—from 13 % in Jordan to 28 % in Tunisia—although this is partly due to the low sample size of only five countries. Noticeable, however, is the low female shares in PhD-holders in the three Middle Eastern countries in the sample: Women accounted for only 5 % of the total number of researchers with PhD degrees in Iran, Jordan, and Syria. The share of female researchers in government agricultural research agencies within Europe varied as well, ranging from 16 % in Cyprus to more than 50 % in Bulgaria, Estonia, Lithuania, Malta, and Portugal.

16.4 Female Participation Trends in Sub-Saharan Africa

More detailed information on women's participation in agricultural research is available for Sub-Saharan Africa, linking overall survey data for benchmark years 2000–01 and 2008 for 23 and 32 countries, respectively, and more detailed

---

sex-disaggregated science indicators. The group has published a series of reports under the title *She Figures* (EC 2009b).

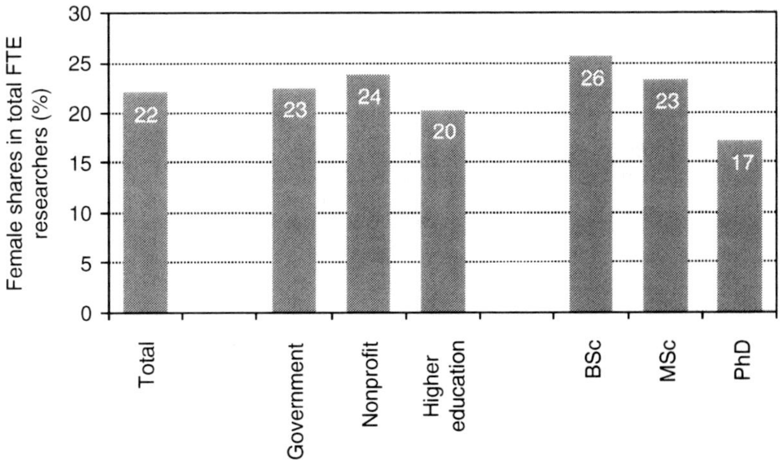

Fig. 16.3 Share of women in total agricultural R&D staff for 32 countries, 2008 (Source: Calculated by author from ASTI datasets. Note: See Fig. 16.5 for country coverage)

sex-disaggregated capacity indicators covering 15 African countries, which was conducted for the year 2007–08 in close collaboration with the African Women in Agricultural Research and Development (AWARD) program.[5]

In 2008, an average of 22 % of the total agricultural researchers, in FTE terms, in a sample of 32 countries were female (Fig. 16.3). Surprisingly, the average female shares were lower at the higher education agencies involved in agricultural research compared to the government and nonprofit sectors. Logically, these averages conceal a wide variation of shares across agencies within the three qualification levels. Unsurprisingly, the share of women was higher for total agricultural researchers that were trained up to the BSc level (26 %) compared to MSc- and PhD-holders (23 % and 17 %, respectively).

Female scientists are also consistently less likely to have advanced degrees than their male counterparts. In 2008, of the total female scientists in the 32 African countries, an average of 24 % held PhD degrees, while 32 % of the male researchers were trained to the PhD level.

The share of female scientists in agricultural research differed substantially from one country to the other (Fig. 16.4), but, in general, countries in Southern Africa have relative higher female shares than those in West and Central Africa. In 2008, female researchers represented at least 30 % of all agricultural research staff in Eritrea, Madagascar, Mauritius, South Africa, Sudan, and Zimbabwe. In contrast, of the total agricultural researchers employed in Ethiopia, Guinea, Mauritius, Niger, and Sierra Leone that year, only 3–8 % were female.

[5] AWARD, managed by the Gender and Diversity (G&D) Program of the Consultative Group on International Agricultural Research (CGIAR), offers competitive two-year fellowships focusing on building capacity in science, mentoring, and leadership to high-performing female African scientists (see http://awardfellowships.org/).

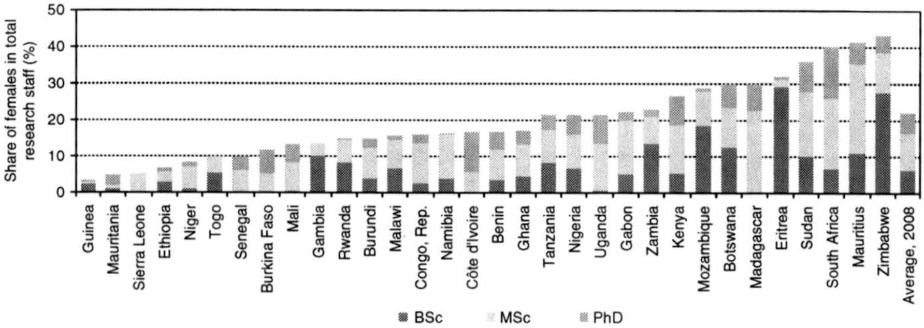

Fig. 16.4 Female scientist shares, by country and degree level, 2008 (Source: Calculated by author from ASTI datasets)

With respectively two-thirds and one-half of the total (that is female and male) agricultural research staff qualified to the BSc level, Eritrea and Zimbabwe are among the African countries with relatively the lowest qualification levels (Beintema and Stads 2011). As a result, most women scientists in these two countries were only trained to the BSc levels, despite their overall high female-scientist share.

In 2008, Sub-Saharan Africa as a whole employed more than 12,120 FTE researchers, compared with 9,824 FTEs in 2001. Unsurprisingly, absolute levels of public agricultural R&D staffing varied. In 2008, Nigeria employed more than 2,000 FTE researchers, followed by Ethiopia, Sudan, and Kenya with each employing more than 1,000 FTEs. In contrast, 9 of the 32 countries in the ASTI sample had less than 100 FTE researchers that same year. The capacity growth during 2001–2008 was largely driven by only a handful of countries. Nigeria was the main driver of the regional growth, accounting for 724 of the region's 2,285 additional FTE researchers. Ethiopia, Kenya, and Sudan also reported significant increases in staff. In contrast, the total number of agricultural research staff declined in South Africa, with 140 FTEs during the same period (Beintema and Stads 2011).

Figure 16.5 further illustrates the increased capacity in absolute terms and as annual growth rates for the 23 countries for which both 2001 and 2008 sex-disaggregated researcher data were available. In absolute terms, the pool of male agricultural FTE researchers continued to increase faster than the pool of female researchers. But during this period, far more women with MSc degrees were added compared to male researchers with MSc degrees. In fact, in quite a number of countries, the total number of male FTE researchers with MSc degrees declined from 2001 to 2008. Some, although not all, of these declines can be attributed to male staff obtaining PhD degrees or leaving their organizations. Although in absolute terms, the increase of female scientists was lower, in relative terms, the growth outpaced that of the male researchers. The total number of female FTE researchers increased by 5 % per year from 2001 to 2008—more than twice the annual rate for the male population (2 %).

The growing shares of professional women employed in agriculture indicate that the gender gap in African agricultural sciences may be narrowing, especially in

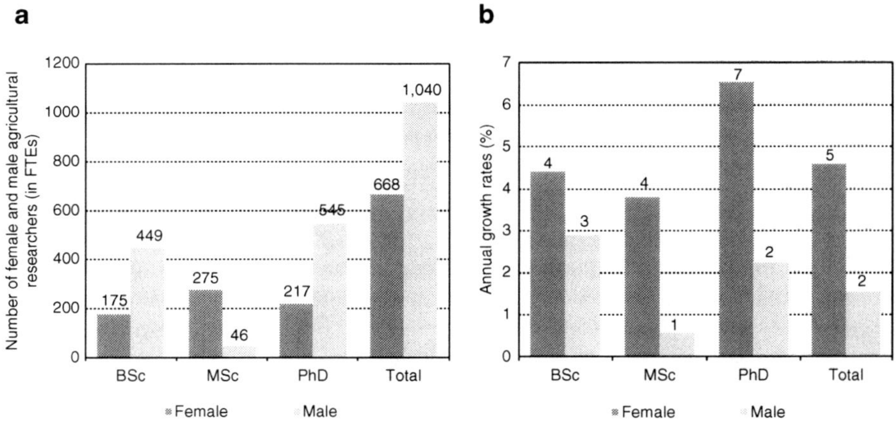

Fig. 16.5 Growth in total female and male scientists for 23 African countries, in 2001–2008. (**a**) Growth in absolute numbers. (**b**) Annual growth rates (Source: Calculated by author from ASTI datasets. Notes: Data for a number of countries are for 2000. The trends in the figure are different from those reported in Beintema and Di Marcantonio (2010) and are the combined results of a larger agency/country sample and the exclusion of professional staff with degrees holding support staff positions (e.g., research assistants and technicians))

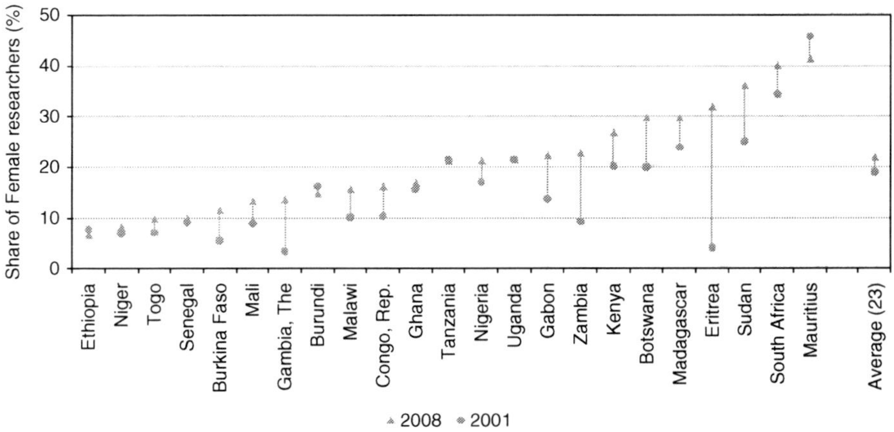

Fig. 16.6 Growth in female scientist shares, by country for 23 African countries, 2001–2008 (Source: Calculated by author from ASTI datasets. Note: See Fig. 16.5)

southern Africa. In relative terms, the share of women in total professional staff increased from 19 % in 2001 to 22 % in 2008. But this increase varied considerably across the 23 countries under study (Fig. 16.6). It is noteworthy that in most countries with relatively higher shares, the percentage of women in agricultural research continued to grow. For example, Sudan and South Africa's shares increased from

24 % and 34 % to 36 % and 40 %, respectively. In contrast, Ethiopia, Niger, Senegal, and Togo, which had relatively low shares of women in total research capacity, also had little growth.

Despite the overall growth in agricultural R&D capacity across the region, the average levels of staff qualifications have deteriorated since 2001. In 2008, 27 % of total agricultural researchers held BSc degrees, an increase from 24 % in 2001. In a number of countries, such as Zambia, Nigeria, and Botswana, this increase was more significant than in others.

Future trends in female participation in agricultural research are influenced by current female student enrollment and graduation levels. An increasing number of women have been enrolling in higher education, not only in Sub-Saharan Africa, but also in other regions of the world (UIS 2006). This also appears to be the case in agricultural sciences, but unfortunately no sex-disaggregated data are available on student enrollments and graduations in agricultural sciences over a longer period. Beintema and Di Marcantonio (2010) presented sex-disaggregated student population data for 28 higher education agencies in agricultural sciences. On average, about one-third of the students enrolled in 2007 were female and most of these were enrolled in BSc studies (83 %), whereas only 13 % and 4 % of women were enrolled in MSc and PhD studies, respectively. This distribution was similar for male students and reflects the reality that many faculties and schools lack, or have only small, PhD programs. There were relatively more women enrolled in agricultural sciences in three major agricultural faculties in South Africa that were included in the sample. In contrast, less than one-fifth of the agricultural student populations in Ethiopia, Ghana, Senegal, and Burundi comprised female students.

Although the share of female students in the total student population in Ethiopia was relatively low, it was much higher than the proportion of females in agricultural sciences (18 % compared with 7 %). In South Africa, Malawi, and Nigeria, the number of female students was also substantially higher than the number of female researchers, indicating potential for these countries to increase their female capacity in the near future, given appropriate incentives to attract new graduates into the workforce. Of course, this also depends on the role of women in the society, as well as the institutional environment for female researchers.

16.5 Age Structure and Seniority Levels of Female Scientists in Sub-Saharan Africa

Many research agencies in Sub-Saharan Africa have difficulties in retaining staff as a result of the relatively poor remuneration and conditions of service. Even in countries with better training opportunities, staff that obtain higher degrees often leave the research agencies for other employment opportunities with better salaries and conditions of services in the private sector, international organizations, or even abroad. A major and growing concern in many countries, particularly in

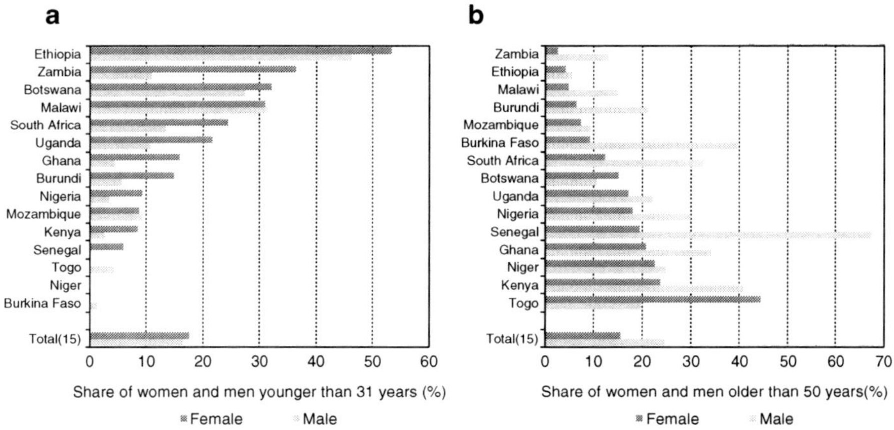

Fig. 16.7 Sex disaggregated proportions of professional staff aged younger than 31 and older than 50 years, 2008. (**a**) Share of women and men younger than 31 years. (**b**) Share of women and men older than 50 years (Sources: Calculated by author from ASTI-AWARD datasets. Notes: Data for a number of countries are for 2008. The pool of professional staff in some of these age groups is extremely small. For example, neither the two young staff in Burkina Faso nor the three in Togo were female, and no agricultural staff in Niger were aged 30 years or younger)

West and Central Africa, is a rapidly aging researcher pool, many of whom will retire within the next 10 years. Senegal, for example, has one of the oldest pools of scientists in West Africa. In 2007, about 60 % of the total researchers were 51 years or older. In Ghana, Burkina Faso, and Kenya, more than one-third of the research staff was 50 years or older. In contrast, other countries are experiencing a large influx of young scientists (many holding only BSc degrees), often after prolonged recruitment restrictions were lifted. One such example is Ethiopia, where close to one-half of the research staff was 30 years or younger (Beintema and Stads 2011; ASTI-AWARD 2008).

On average, the majority of women (as well as men) were aged between 31 and 50 years (Fig. 16.7). Less than one-fifth of the female and male researchers in a 15-country sample were 30 years or younger. As usual, there is a wide variation across countries. In general, however, women are relatively more represented in the younger age group, and underrepresented in the older age group. This is particularly true in Ethiopia, where more than half the female researchers were 30 years or younger. The case is similar for the male staff, although to a slightly lesser extent (44 %). More than one-fifth of the female researchers in Botswana, Malawi, Zambia, South Africa, and Uganda were younger than 31 years old. With the exception of Malawi, these proportions were considerably higher than the corresponding figures for men in this age group. In contrast, about two-thirds of the male agricultural researchers in Senegal were 51 years or older, while female researchers in the age group accounted for less than 20 % of the country's total female research pool.

Huyer and Westholm (2007) point out that women entering the R&D system encounter two levels of segregation. First, "horizontal segregation" measures the concentration of women in certain disciplines and sectors. Evidence shows that female participation levels are higher in areas such as biology and other life and social sciences ("soft science") and much lower in areas such as physics and engineering ("hard science"). It is difficult to make a clear distinction between "soft" and "hard" fields of science within agricultural R&D, although some fields are clearly related to engineering and others are clearly related to life or social sciences. The proportion of women in total agricultural professional staff for the 15-country sample that were engaged in food and nutritional science was very high at 44 %. There were also relatively large numbers of female professional staff trained in biodiversity (30 %), molecular biology (31 %), and agricultural economics (24 %). In contrast, female shares were particular low for disciplines such as water and irrigation (8 %), forestry (15 %), and soil science (13 %).

The other level of segregation is "vertical segregation" and involves the over-representation of women at lower levels of the professional hierarchy; this is true in industry and education, as well as in S&T. Hence, women are less represented in high-level research and management positions compared with their male colleagues. As a result, women have less influence in policy- and decisionmaking processes (Huyer and Westholm (2007). In 2008, 14 % of the total research staff in management positions were female for the 15-country average. This is considerably lower than the overall proportion of women research staff (24 %) and indicates a high level of vertical segregation in African agricultural research. Only in Ethiopia was this share about equal to the share of female professional staff, but Ethiopia has one of the lowest shares of female scientists in Sub-Saharan Africa. Although South Africa has one of the highest, the share of female staff in management positions was similar to the 15-country average of 14 %. As in other sectors, the low proportion of women in management means that women have less influence in policy- and decisionmaking processes, which can further result in biased decisionmaking and priority setting, which may result in further neglect of the needs of women farmers.

A "scissor" diagram is commonly being used to illustrate the way the gender gap progresses along various points of an R&D career path, starting with higher education (EC 2009b). This scissor pattern occurs because generally at BSc and MSc levels, more women than men enroll and graduate, but from PhD level onward, this trend reverses. In the following stages in the career path, the proportion of women further declines, reaching its lowest level in management. These general figures include all academic fields, even when women are underrepresented. But female participation in BSc and MSc agricultural courses at higher education agencies for a 10-country sample is still considerably lower than that of men in Sub-Saharan Africa. As a result, the diagram does not follow the classic scissor trend in the case of female participation in agricultural research (Fig. 16.8).

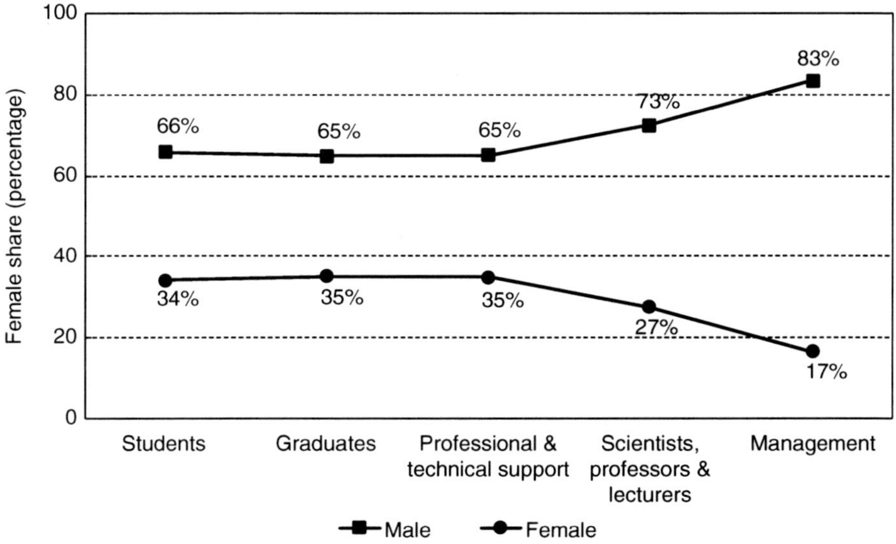

Fig. 16.8 Shares of women and men at different points along the career path, 2008 (Sources: Calculated by author from ASTI-AWARD datasets. Notes: Data for a number of countries are for 2008. Burkina Faso, Burundi, Ethiopia, Niger, and Togo were excluded because data on student enrollments and graduations were either unavailable or incomplete. Scientists, professors, and lectures also include assistant professors and senior lecturers; management includes directors, deans, and department heads. When all 15 countries are included, the share of women in management positions falls to 14 %)

16.6 Human Resource Challenges in Agricultural R&D, Particularly Faced by Women

16.6.1 General Human Resource Challenges

Agricultural R&D systems in developing countries are facing a number of general human resource challenges, particularly so in Sub-Saharan Africa:

- Despite the overall growth in the total pool of agricultural researchers, average qualifications levels have deteriorated in a number of African countries since the turn of the millennium. This was mostly the result of a large influx of young scientists with bachelor degrees, often after prolonged recruitment restrictions were lifted. As a result, many institutions are unable to provide adequate training opportunities and mentoring by senior researchers.
- Salaries and conditions of service at the main agricultural research agencies are often poor. Many agencies have lost a large number of researchers to the private

sector and, occasionally, to universities, because they offer higher-paid jobs with better conditions.

- Another key challenge is the rapidly aging pool of scientists, many of whom will approach retirement age within the next decade. This situation is particularly grave in West Africa, but also a concern in a number of countries in East and Southern Africa as well.
- In addition, student enrollments in agricultural sciences (and science and technology in general) appear to have declined in a number of countries since the early 1990s.

16.6.2 Challenges Specific to Female Scientists

The integration of women in S&T, which have been traditionally staffed by men, poses significant workplace and societal challenges, which are a combined effect of the institutional, social, and cultural environments of women. Based on Meinzen-Dick et al. (2011) and Andres (2011), these key challenges are

- Girls have unequal access to basic education in developing countries, which, in turn, results in low representation of young women in universities.
- Because of the traditional beliefs on the role of females in the societies of many developing countries, women are often seen as unsuited for sciences.
- When women do overcome these barriers at the secondary and higher education levels, they often start their science career at the time of childbearing. Balancing work and a family is one of the more substantial challenges female scientists face. The perception in many countries that the role of a woman is that of a mother and her domain is the family limits female scientists' advancements in research and higher education agencies.
- Female scientists potentially face gender discrimination in various levels of their employment and career. Formal and informal networks, for example, are mostly dominated by men, which makes entering these networks difficult for women. Networks are important to integrate into collaborative research programs, lobby for research grants, and improve publication records. Recruitment and promotion committees are also mostly populated by senior researchers and, are, therefore, often not gender balanced. This can potentially negatively affect their performance evaluation and furthermore their promotion chances to more senior positions. Finally, women, in general, receive less payment for the same work than their male colleagues with similar qualifications and responsibilities.

In many countries, various institutional reforms and policies have been initiated that promote gender equality. This was further facilitated by the inclusion of gender equality in the Millennium Development Goals (Andres 2011). A detailed assessment of these policies and reforms is, however, only available for a handful of countries. The governments of Ghana and Nigeria, for example, established a Ministry of Women's Affairs as well as national councils for women. Specifically related to S&T,

Ghana put in place affirmative action to facilitate both women's admission to universities and the participation of girls in science and mathematics clinics that are set up at the high-school levels. Other African countries have also made significant progress in promoting gender equality (Beintema and Di Marcantonio 2010).

16.7 Conclusion

Evidence, although scare and spotty, show that participation by women in overall S&T has increased in recent years; this has been the case for agricultural R&D as well. In Sub-Saharan Africa, for example, growth in total number of female agricultural researchers was more than twice the growth in total male research staff during 2001–2008. But because of the growing representation of women in entry-level positions, women in agricultural research and higher education are typically younger, have less advanced degrees, and are by definition overrepresented in lower positions and underrepresented in management positions compared with men. Furthermore, the overall quality of agricultural research staff (male and female) declined over the seven-year period, at least in some of the countries. This is a particularly worrisome trend in light of significant overall challenges that agricultural research capacity in Africa—and elsewhere in the developing world—faces.

The proportion of women studying agricultural sciences is actually larger than the share of female professional staff employed in agriculture, which is a positive indicator for the future, assuming that appropriate incentives can be provided to encourage these students to pursue careers in agricultural research, undertake higher degrees, and ultimately attain positions of seniority. Given that a large proportion of the current pool of students is only enrolled in BSc degrees, it is extremely important that MSc and PhD training programs be provided to ensure the quality of the future pool of researchers. National governments as well as donors and development banks seriously need to expand their investment in agricultural R&D. In addition, increased investments in agricultural higher education are needed to allow universities to increase the number and size of PhD and MSc programs and to improve curricula of existing programs (Beintema and Stads 2011).

Female participation is known to diminish with career advancement in S&T systems. Hence, women are less represented in high-level research and management positions compared with their male colleagues and, as a result, will have less influence in policy- and decisionmaking processes. Various actions should be put in place that will cause the gender gap in science in general, and agriculture specifically, to decline. At the family level, this will require a culture change away from the current stereotyping that science is not a career for females. More girls will need to stay in school, but they also need to be encouraged to choose science careers. This can only be achieved by continuous awareness campaigns of the benefits of providing education to girls and women at national and community levels. At the institutional level, access to role models and mentors, as well as building networks and support systems that support women, have been proven to be powerful tools to inspire (young) female scientists to remain in science. Finally, eliminating or at least lessening the payment gap may also attract more women into science, and agricultural sciences in particular (Andres 2011).

Although this chapter provides new insights into existing female and male capacity in African agricultural research and higher education, more research is needed to improve our understanding of underlying factors such as staff mobility, career paths, and the relationship between age distribution and professional levels of women and men. Furthermore, the sex-disaggregated capacity indicators collected for this study only reflect a certain point in time and are subject to fluctuations. Ongoing survey rounds, at least every 2–3 years, are necessary to maintain an accurate picture not only of women's participation in agricultural research and higher education in Africa, but also of the region's overall capacity (Beintema and Di Marcantonio 2010). Collecting and analyzing quantitative and qualitative information disaggregated by gender is important for policymakers, R&D managers, and development agencies to better understand gender issues within S&T and formulate policies that increase women's participation in agricultural R&D (Huyer and Westholm 2007; Andres 2011). A greater representation of women in agricultural research will offer a wider diversity of insights and perspectives to help research institutes to more fully address the unique and pressing challenges of both female and male farmers. Furthermore, attracting more women into science and engineering would be a highly beneficial way of reducing the capacity shortages (Huyer and Westholm 2007; Beintema and Di Marcantonio 2010).

Acknowledgments Parts of this chapter are revised sections from Beintema and Di Marcantonio (2010). The underlying datasets on which these reports were based were developed through (1) a comprehensive survey on sex-disaggregated capacity indicators conducted in 19 countries during 2008 as part as a partnership between ASTI and the African Women for Agricultural Research and Development (AWARD) program; (2) primary survey rounds on investment and human resources trends datasets conducted in 32 African countries during 2001–2003 and 2009–2010; and (3) survey results for various Asian and Latin American countries for earlier years. "The data for Sub-Saharan Africa have been updated to 2011 and will be released around mid-2014. The main findings and conclusions of this chapter, however, remained valid."

References

Andres JTH (2011) Overcoming gender barriers in science: facts and figures. Science and Development Network, 22 June. Available at http://www.scidev.net/en/editorials/women-in-science-still-an-untapped-resource-1.html?utm_source=link&utm_medium=rss&utm_campaign=en_editorials

ASTI–AWARD (Agricultural Science and Technology Indicators–African Women in Agricultural Research and Development) (2008) Gender-disaggregated capacity indicators in Sub-Saharan Africa, 2007/08. Unpublished surveys. International Food Policy Research Institute, Washington, DC, and CGIAR Gender and Diversity Program, Nairobi

Beintema NM, Di Marcantonio F (2010) Female participation in African agricultural research and higher education: new insights. Synthesis of the ASTI–AWARD Benchmarking Survey on gender-disaggregated capacity indicators. IFPRI discussion paper 00957. International Food Policy Research Institute, Washington, DC and CGIAR Gender Diversity Program, Nairobi

Beintema NM, Stads GJ (2008) Diversity in agricultural research resources in the Asia-Pacific region. International Food Policy Research Institute, Washington, DC and Asia-Pacific Association of Agricultural Research Institutions, Bangkok

Beintema NM, Stads GJ (2011) African agricultural R&D in the new millennium: progress for some, challenges for many. IFPRI food policy report. International Food Policy Research Institute, Washington, DC

EC (European Commission) (2008) Mapping the maze: getting more women to the top in research. EC (European Commission), Brussels

EC (European Commission) (2009a) She figures 2009. EC (European Commission), Brussels

EC (European Commission) (2009b) Women and science: gender difference, gender equality. http://ec.europa.eu/research/science-society/index.cfm?fuseaction=public.topic&id=27. Accessed May 2009

Goh A, Recke H, Hahn-Rollins D, Guyer-Miller L (2008) Successful women, successful science. Gender & diversity working paper 48. CGIAR Gender & Diversity Program, Nairobi

Huyer S, Westholm G (2007) Gender indicators in science, engineering, and technology: an information toolkit. Science and technology for development series. United Nations Educational, Scientific and Cultural Organization, Paris

IAC (InterAcademy Council) (2004a) Inventing a better future: a strategy for building worldwide capacities in science and technology. IAC (InterAcademy Council), Amsterdam

IAC (InterAcademy Council) (2004b) Realizing the promise and potential of African agriculture. Science and technology strategies for improving agricultural productivity and food security in Africa. IAC (InterAcademy Council), Amsterdam

Meinzen Dick R, Quisumbing A, Behrman J, Biermayr-Jenzano P, Wilde V, Noordeloos M, Ragasa C, Beintema N (2011) Engendering agricultural research, development and extension. Research monograph. International Food Policy Research Institute, Washington, DC

Otchet A (2007) Head counts and headaches: measuring women in science. A World of Science 5(2):21–23

Stads GJ, Beintema NM (2009) Public agricultural research in Latin America and the Caribbean: investment and capacity trends. ASTI synthesis report. International Food Policy Research Institute and Inter-American Development Bank, Washington, DC

UIS (UNESCO Institute of Statistics) (2006) Women in science: under-represented and under-measured. UIS Bulletin on Science and Technology Statistics Issue No. 3 (November). New York

World Bank (2007) Cultivating knowledge and skills to grow African agriculture. World Bank, Washington, DC

Chapter 17
Improving Gender Responsiveness of Agricultural Extension

Catherine Ragasa

Abstract This chapter makes the case for improving the gender-responsiveness of agricultural extension systems through consideration of (1) whether the gender of the extension agent affects the effectiveness of extension services; (2) whether both men and women receive extension advice; and (3) how extension services are delivered. As part of this, the author looks at issues surrounding quality and emphasis of extension services. The author then reviews the evidence on gender differences in access to formal extension agent visits and to other sources of extension information, and the factors that lead to women having lower access to extension services. At this point, the chapter examines the experience of programs and projects that aim to increase women's access to extension, with more detailed analysis of extension system reforms in India (ATMA model), Uganda (NAADS), Venezuela (privatization and decentralization), and Ethiopia (sectoral policies). Subsequently, the chapter reviews innovative literature on the use of Information Communication Technologies (ICTs) in extension. The concluding section examines lessons learned and key messages.

Keywords Gender • Agricultural extension services • Rural advisory services • Information and communication technologies (ICTs) • Extension agents

17.1 Introduction

Agriculture extension services (also known as agricultural advisory services) refer to the range of information, advice, training, and knowledge related to agriculture or livestock production, processing, and marketing provided by governments,

---

C. Ragasa (✉)
Development Strategy and Governance Division, International Food Policy Research Institute, 2033 K Street, N.W, Washington, DC 20006, USA
e-mail: c.ragasa@cgiar.org

A.R. Quisumbing et al. (eds.), *Gender in Agriculture: Closing the Knowledge Gap*, DOI 10.1007/978-94-017-8616-4_17,

nongovernmental organizations (NGOs), and other sources that increase farmers' ability to improve productivity and income. Agriculture extension is recognized by many governments and experts as a crucial element in increasing agricultural productivity (Feder et al. 2010; Davis et al. 2010). Extension systems have contributed to more rapid dissemination and adoption of Green Revolution technologies, particularly in irrigated areas. Various meta-analyses have shown that investments in extension have substantial short-term pay-offs and high returns (Swanson and Rajalahti 2010; Alston et al. 2000; Evenson 1997). However, a vast literature has emphasized many shortcomings of past approaches and delivery methods, including weak linkages among extension, research, and farmers; weak accountability to farmers and lack of incentives for extension workers to perform; financial sustainability problems; elite capture; and limited established impact on remote areas, female farmers, and the poor (Feder et al. 2010; Davis et al. 2010).

Agricultural extension systems use various forms of delivery, including individual or group visits; organized meetings; use of Information and Communication Technologies (ICT); or learning through demonstration plots, model farms, or farmer field schools (FFSs). The emphasis of agricultural extension has changed recently from disseminating new technologies and providing education to facilitating knowledge exchange, which involves considerable challenges for the empirical analysis of advisory methods since the facilitation role is difficult to measure. The variety of delivery mechanisms and approaches pose challenges for empirical impact assessment due to attribution issues. Moreover, available impact assessment studies often leave unclear what main factors constrain the attainment of better outcomes: inappropriate advisory methods, low levels of training of advisory services agents, excessive centralization or poor management of the extension system, or other factors (Birner et al. 2006).

Although impact evaluation of agricultural extension services can be improved, a consistent message from reviews across different types of delivery methods and advisory approaches is that remote rural areas and poor rural women tend to be underserved. Evidence of gender bias in access to extension services and gender-specific constraints in adoption of new technologies is voluminous.[1] There are pockets of success, but most of them are boutique cases where replicability and scaling-up are major challenges. For those who have access, the quality of service and the appropriateness of information are often cited as issues. Moreover, resource-constrained farmers, particularly poor women farmers, often have limited land, cash, or labor time required to apply the new knowledge and information acquired.

This chapter makes the case for improving the gender-responsiveness of agricultural extension systems through consideration of (1) whether the gender of the extension agent affects the effectiveness of extension services; (2) whether both men and women receive extension advice; and (3) how extension services are delivered.

[1] Empirical studies reviewed here have looked at a wide range of technologies being promoted, from planting techniques, improved seed varieties, use of fertilizer, and disease control to postharvest techniques, group formation, marketing, sanitation, and hygiene.

After reviewing the evidence on gender differences in access to formal extension agent visits and to other sources of extension information, the chapter discusses the factors that lead to women having lower access to extension services. The chapter then examines the experience of programs and projects that aim to increase women's access to extension, with more detailed analysis of extension system reforms in India, Uganda, Venezuela, and Ethiopia, and the use of Information Communication Technologies (ICTs) in extension. The concluding section examines lessons learned and key messages.

17.2 Access to Extension Agent Visits

Although private-sector, nongovernmental organizations and community-based extension service delivery are starting to play a role in technology dissemination and extension in many contexts, public extension visits will remain a key medium in bringing in information and knowledge to farmers (World Bank/IFPRI 2010). Thus, the commonly used measure of households' access to extension services is contact with or a visit from agricultural extension agents or livestock officers. Studies show that access to extension services is consistently (and statistically in the majority of cases) lower for women as compared with men: 19 % for women versus 81 % for men in Malawi (Gilbert et al. 2002); 1.13 versus 2.03 contacts in Uganda (Katungi et al. 2008); 20 % versus 27 % in Ethiopia; 8–19 % of female-headed households versus 29 % of male-headed households in Karnataka, India; and the lowest figures in the recent empirical evidence, in Ghana, 0–2 % among female-headed households and 0.5–2 % of female spouses of male-headed households have access to extension agent visits, versus 11–12 % among male-headed households (World Bank/IFPRI 2010). For livestock-related extension services, results are slightly better: 0–24 % of female-headed households and 0–15 % of female spouses have access, versus 5–34 % for male-headed households in Ghana; and 71 % of female-headed households versus 78 % of male-headed households in Karnataka, India (World Bank/IFPRI 2010). In India, the role of dairy cooperatives as an important medium for providing and accessing livestock services accounts for the higher coverage for both women and men in the livestock sector. Dairy cooperative are slightly more gender-inclusive than other farmer organizations; for example, the male–female ratio of membership in dairy cooperative is 2.6 compared with 4.2 in other farmer-based organizations (FBOs) in Karnataka, India.

Findings by FAO's global survey on extension in 115 countries (Swanson et al. 1990) showed that women received only between 2 and 10 % of all extension contacts and a mere 5 % of extension resources worldwide. More recent evidence does not show any substantial improvements in gender equality in extension service delivery, despite decades of gender-mainstreaming efforts. Figures suggest that extension provision is low for both men and women, but more so for the latter, and this has major implications for attaining higher productivity and agricultural development.

When extension service access is more broadly defined to include community meetings or group meetings held by extension agents, gender differences persist: 0–6 % for female-headed households and 5–9 % for female spouses, versus 11–24 % for male-headed households in Ghana; and 11 % of women versus 28 % of men in Ethiopia (World Bank/IFPRI 2010). The differences between female-headed and male-headed households are statistically significant in both countries.

17.3 Access to Other Sources of Extension Services

17.3.1 Farmer Field Schools

Farmer field schools (FFSs) that use experiential, group-based learning processes are an increasingly used approach to education and extension, but existing studies show mixed impacts on women. In Kenya, Tanzania, and Uganda, female membership was 50 % in FFSs, and gender of the household head did not significantly affect participation in FFSs (Davis et al. 2010). Adoption of nearly all major technologies was significantly higher among the FFS farmers. Participation in FFSs increased income by 61 % in an analysis that pooled the three countries; and female-headed households benefited significantly more than male-headed households in Uganda. In Vietnam, Braun et al. (2006) show that women's participation has positive effects on women's leadership, incomes, and livelihoods. However, the authors acknowledged that other countries have lagged behind, which remains a major concern.

Van den Berg (2004) synthesizes 25 evaluation studies of integrated pest management (IPM) FFSs. Most studies focused on rice and measured immediate impact of the FFSs in terms of reduced pesticide use and changes in yields, reporting considerable reductions in pesticide use, with some studies also showing an increase in yields. Van den Berg and Jiggins (2007) review studies evaluating FFS and pest management, finding that FFSs have had additional benefits in addition to IPM, including facilitating collective action, leadership, organization, and improved problem-solving skills. Noting that discussions on the fiscal sustainability of FFSs should include considerations of who will pay for the externalities of pesticide use, they conclude that FFSs can be a cost-effective way of increasing farmers' skills and may contribute toward escaping poverty.

However, these impacts have not translated into changes beyond the local level; several studies suggest that FFSs are having limited or no effect on the agricultural sector's economic performance, environmental sustainability, and on dissemination of information by FFS participants to other farmers. There are also many questions about their sustainability (Davis 2006); cost-effectiveness (Quizon et al. 2001); and scaling up their impacts beyond the relatively small numbers that can be reached directly (Braun et al. 2006). For example, in Uganda, Isubikalu (2007) relates that although women outweighed men in number, men dominated most discussions and activities in FFSs. In Indonesia, the World Bank (2000) concludes that despite substantial increases in women's participation in FFSs (the average percentage of women trained in IPM field schools increased from 5.6 % at the beginning to 21.5 % in the last year of the project), findings of the gender studies were not always applied correctly and excessive pressure to meet gender targets resulted in participation of non-farmers (i.e., farmers' daughters who were students) in field schools in some provinces where women do not play an active role in farming.

17.3.2 Community-Based Organizations

Other potentially important sources of extension service are community-based organizations (CBOs). Despite their huge potential, the involvement of CBOs in extension services remains low in Ghana, Ethiopia, India, and Kenya; and these organizations are not typically inclusive (World Bank/IFPRI 2010; Davis and Negash 2007). There are considerable gender differences in participation and membership in CBOs. For instance, in Ghana, typically the male head was a member, while only 2–5 % of female spouses and only 3–7 % among female-headed households said they belonged to a CBO. In Ethiopia, gender differences exist in services by cooperatives: 24 % of men and 4 % of women belonged to some kind of cooperative; 13 % of men and 2 % of women belonged to agricultural cooperatives (World Bank/IFPRI 2010). In India, there are gender differences in participation in farmers' and dairy cooperatives, as cited above, but multivariate analysis suggests that gender of the household head does not play a significant role in determining the number of memberships in CBOs.

However, the type of group joined varied along gender lines. Women in India mainly joined self-help groups or women's groups, and men primarily joined forest groups, cooperative societies, and caste associations. Church groups, parent-teacher associations, and women's groups were the most important forms of social organization in all zones in Ghana, especially for women. Women in Kenya tended to participate more in revolving savings and loan associations, church groups, and women's groups, while men participated more in clan and water groups. These figures suggest that certain groups might be a better vehicle than farmer-based organizations (FBOs) for reaching women in rural areas. However, it may be best to work through traditional types of groups or institutions where women and the poor may feel more comfortable participating.

In Mozambique, it has been reported that the husband would typically become a member in a group to represent the household. Consequently, he would sell "his" crops through the group, even if the crops had been grown by the family or by his wife. This is particularly true among groups that allow only one member of the household to become a member. Even where membership is open to both spouses, only one may join, because most groups require members to pay monthly membership fees. In female-headed households, the female head can join the group in her own name, register, and sell the products as her own. Membership therefore has an inherently gendered subtext: the participation of women in farmer groups depends on their personal circumstances, such as age, education, and civil status.

Because of household and reproductive responsibilities, a wife rarely participates in a group in her own name or has her own voice in a mixed group; rather, she is likely to be represented by her husband who assumes membership as the head of the household, sells the joint production, and takes leadership responsibilities (Gotschi et al. 2009). Therefore, attention to the rules governing membership, and even to details such as the timing of meetings, will affect the extent of women's participation (also see Meinzen-Dick et al., Chap. 10).

17.3.3 Information and Communications Technologies

Both innovative ICT applications, such as mobile phones, print media, and Internet, as well as more traditional technologies, such as radio, satellite radio, and television, have been employed in extension service delivery. While comprehensive sex disaggregated data on ICT usage in developing countries do not exist, available data show that female participation in most aspects of ICT use falls behind men in most developing countries (Huyer et al. 2005). Due to unequal access to the factors that appear to enhance ICT access and use, such as income and education, women generally have less access to ICTs, and this pattern increases as the technologies become more sophisticated and expensive. A study by Gillwald et al. (2010) using empirical data across 17 African countries suggests that there are statistically significant gender differences in access to and use of Internet services for almost all the countries. No clear pattern of television usage can be seen for women or men, although the process and method are quite different for the two; women are as likely as men to watch television at friends', relatives', or neighbors' houses (Gillwald et al. 2010).

There are more promising patterns for rural women's access to radio, although men still are more likely to access and use it. Across 17 African countries, average hours listened to the radio per day were higher for men than women (Gillwald et al. 2010). In terms of mobile phones, the relative difference between men's and women's access and use is diminishing, especially in urban areas (Sorensen 2002). Zainudeen et al. (2008) find a strong gender divide in access to ICTs in India and Pakistan, a smaller gender gap in Sri Lanka, and none in the Philippines and Thailand, where mobiles are most pervasive. In rural areas, however, men are still more likely to own and have access to phones than women. This could be attributed to a number of factors, including illiteracy, cultural barriers, lack of availability of cash, and the age structure of inhabitants in rural areas as opposed to urban areas.

The perception that ICTs are a man's domain has often had an impact on the distribution of these technologies in the rural areas, making ICT-based programs highly gender biased or even gender blind. Thus, even though a lot of developing countries are making efforts to reach communication technologies to the rural poor, the women in these areas are still not benefitting from them. Even programs with good intentions miss reaching out to women because of lack of gender-sensitive planning (Awoyemi and Oluwatayo 2010).

17.3.4 Quality and Emphasis of Extension Services

Gender differences also appear in the type of technology and information disseminated to farmers. In Ghana, the outreach to women farmers is significantly lower than to men (from 0.33 to 0.89 women-to-men ratio) for almost all technology and information disseminated, except on livestock diseases and sanitation and hygiene (1.07–1.69 women-to-men ratio) (World Bank/IFPRI 2010). The gender gap is wider for production-related technologies and narrower for information related to marketing, livelihood strategies, and HIV/AIDS. In Senegal, researchers found that female extension agents have a significant positive impact on the dissemination of natural resource management knowledge among both men and women (Moore et al. 2001).

There is no conclusive evidence about gender differences in the quality of extension services or satisfaction derived from them. In India, households with assets and households with a female head were more likely than other households to report being satisfied with extension service delivery. In Ethiopia, individuals reported being satisfied with extension advice at staggering rates: 92 % of men and 94 % of women. Surprisingly, given these satisfaction rates, only 8 % of respondents stated that they had tried something new in the past 2 years, making it unclear what these farmers consider *satisfaction with extension agents* (World Bank/IFPRI 2010). Studies and expert opinion (Davis, personal communication) suggest that it is the lack of complementary inputs and credit that constrained farmers to try new technologies in Ethiopia, despite being satisfied with the extension services provided or believing that the extension agent knows all the best technologies and farming practices. In Uganda, the lack of resources (especially land and cash) to take advantage of information provided was said to have made information conveyed through the NAADS program in Uganda of limited use to poorer farmers, and in particular to women (Driciru 2008; Bukenya 2010). Moreover, the gap between farmers' priorities and the priorities of the NAADS program (which often were imposed on farmers) led to dissatisfaction among many farmers (Parkinson 2008).

17.4 Factors Contributing to Women's Low Access to Extension Services

Studies on agricultural extension have highlighted a number of challenges in reaching rural women. First, the perception bias that "women are not farmers" persists even though women are engaged in a wide range of agricultural activities. A recent report the World Bank and IFPRI (2010) finds strong evidence for a cultural perception that "women don't farm." In Ethiopia, because extension agents were evaluated on how many farmers they could get to adopt technology packages, they preferred to work with the household decisionmaker, who in a husband-wife household was always the male. Second, there are also perceptions that if extension services are given to a member of the family, then the information will trickle down to the rest of the household, including female members. However, men do not necessarily discuss production decisions with their wives or transfer extension knowledge to them, and if the extension information is tailored to men's crops or priorities, the information may not help women. Third, most extension services have been traditionally devoted to farmers who own land and who are willing and able to obtain credit and invest it in inputs and technological innovations. Since women often lack access to land or to other collateral with which to obtain credit, extension services unintentionally bypass women. Also, women have lower levels of formal education, and this hampers their ability to take part in extension activities requiring reading and arithmetic skills. Women farmers may also not be comfortable dealing with male extension workers or with the time and location of training, but this is culturally-specific. Due et al. (1997) found that in Tanzania, 40 % of women farmers preferred to work with female extension agents (compared to 26 % who preferred male extension agents; while the remaining 34 % had no preference). Female farmers stated that they preferred female extension agents as they were freer to discuss problems with them and they can better accommodate their time preferences for meetings than with male extension agents.

Relatively lower provision of extension services to women is also a reflection of the gender-sensitive policies, or lack thereof, at the ministry or service-provider level. In Ghana, a World Bank and IFPRI (2010) study shows that of the 70 agricultural extension agents surveyed, only 10 were female. Although about two-thirds of all extension agents stated that they received training that had a gender component, only 7 % reported receiving training that was entirely focused on gender issues. In Ethiopia, extension agents were overwhelmingly male, and cultural taboos restricted their interaction with women. In Karnataka, India, a survey conducted with front-line professionals responsible for extension service provision show a limited number of women (no female among 41 agricultural extension workers, 1 female out of 41 junior engineers, and 4 females out of 40 veterinary assistants). Lack of staff is the most frequently cited constraint and main problem: more than 70 % of agricultural extension workers serving the surveyed blocks are over 50 years old.[2]

In all three countries, there were no statistically significant differences between the types of technologies promoted by female and male extension agents, and female agents were just as likely to establish and run demonstration plots as their male counterparts. Female extension workers serve a higher proportion of female farmers than male agents (the average ratio of women to men is 1.3 for female agents and 0.53 for male agents) (World Bank/IFPRI 2010). This suggests that extension

[2] Conversely, with real commitment, it is possible to increase women's involvement in extension. In The Gambia, the proportion of female agricultural extension workers has increased from 5 % in 1989 to more than 60 % today, paying increased attention to women's livestock information needs, especially regarding small ruminant and poultry production (FAO 2003).

services from female extension agents are better targeted to female farmers. Interestingly, the World Bank/IFPRI (2010) study found that in Ghana, when male and female agents were asked about their biggest constraint to achieving their missions, most male agents mentioned transportation, while female agents cited farmer-related problems, such as lack of access to credit. This difference may partly reflect the fact that female extension agents are more likely to work with female workers, who may face more obstacles to adopting new practices.

Although the presence of women professionals in extension organizations and their representation in decisionmaking is critical, the predominant practice continues to direct training and resources to men only. For example, the Sasakawa Africa Fund for Extension Education (SAFE) training program in West Africa, which set up in universities for mid-career agricultural extension staff of the Ministry of Agriculture, has recorded only 21 % female participation (Akeredolu 2009). The reasons behind low female participation include (1) perception bias—the community's low perception of women's talents and potentials and perception of agriculture as a domain of men, (2) limited access to information about opportunities for further education, (3) limited opportunities that target professional women, (4) family concerns and time constraints, and (5) other social, cultural, and religious barriers.

17.5 Lessons from Reforms and Recent Approaches

Alternative organizational and institutional arrangements for extension programs are being explored, including restructuring of current systems to be more inclusive, farmer-led, market-driven, decentralized, and cost-effective. Reforms in agricultural extension systems include decentralization (such as the Agricultural Technology Management Agency [ATMA] model in India), privatization (e.g., in Chile, Ecuador, Estonia, Pakistan, South Africa, and Uganda), contracting for extension service delivery (Honduras and Venezuela), private market-oriented extension services, and different forms of public–private partnerships. However, a continuous challenge has been how to increase farmers' engagement, particularly women's, into program planning and resource allocation and to increase accountability to stakeholders.

There is a dearth of research that looks at how gender issues are integrated and affected by these reform strategies. In some cases, it is assumed that more attention and special provisions to small farmers will automatically and equally reach and benefit women farmers. In other cases, special provisions for women farmers have been explicitly incorporated into reform policies and strategies. Numerous gender-responsive strategies have been adopted and can be grouped into the following:

- strategies that specifically target female household members and CBOs, such as creating and strengthening self-help groups and women's associations; adopting affirmative action in user group associations or farmer-based organizations; and promoting political awareness, leadership, and advocacy abilities for women.

- strategies that target service providers, such as recruiting and training women extension agents and designing, implementing, and monitoring projects in a gender-sensitive manner.
- strategies that target public administration, elected representatives, and political parties, such as reserving seats for women representatives in local councils or committees, gender machineries, sectoral gender focal points, and gender-sensitive training for staff.

While a number of these initiatives have some pockets of success, scaling-up remains difficult, and major gaps in policy and actual implementation persist. Many of these initiatives are superficially done (e.g., to satisfy donors' requirements) and remain supply-driven and far from being transformative. Experiences from a number of reform processes, described below, are indicative of the challenges in reaching out to women farmers and being more responsive to their needs.

17.6 ATMA Model in India

The ATMA model in India is often cited as an innovative model for a decentralized extension service delivery. The ATMA is a semiautonomous organization composed of many key stakeholders involved in agricultural activities and is responsible for technology dissemination at the district level. Several gender-specific provisions are included, namely, (1) mandating 30 % of the resources on beneficiary-oriented programs and activities to be allocated for women farmers and women extension functionaries across 252 ATMAs set up or to be set up in all the major states of the country; (2) introducing gender-sensitization aspects in the training of trainers; and (3) mandating representation of women in all committees and groups at the district level. Although evidence on the impact of ATMA on women farmers is limited, some reports indicate gaps in implementation where the actual allocation and spending is lower than the stipulated. In Bihar, staff shortage prevented agricultural extension workers from promoting the formation of farmer interest groups. While ATMA guidelines did induce agricultural extension workers to seek women's participation in ATMA-sponsored programs, such efforts or programs were not always geared toward improving agricultural production or the marketing practices of women (World Bank/IFPRI 2010). For example, in a case study in Bihar, a group of landless female members of a self-help group was selected for an exposure visit to West Bengal to learn new agricultural technologies. It turned out that the new technologies required access to land, and instead of sending five women for the exposure visit, as specified, only three were sent and the remaining two positions were filled by men from a dominant caste group (World Bank/IFPRI 2010).

Training programs targeting women's groups provide indications of positive impact, such as increases in general awareness among women farmers, increase in income, and visible impact on women's socioeconomic status and food security (Planning Commission 2007). An estimated 1.4 million women farmers benefited from women-specific programs in India over more than 23 years at a cost of INR 2.3 billion (US$50 million), or approximately INR 1,700 (US$37) per woman farmer (Planning Commission 2007). The schemes have covered about 143 districts in most of the states and have promoted 28,000 self-help groups. Concerns expressed relate to limited coverage, especially in terms of direct beneficiaries, despite large coverage of states and districts and seemingly low impact in terms of economic and overall empowerment (Planning Commission 2007). Recently, the ATMA model has moved from a World Bank-funded pilot phase to a domestically-funded national scale. Questions regarding the effectiveness and sustainability of this national program have been raised by a number of authors. For example, Sulaiman and Hall (2002) predict that ATMA may suffer the same fate as the failed T&V extension system. Substantial gaps exist in most of the aspects, like access to technology, markets, and credit, and this mismatch between people trained and their access to resources often results in poor outcomes.

17.7 NAADS in Uganda

Uganda's National Agricultural Advisory Services (NAADS) program focuses on farmers' groups as the lead players in extension service delivery, where government provides services through private service providers in line with farmers' needs. The strategy of NAADS stipulates sensitizing district staff in gender issues and concerns, and identifying indicators to address gender issues at the district and subcounty levels. The NAADS program is acknowledged to have enabled people from remote villages to obtain information on agriculture production; the demonstration farms were also highly appreciated for their practical training components, for being close to the farmers, and for farmers' control of the proceeds from the demonstrations. However, its gender impact seems to be limited. First, individual farmers generally lack the resources to apply the ideas acquired, and while NAADS provides a lot of training, farmers, particularly women, do not necessarily adopt the techniques because of low literacy rates and lack of capital to access the required inputs and technology. Many women do not apply the agricultural technologies for which they are trained owing to limited education, lack of control over land, and cultural factors that limit women from using some technologies (like sitting on tractors) in some communities. While women and youth, in particular, were eager to engage with NAADS, they expressed concern about the limited nature and scope of the enterprises that NAADS made available (Bukenya 2010). The enterprise approach embraced by NAADS had tended to favor farming enterprises requiring substantial amounts of land or capital, thus putting women and youth at a disadvantage compared to men (Bukenya 2010).

Second, although many elderly women belong to the groups, they are not empowered to influence group decisions in the groups, and very few have leadership positions. Despite the overwhelming participation of women in farmer groups, men still retain control over NAADS processes and actual decisionmaking, even in supposedly women-only groups. Some of the factors found to undermine women's control and influence over NAADS processes include (1) low literacy rates for women (as a result, even in many women-only groups, men advisors or secretaries are co-opted to provide linkage to the literate outside world); (2) time burden due to women's triple roles (productive, reproductive, and community service); and (3) weak ownership and control over resources, especially land (where the level of influence or control over group activities relates to the resources at one's disposal) (CEEWA n.d.; NAADS Secretariat 2004).

The NAADS gender analysis conducted by CEEWA-Uganda highlights a few important points. First, it highlights the importance of sex-disaggregated data to identify sources of inequalities and biases to better inform policy and project design. Second, it identifies the need for more capacity building of service providers in gender analysis to enable them to identify the gender-specific needs of the farmers in the program. Many seem to view gender wrongly, as numbers of women in the program, not in terms of factors that make women socially and culturally different from men. Third, service providers need to promote women's ability to identify their own needs and implement their own solutions. In addition, women should be given responsibility at all levels, such as farmer fora, parish, and group levels (Driciru 2008). Fourth, a wide range of channels for communication, like drama, pictorials, and use of local languages, needs to be used in message delivery. Women need to be consulted on timing of radio messages to meet their time constraints and also on location of messages, for example, at water points or health centers where the majority of women converge. When selecting a technology, efforts should be made to assess the impact of the technology on time use, its cost, availability of credit to purchase it, and its appropriateness to the level of education of women.

17.8 Privatization and Decentralization in Venezuela

The third case is the Venezuelan reform initiative, which combined decentralization, privatization, and cost sharing by different government levels, agencies, and beneficiaries. Extension service provision was to be the responsibility of the Foundation for Training and Innovation for Rural Development (CIARA), which contracts private service providers. State and municipal participation is also established through contracts, and cost sharing was introduced through municipal Civil Extension Associations (Spanish acronym, ACEs). The increased program focus on gender and the environment has heightened recognition of the productive role of women and youth and promoted an enhanced awareness of environmental conservation issues. Programs show a positive trend toward a greater participation of women and youth in income-generating activities. The access of rural women and youth to extension services is enhanced by mainstreaming women's programs, identifying potentially differentiated needs for these groups, conducting additional gender-sensitizing programs for policymakers and implementers, and maintaining program flexibility in dealing with situational specificities. The program's achievements include a

42 % increase in annual farm income; a 54 % increase in average crop productivity in relation to the base year; a 127 % increase in average livestock productivity in relation to the base year; a 21 % share for women in program participation in productive activities; and a total of 68 organizations created or strengthened by the program (World Bank et al. 2009).

The Venezuela privatization experience raises two important lessons. The first is the crucial need to focus strongly on the social and human capital development needs of resource-poor smallholder farmers. To respond adequately to the complex needs of those groups, extension service providers need public-sector program managers and field advisers with greatly enhanced competencies to plan and provide services using facilitation and problem-solving approaches in the context of wider community needs. This implies a significant shift from the traditional paradigm of technical expertise alone to the broader competencies needed for effective responses to the new "social" challenges faced by extension personnel, including (1) the practice of participatory extension approaches and (2) local farmer organization development. Second, other extension system reforms have been initiated but were silent in terms of gender strategy and gender-differentiated impact. Experience indicates that these reforms do not guarantee greater outreach to women farmers. This requires accompanying earmarked funding and provisions or conditions for gender in terms of more capacity building, literacy, and consideration of women's time constraints. Proponents of extension system reforms need to take a broad view of extension services, and as Rivera and Alex emphasized, "the client base goes beyond that of the 'male-head-of-household' and the agenda goes beyond the traditional agricultural production focus" (2004, 79).

17.9 Sectoral Policies in Ethiopia

To ensure that gender is taken into account in the agricultural planning process, many districts in Ethiopia have established a system of gender desks or focal points within sectoral agencies and are supposed to guarantee that the gender focal point reviews budgets, plans, and operations. However, there are considerable variations in the effectiveness of this policy. All of the gender focal points in the World Bank/IFPRI (2010) study were junior staff members, both women and men, but qualification or match for the job has been cited as a major issue. Gender focal points in some districts have conducted training in gender analysis for all the extension agents in those districts; however, some feel that the sectoral focal point system is somewhat redundant, because the district offices of women's affairs are already responsible for mainstreaming gender issues in planning activities. In some districts, neither a women's affairs officer nor a focal point system is present. Lack of funds and/or timeliness of fund release have been cited as issues in some districts.

Extension agents interviewed under the World Bank/IFPRI (2010) study were aware of the gender bias in the extension system and had employed strategies to deal with it. Because of cultural barriers to male extension agents interacting with women alone, extension agents have employed different approaches to reaching women farmers, such as contacting their husbands first and explaining the purpose of the visit; meeting women in groups; addressing women in public meetings; and seeking support of women's affairs offices. The district governments also carried out gender analysis as part of a comprehensive needs assessment, and all district government staff received in-service training on gender issues. Awards and recognitions for high-quality work among women extension agents are also being provided in some districts. The Ministry of Agriculture has also developed a broader variety of extension packages, recognizing that one size does not fit all farmers. This includes the "women's development package," which emphasizes support for women's agricultural activities (poultry, small ruminants, and home gardens). However, the women's package does not recognize women's engagement in different tasks and, in particular, does not distinguish between the needs of female household heads and female spouses. For example, a case study by the World Bank/IFPRI (2010) indicates that it is difficult for female household heads to raise chickens, because they have to provide weeding services to male farmers to earn income. To the extent that the women's packages emphasize poultry, it is really a "married women's development package" (177).

17.10 Information and Communication Technologies (ICTs)

Developing content targeted to women's needs would help to increase women's use of this new medium. Several organizations in developing countries are actively working to increase women's understanding and usage of various ICTs for agriculture and related activities. In a review for the Technical Centre for Agricultural and Rural Cooperation ACP-EU (CTA), Hafkin and Odame (2002) find that Africa still lags behind in such projects. However, a start has been made by several organizations.

Since infrastructure has been a leading constraint in women's access to ICTs, some projects were implemented using the ICT tool most familiar and readily available to women: radio is almost universally owned by households, even in developing countries, and is a low cost medium to reach the rural poor, who may not have the infrastructure to access any other technologies (Bates and Tony 2000). Although radio is an older technology, it can be used along with newer forms of ICT to provide agricultural knowledge and information to rural women (Giles 2004). Both Wambui (2002) and Hafkin and Odame (2002) discussed how digital radio can be used to deliver local language programs through links with the internet and mobile phones. Radio broadcasts can also be used in distance education to air both formal and informal learning content for rural women (Maskow 2000). The Kenya AIDS prevention Project Group conducted nutritional field schools in six projects sites in western Kenya. Similarly, the radio was used in rural Ghana to conduct panel discussions featuring women who could talk on a variety of topics affecting women on local FM. While training the rural women to serve as panelists, the project also increased their capacity and knowledge in these areas. However, radio is a one-way medium and needs to be complemented by some other forms of ICT to ensure maximum two-way learning.

Projects using mobile phones to deliver messages to women have been based on the premise that phones are now more readily available in rural areas, due to increasing upgrades to infrastructure in developing countries. Men and women view and use mobile phones differently. While mobile phones symbolize social status for men, for women they are instruments of expression and sociability (Plant 2003). Mobile phones have been used by fishermen in India to check prices in various markets before selling their products, thereby increasing their profits and reducing wastage (Jensen 2007). The Grameen Phone project (2005) specifically targeted women in Bangladesh. Learning the skills required to be a village phone operator not only augmented a woman's earnings, but also enhanced her social prestige. Learning modules related to sheep and goat enterprises delivered to women through their mobile phones while they tended to the animals overcame the barrier imposed by time constraints (Balasubramanian et al. 2010). Five hundred one-minute audio messages on topics like buying goats, feed, disease and health management, and marketing management were sent as voicemails. These were followed up with weekly group meetings to share experiences and recall information. The project not only increased women's confidence from the information sharing, but it also linked them to information sources. The flow of communication within the self-help groups, relatives, and friends enabled the women to learn to use the phones as well as benefit from the information shared. It also enhanced their self-respect and status within the family.

The internet is, of course, the most modern form of ICT and has the capacity to be both a one-way and an interactive learning medium. The SISSI project in Uvira, Congo, created a support group of women accessing agricultural information. The information center not only provided internet access to the 60 women farmers involved in the project, but also matched them with mentors from other local communities to reinforce their support networks (GenARDIS 2010). In another project in Uvira, 48 women and 18 men from nine different women's groups received information on production and disease management for cassava crops. They were provided with mobile phones to contact their potential buyers. A radio show was also created on topics related to gender and agriculture. Similarly, the Ndola resource center in Zambia trained 115 women in open-source software (GenARDIS 2010).

The outcomes of the projects illustrate that ICTs have the capacity to empower women and overcome the socially constructed digital divide that placed such technologies within the man's domain. Given proper training, women have been successfully using various forms of communication technologies, even if both the use and social meaning of the technology varies between men and women. While most projects find that the ownership and control of costlier communication technologies like mobile phones or computers are still with men, women control the usage of the devices, especially mobile phones, in the house. Using these technologies as a form of e-learning, women have increased their knowledge of agro production and animal husbandry. At the same time, they have increased their social status within the family and community.

Since women's literacy rates lag behind men in the rural areas in most developing countries, software in local languages will go a long way to improve the benefits of internet learning for women (Kwapong 2009). Financial constraints are still a deciding factor that determines whether a woman can afford a phone or pay for internet access.

17.11 Conclusions

Problems with access to extension services, and the priorities to address them, vary from country to country. Thus, analysis of these problems and the design of programs to overcome them should cater to the heterogeneity of farmers and the variability of local contexts. For example, in India, the issue is low government extension capacity overall; in Ghana, it is the lack of focus on measurable development targets and low access to extension services in rural areas, particularly among women, despite the comparatively high ratio of extension agents to farmers; in Ethiopia, the problem lies in overreliance on fixed technology packages that give less discretion to extension agents and are irresponsive to farmers' demands, especially women's needs. However, despite differences, programs have common needs.

17.11.1 Project Interventions

From the program or project perspective, there is a need for increased earmarked funding and explicit targets for reaching poor women farmers. Reforms in extension systems such as privatization and decentralization do not automatically guarantee greater attention to women's needs, due to persistent social and cultural norms and perception bias that often prevent equal access and opportunities for women and men. The above-mentioned case studies highlight several entry points from both supply and demand sides. From the supply side, options include affirmative action and policy shifts to enable research and extension to focus more on women; engaging role models within the agricultural extension service systems to make the contribution of women visible at every opportunity; leadership training to increase women's capacity to leverage and negotiate; educational opportunities for women who wish to study in the field of agriculture and for mid-career women to improve their skills and competencies. Extension organizations must encourage and recruit more female extension agents, who were found to be more effective than male extension agents in reaching female farmers. Creating incentives for reaching female farmers by, for example, rewarding such outreach in performance reviews, would be important. At the same time, given the low current staffing levels of female extension agents, there is a need to develop strategies that will help male agents to work better with women farmers. From the demand side, options include strengthening women's groups to better articulate their needs and demands; formulating messages and designing training packages that are simple and clearly understood, especially in areas with many illiterate farmers; understanding and nurturing social networks; exploring the potential of ICT; and pilot-testing community-based extension approaches that provide more accountability to women farmers.

In most cases, information from extension services and training is not applied and does not create sustainable business enterprises because of lack of complementary inputs and resources. Because women have disproportionately fewer resources than men, programs that specifically target female household members will be important. Some initiatives include creating policies that strengthen poor women's rights to assets, strengthening group-based approaches that target women, and piloting voucher programs or grants to ensure women smallholders' access to resources. There is a need to scale up pockets of success from gender-responsive strategies and approaches, which include creating and strengthening women's groups, innovative forms of extension and education such as FFSs and radio, and women-friendly forms of information technologies. Strategies and approaches need to be designed that address women's needs, but more important, that pay special attention to the implementation and monitoring.

17.11.2 Research

From the research perspective, more sex-disaggregated data collection and rigorous impact assessments are needed. These play a crucial role in identifying sources of bias and inequality and bottlenecks in furthering food security and agricultural development to inform policy and project design. Topics that need further research include (1) understanding the constraints, bottlenecks, and opportunities for scaling-up and rolling out successful gender-responsive actions; (2) rigorous methodologies for assessing quality or satisfaction from extension services as current methods and studies show conflicting results; (3) studies exploring the supply side, including motivations, incentives, and constraints faced by women in becoming extension agents; and (4) gender-sensitive impact assessment of reforms in extension systems.

The role of agricultural extension in empowering women and increasing their incomes and their family's food security is potentially large. If done well and context-specific delivery methods are adopted, agricultural extension can bring knowledge and skills to those who need them the most. Finding the right combination of approaches will likely remain a major challenge in the near future, but regular sharing of experiences and lessons learned will help identify under what conditions a particular delivery method or advisory approach works or does not work. Putting poor women and men at the center of project and program design and focusing on impact on them in every step of the project cycle will be an important guide toward the right direction.

Acknowledgments This chapter draws from the discussion on extension in Meinzen-Dick et al. (2010).

References

Akeredolu M (2009) Female students' participation in the University Mid-Career Agricultural Extension Training Programme in West Africa: constraints and challenges. In: Proceedings of the 25th annual meeting of the association for international agricultural and extension education (AIAEE) at the InterContinental San Juan Resort in Puerto Rico, 24–27 May 2009

Alston JM, Wyatt TJ, Pardey PG, Marra MC, Chan-Kang C (2000) A meta-analysis of rates of return to agricultural R&D—Ex Pede Herculem. International Food Policy Research Institute, Washington, DC

Awoyemi TT, Oluwatayo IB (2010) Gender mainstreaming of ICT projects: lessons from rural South West Nigeria. Libyan Agric Res Center J Int 1(2):65–69

Balasubramanian K, Thamizoli P, Umar A, Kanwar A (2010) Using mobile phones to enable lifelong learning among rural women in southern India. Distance Educ 31(2):193–209

Bates A, Tony W (2000) Managing technological change: Strategies for college and university leaders. Jossey-Bass, San Francisco

Birner R, Davis K, Pender J, Nkonya E, Anandajayasekeram P, Ekboir J, Mbabu A, Spielman D, Horna D, Benin S, Cohen M (2006) From "best practice" to "best fit": a framework for analyzing pluralistic agricultural advisory services worldwide. Development strategy and governance division discussion paper 37. International Food Policy Research Institute, Washington, DC

Braun A, Jiggins J, Röling N, van den Berg H, Snijders P (2006) A global survey and review of farmer field school experiences. Report prepared for the International Livestock Research Institute. Final Report, June 12. Available at http://www.share4dev.info/kb/documents/1880.pdf

Bukenya C (2010) Meeting farmers demand? An assessment of extension reform in Uganda. Unpublished PhD dissertation, Wageningen University, the Netherlands. Available at http://edepot.wur.nl/15663

CEEWA (Council for Economic Empowerment of Women of Africa, Uganda) (n.d.) Women should be the driving force in NAADPS program. http://www.ceewauganda.org/. Accessed 3 Feb 2010

Davis K (2006) Farmer field schools: a boon or bust for extension in Africa? J Int Agric Ext Educ 13(1):91–97

Davis K, Negash M (2007) Gender, wealth, and participation in community groups in Meru Central District, Kenya. CAPRi working paper series 65. Consultative Group on International Agricultural Research, Washington, DC

Davis K, Nkonya E, Kato E, Ayalew Mekonnen D, Odendo M, Miiro R, Nkuba J (2010) Impact of farmer field schools on agricultural productivity and poverty in East Africa. IFPRI discussion paper 00992. International Food Policy Research Institute, Washington, DC

Driciru C (2008) Women participation in national agricultural advisory services in Uganda: a case study of Vurra Sub-County, Arua District, 2001—2007. Unpublished master's degree thesis, Public Administration and Management, Makerere University, Uganda

Due JF, Magayane AA, Temu A (1997) Gender again: views of female agricultural extension officers by smallholder farmers in Tanzania. World Dev 25(5):713–725

Evenson R (1997) The economic contributions of agricultural extension to agricultural and rural development. In: Swanson B, Bentz R, Sofranko A (eds) Improving agricultural extension: a reference manual. Food and Agriculture Organization of the United Nations, Rome, pp 27–36

FAO (Food and Agriculture Organization of the United Nations) (2003) Gender: key to sustainability and food security. Gender and Development, Rome

Feder G, Anderson JR, Birner R, Deininger K (2010) Promises and realities of community-based agricultural extension. IFPRI discussion paper 959. International Food Policy Research Institute, Washington, DC

GenARDIS (2010) Small grants that make big changes for women in agriculture. Association for Progressive Communications, Melville

Gilbert RA, Sakala WD, Benson TD (2002) Gender analysis of a nationwide cropping system trial survey in Malawi. Afr Stud Q 6(1 & 2):223–243. http://web.africa.ufl.edu/asq/v6/v6i1a9.htm. Accessed 23 Feb 2010

Giles L (2004) ICT to reduce gender disparity in education in Sub-Saharan Africa. Capstone project: master of distant education. University of Maryland, University College, College Park, MD, US

Gillwald A, Milek A, Stork C (2010) Gender assessment of ICT access and usage in Africa, vol 1, Policy paper 5. Research ICT Africa, Johannesburg

Gotschi E, Njuki J, Delve R (2009) Equal numbers, equal chances? A case study of gender differences in the distribution of social capital in smallholder farmer groups in Búzi District, Mozambique. Eur J Dev Res 21(2):264–282

Grameen Phone (2005) Annual report. Retrieved from http://investor-relations.grameenphone.com/irportal/annualreport/gp_ar_05.pdf

Hafkin NJ, Odame HH (2002) Gender, ICTs, and agriculture. Situation analysis for the 5th consultative expert meeting of the technical centre for agricultural and rural cooperation's (CTA) ICT observatory meeting on Gender and Agriculture in the Information Society, Sageningen, The Netherlands, 11–13 September 2002

Huyer S, Hafkin N, Ertl H, Dryburgh H (2005) Women in the information society. In: Sciadas G (ed) From the digital divide to digital opportunities: measuring infostates for development. Orbicom/ITU, Ottawa, pp 135–195

Isubikalu P (2007) Stepping-stones to improve upon functioning of participatory agricultural extension programmes: farmer field schools in Uganda. PhD dissertation, Wageningen University, Netherlands. Available at http://edepot.wur.nl/42679

Jensen R (2007) The digital provide: information (technology), market performance, and welfare in the South Indian Fisheries Sector. Q J Econ 122(3):879–924

Katungi E, Edmeades S, Smale M (2008) Gender, social capital, and information exchange in rural Uganda. J Int Dev 20(1):35–52

Kwapong OATF (2009) A comparison of ICT knowledge and usage among female distance learners in endowed and deprived communities in developing countries. E-Learn Digit Media 6(2):164–174

Maskow M (2000) Radio as a learning technology. In: Burge EJ (ed) New directions for adult and continuing education: the strategic use of learning technologies. Jossey-Bass, San Francisco, pp 59–68

Meinzen-Dick R, Quisumbing A, Behrman J, Biermayr-Jenzano P, Wilde V, Noordeloos M, Ragasa C, Beintema N (2010) Engendering agricultural research. Discussion paper 00973. International Food Policy Research Institute, Washington, DC

Moore KM, Hamilton S, Sarr P, Thiongane S (2001) Access to technical information and gendered NRM practices: men and women in rural Senegal. Agric Hum Values 18(1):95–105

NAADS (National Agricultural Advisory Services) Secretariat) (2004) NAADS poverty and gender fact sheet No. 6 (June). http://www.naads.or.ug/

Parkinson S (2008) Learning participation in rural development: a study of Uganda's national agricultural advisory services. Unpublished PhD dissertation. University of Guelph, Ontario

Planning Commission (India) (2007) Report. Report of sub-group on gender and agriculture submitted to the working group on gender issues, Panchayat Raj Institutions, Public Private Partnership, Innovative Finance and Micro Finance in Agriculture for the Eleventh Five Year Plan (2007–2012). New Delhi

Plant S (2003) On the mobile: the effects of mobile telephones on social and individual life. http://www.motorola.com/mot/doc/0/234_MotDoc.pdf

Quizon J, Feder G, Murgai R (2001) Fiscal sustainability of agricultural extension: the case of the farmer field school approach. Working paper. Development Research Group, World Bank, Washington, DC

Rivera W, Alex G (eds) (2004) Privatization of extension systems: case studies of international initiatives. Volume 2. Agriculture and rural development discussion paper 9. World Bank, Washington, DC

Sorensen K (2002) Love, duty and the S-curve: an overview of some current literature on gender and ICT. SIGIS report. University of Edinburgh, Edinburgh

Sulaiman RV, Hall A (2002) Beyond technology dissemination—can Indian agricultural extension re-invent itself? Policy brief 16. Indian Council of Agricultural Research, New Delhi

Swanson B, Rajalahti R (2010) Strengthening agriculture extension and advisory systems: procedures for assessing, transforming, and evaluating extension systems. Agricultural and rural development discussion paper 45. World Bank, Washington, DC

Swanson B, Farner B, Bahal R (1990) The current status of agricultural extension worldwide. In: Report of the global consultation on agriculture extension. Food and Agricultural Organization of the United Nations, Rome

Van den Berg H (2004) IPM farmer field schools: a synthesis of 25 impact evaluations. Prepared for the Global IPM Facility, Wageningen University, the Netherlands. ftp://ftp.fao.org/docrep/fao/006/ad487E/ad487E00.pdf

Van den Berg H, Jiggins J (2007) Investing in farmers—the impacts of farmer field schools in relation to integrated pest management. World Dev 35(4):663–686

Wambui M (2002) Meet the champions: DG interview with Dr. Nancy Hafkin on ICT and Gender in Africa. Development Gateway, ICT for Development. Retrieved November 10, 2004 from: http://www.developmentgateway.org/

World Bank (2000) Implementation completion report for Indonesia: Integrated Pest Management training project. Report No. 20212, June 6. Available at http://documents.worldbank.org/curated/en/2000/06/728922/indonesia-integrated-pest-management-training-project

World Bank/IFAD/FAO (International Fund for Agricultural Development/Food and Agriculture Organization for the United Nations) (2009) Gender in agriculture sourcebook. World Bank, Washington, DC. http://siteresources.worldbank.org/INTGENAGRLIVSOUBOOK/Resources/CompleteBook.pdf. Accessed 20 July 2009

World Bank/IFPRI (International Food Policy Research Institute) (2010) Gender and governance in rural services: insights from India, Ghana, and Ethiopia. Gender and governance author team. International Bank for Reconstruction and Development/World Bank, Washington, DC

Zainudeen A, Iqbal T, Samarajiva R, Ratnadiwakara D (2008) Who's got the phone? The gendered use of telephones at the bottom of the Pyramid. In: Paper presented at the conference of the International Communication Association, Montreal, 22–26 May

Index

A.R. Quisumbing et al. (eds.), *Gender in Agriculture: Closing the Knowledge Gap*, DOI 10.1007/978-94-017-8616-4,

J

K

L

M

Q

R

S

Y

Z